Mathematical Logic and Its Applications 2020

Mathematical Logic and Its Applications 2020

Editors

Vassily Lyubetsky
Vladimir Kanovei

MDPI • Basel • Beijing • Wuhan • Barcelona • Belgrade • Manchester • Tokyo • Cluj • Tianjin

Editors
Vassily Lyubetsky
Department of Mechanics and Mathematics of Moscow Lomonosov State University
Russia

Vladimir Kanovei
Institute for Information Transmission Problems of the Russian Academy of Sciences
Russia

Editorial Office
MDPI
St. Alban-Anlage 66
4052 Basel, Switzerland

This is a reprint of articles from the Special Issue published online in the open access journal *Mathematics* (ISSN 2227-7390) (available at: https://www.mdpi.com/journal/mathematics/special_issues/math-logic-2020).

For citation purposes, cite each article independently as indicated on the article page online and as indicated below:

LastName, A.A.; LastName, B.B.; LastName, C.C. Article Title. *Journal Name* **Year**, *Volume Number*, Page Range.

ISBN 978-3-0365-0778-1 (Hbk)
ISBN 978-3-0365-0779-8 (PDF)

Contents

About the Editors

Vassily Lyubetsky is a Doctor of physical and mathematical sciences, professor, and a principal researcher. He is also head of the Laboratory of Mathematical Methods and Models in Bioinformatics at the Institute for Information Transmission Problems of the Russian Academy of Sciences (IITP, Moscow), and Professor of the Faculty of Mechanics and Mathematics of the Moscow State University of the Department of Mathematical Logic and Theory of Algorithms. He has published more than 200 scientific papers and 9 books, and the Guest Invited Editor for the Regular Special Issue "Molecular Phylogenomics" of the "Biomed Research International journal (Molecular Biology)".

Vladimir Kanovei is a principal researcher at the Institute for Information Transmission Problems of the Russian Academy of Sciences (IITP, Moscow). He was awarded his Ph.D. by the Moscow State University, and his D.Sc. by the Steklov Mathematical Institute of the Russian Academy of Sciences. His research interests in mathematics include mathematical logic, descriptive set theory, forcing, and nonstandard analysis. He is an author of about 300 scientific publications, including 7 monographs.

Preface to "Mathematical Logic and Its Applications 2020"

This Special Issue contains articles representing three directions: Descriptive set theory (DTM), exact polynomial complexity algorithms (EPA), and applications of mathematical logic and algorithm theory (Appl). We will say a few words about each of the directions.

In accordance with the classical description of Nicolas Luzin, DTM considers simple properties of simple sets of real numbers $\mathbb{R}$. "Simple" sets are Borel sets (the smallest family containing open and closed sets in $\mathbb{R}^n$ and closed with respect to the operations of countable union and countable intersection) and projective sets (the smallest family containing Borel sets and closed with respect to the operations of projecting from $\mathbb{R}^n$ to $\mathbb{R}^m$, $m < n$, and the complement to the whole space). The question of what is a "simple" property is more complicated, but it is not important, since in fact we study a small list of individual properties, including the Lebesgue measurability, Baire property[1], and the individual definability of a set, function, or real. The latest means that there is a formula that holds for a given real number and for no others. This depends on the class of formulas allowed. Such a natural class consists of formulas of the form $\forall x_1 \exists y_1 \forall x_2 \exists y_2 \ldots \forall x_n \exists y_n \, \psi(x_1, y_1, \ldots, x_n, y_n, x)$, where the variables $x_1, y_1, \ldots, x_n, y_n, x$ run through the whole $\mathbb{R}$, and the elementary part $\psi(x_1, y_1, \ldots, x_n, y_n, x)$ is any arithmetic formula (which contains any quantifiers over the natural numbers, as well as equalities and inequalities that connect the superpositions of operations from the semiring of natural numbers). To date, the development of DTM leads to a non-trivial general cultural conclusion: every real number is *definable* (using countable ordinals[2]) or *random*; in the latter case it does not possess any non-trivial properties. This implies that there are absolutely undecidable statements[3]; as well as surprising connections between seemingly very different absolutely undecidable ones. For example, the measurability implies the Baire property for a wide class of sets. The first three articles belong to this direction. In particular, they solve the well-known problem (1948) of A. Tarski on the definability of the notion of definability itself, and prove the statement (1975) of H. Friedman.

The EPA section contains an article contributing a solution for the meaningful combinatorial and, at first glance, complicated algorithmic problem of optimization of the functional given on paths of passing from one graph to another. It is solved by an algorithm of linear complexity, being at the same time exact. The latter means that for any input data, that is for any ordered pair of graphs A and B, accompanied by costs of elementary graph transformations, the algorithm produces exactly the minimal value of the above functional (i.e., the minimum distance between A and B and the minimum path itself from A to B).

Here the complexity of the problem turned into the logical complexity of this, albeit linear, algorithm. Our goal was to draw attention to the search for, and possible discussion of, algorithmic problems that seem to require exhaustive search but are actually solved by exact algorithms of low polynomial complexity. This ensures their practical significance when working with large data (terabyte and larger sizes).

The Appl section contains two articles. First of them is devoted to the application of non-standard analysis (and other logical methods) to the problems of isomorphism in algebra and mathematical physics (the Jacobian and M. Kontsevich's conjectures, and algorithmic undecidability). The second is devoted to the application of logical and algorithmic approaches to the problem of theoretical medicine — a quantitative description of the balance and the adaptive resource of a human that determines his resistance to external influences. Applied problems in which logic and theory of algorithms have shown their usefulness could be of interest.

The Editorial Board of *Mathematics* (WoS: Q1) has announced the preparation of the issue "Mathematical Logic and Its Applications 2021"; contributions in these directions and especially in other ones of this huge mathematical area, including various applications, are invited.

1) The Baire property of a set X says that there is an open set U such that the symmetric difference $X \Delta U$ is a meager set (the union of a countable number of nowhere-dense sets).

2) Countable ordinals are the natural numbers themselves and their natural extension: taking the limit over all natural numbers we get ω, adding $+1$ to it consecutively and taking the limit yet again we get $\omega + \omega = \omega \cdot 2$, and so on. Each time, the limit is taken over a countable sequence.

3) This means that a natural statement about measurability (or other simple subjects) cannot be proved or disproved in the natural set theory of ZFC, which seems to contain all the mathematics used in physics, biology, computer science, and engineering.

Vassily Lyubetsky, Vladimir Kanovei

Editors

mathematics

MDPI

Article

Models of Set Theory in which Nonconstructible Reals First Appear at a Given Projective Level

Vladimir Kanovei *,† and Vassily Lyubetsky †

Institute for Information Transmission Problems of the Russian Academy of Sciences, 127051 Moscow, Russia; lyubetsk@iitp.ru

* Correspondence: kanovei@iitp.ru

† These authors contributed equally to this work.

Received: 11 May 2020; Accepted: 27 May 2020; Published: 3 June 2020

Abstract: Models of set theory are defined, in which nonconstructible reals first appear on a given level of the projective hierarchy. Our main results are as follows. Suppose that $n \geq 2$. Then: 1. If it holds in the constructible universe $\mathbf{L}$ that $a \subseteq \omega$ and $a \notin \Sigma^1_n \cup \Pi^1_n$, then there is a generic extension of $\mathbf{L}$ in which $a \in \Delta^1_{n+1}$ but still $a \notin \Sigma^1_n \cup \Pi^1_n$, and moreover, any set $x \subseteq \omega$, $x \in \Sigma^1_n$, is constructible and Σ^1_n in $\mathbf{L}$. 2. There exists a generic extension $\mathbf{L}$ in which it is true that there is a nonconstructible Δ^1_{n+1} set $a \subseteq \omega$, but all Σ^1_n sets $x \subseteq \omega$ are constructible and even Σ^1_n in $\mathbf{L}$, and in addition, $\mathbf{V} = \mathbf{L}[a]$ in the extension. 3. There exists an generic extension of $\mathbf{L}$ in which there is a nonconstructible Σ^1_{n+1} set $a \subseteq \omega$, but all Δ^1_{n+1} sets $x \subseteq \omega$ are constructible and Δ^1_{n+1} in $\mathbf{L}$. Thus, nonconstructible reals (here subsets of ω) can first appear at a given lightface projective class strictly higher than Σ^1_2, in an appropriate generic extension of $\mathbf{L}$. The lower limit Σ^1_2 is motivated by the Shoenfield absoluteness theorem, which implies that all Σ^1_2 sets $a \subseteq \omega$ are constructible. Our methods are based on almost-disjoint forcing. We add a sufficient number of generic reals to $\mathbf{L}$, which are very similar at a given projective level n but discernible at the next level $n + 1$.

Keywords: definability; nonconstructible reals; projective hierarchy; generic models; almost disjoint forcing

MSC: 03E15; 03E35

---

1. Introduction

Problems of definability and effective construction of mathematical objects have always been in the focus of attention during the development of mathematical foundations. In particular, Hadamard, Borel, Baire, and Lebesgue, participants of the discussion published in [1], in spite of significant differences in their positions regarding problems of mathematical foundations, emphasized that a pure existence proof and a direct definition (or an effective construction) of a mathematical object required are different mathematical results, and the second one of them does not follow from the first. Problems of definability and effectivity are considered in such contemporary monographs on foundations as [2–5]. Moschovakis, one of founders of modern set theory, pointed in [6] (p. xiv), that

> the central problem of descriptive set theory and definability theory in general [is] to find and study the characteristic properties of definable objects.

The general goal of the research line of this paper is to explore the existence of effectively definable structures in descriptive set theory on specific levels of the projective hierarchy. One of the directions here is the construction of set theoretic models, in which this or another problem is decided, at a predefined projective level n, differently than it is decided in $\mathbf{L}$, Gödel's constructible universe, or, that is equivalent, by adding the axiom of constructibility, dubbed $\mathbf{V} = \mathbf{L}$.

Such set theoretic models are usually defined as generic extensions of $\mathbf{L}$ itself. Any such a generic extension leads to consistency and independence results in set theory, because if a sentence Φ holds in $\mathbf{L}$ or in a generic extension of $\mathbf{L}$ then Φ is consistent with the axioms of **ZFC**, the Zermelo–Fraenkel set theory (with the axiom of choice **AC**).

As a first, and perhaps most immediately interesting problem of this sort, in this paper, we consider the problem of the existence of effectively definable (that is, occurring in one of lightface classes Σ^1_n of the projective hierarchy) but nonconstructible reals. It follows from Shoenfield's absoluteness theorem [7] that every (lightface) Σ^1_2 set $x \subseteq \omega$ belongs to $\mathbf{L}$. Generic models, in which there exist nonconstructible reals on effective levels of the projective hierarchy higher than Σ^1_2, were defined in the early years of forcing; see a brief account in [8]. This culminated in two different generic extensions [9,10] containing a nonconstructible Π^1_2 singleton, hence, a Δ^1_3 set $a \subseteq \omega$. (We are concentrated on generic extensions of $\mathbf{L}$ in this paper, and therefore leave aside another research line, related to models with large cardinals, with many deep and fruitful results connected, in particular, with properties of Π^1_2 singletons, see e.g., [11–13]).

Then it was established in [14] that for any $n \geq 2$ there is a generic extension of $\mathbf{L}$ in which there exists a nonconstructible Δ^1_{n+1} real $a \subseteq \omega$, but all Σ^1_n sets $x \subseteq \omega$ are constructible. Our motivation here is to further extend this research line. The next three theorems are the main results in this paper.

Theorem 1. *If $\mathtt{n} \geq 2$ and $b \subseteq \omega$, $b \notin \Sigma^1_{\mathtt{n}} \cup \Pi^1_{\mathtt{n}}$, then there is a generic extension of $\mathbf{L}$ in which $b \in \Delta^1_{\mathtt{n}+1}$ but still $b \notin \Sigma^1_{\mathtt{n}} \cup \Pi^1_{\mathtt{n}}$, and moreover, any set $x \subseteq \omega$, $x \in \Sigma^1_{\mathtt{n}}$, is constructible and $\Sigma^1_{\mathtt{n}}$ in $\mathbf{L}$.*

Theorem 1 shows that being at a certain lightface projective level is hardly an intrinsic property of a constructible real, unless it is already at that level in $\mathbf{L}$. The theorem definitely fails for $\mathtt{n} = 1$ since being Δ^1_2 is an ablosute property of a real by the Shoenfield absoluteness theorem.

Theorem 2. *If $\mathtt{n} \geq 2$, then there exists a generic extension of the universe $\mathbf{L}$ in which it is true that*

(i) *there is a nonconstructible $\Delta^1_{\mathtt{n}+1}$ set $a \subseteq \omega$, but all $\Sigma^1_{\mathtt{n}}$ sets $x \subseteq \omega$ are constructible and $\Sigma^1_{\mathtt{n}}$ in $\mathbf{L}$;*
(ii) *we can strengthen (i) by the requirement that $\mathbf{V} = \mathbf{L}[a]$ in the extension.*

Theorem 3. *If $\mathtt{n} \geq 2$ then there exists an extension of $\mathbf{L}$ in which there is a nonconstructible $\Sigma^1_{\mathtt{n}+1}$ set $a \subseteq \omega$ but all $\Delta^1_{\mathtt{n}+1}$ sets $x \subseteq \omega$ are constructible and $\Delta^1_{\mathtt{n}+1}$ in $\mathbf{L}$.*

The common denominator of Theorems 2 and 3 is that nonconstructible reals can first appear at a given lightface projective class strictly higher than Σ^1_2, in an appropriate generic extension of $\mathbf{L}$. The lower limit Σ^1_2 is motivated by the Shoenfield absoluteness theorem.

The generic models, which we define to prove the main theorems, make use of modifications of the almost-disjoint forcing by Jensen–Solovay [9].

Some other recent results can be mentioned here, which resemble Theorems 1–3 in that they give models in which a particular property of some kind holds at a certain pre-selected level of the projective hierarchy. Yet they are different in that they use modifications of Jensen's minimal Π^1_2 singleton forcing [10] and its finite-support products first considered by Enayat [15], as well as its collapse-style modification by Abraham [16], rather than the almost-disjoint forcing.

- A model defined in [17], in which, for a given $n \geq 2$, there is a (lightface) Π^1_n Vitali equivalence class in the real line $\mathbb{R}$ (that is, a set of the form $x + \mathbb{Q}$ in $\mathbb{R}$), containing no OD (ordinal definable) elements, and in the same time every countable Σ^1_n set consists of OD elements.
- A model in [18], in which, for a given $n \geq 2$, there is a Π^1_n singleton $\{a\}$, such that a codes a collapse of $\omega_1^{\mathbf{L}}$, and in the same time every Σ^1_n set $a \subseteq \omega$ is still constructible.
- A model defined in [19], in which, for a given $n \geq 2$, there is a Π^1_n non-OD-uniformizable planar set with countable cross-sections, and at the same time, every Σ^1_n set with countable cross-sections is OD-uniformizable.

Organization of the Paper

Our plan of the proofs of the main results will be to construct, in $\mathbf{L}$, a sequence of forcing notions $\mathbb{P}(\nu)$, $\nu < \omega_1$, satisfying the following three key conditions.

1. $\mathbb{P}(\nu)$ are sufficiently homogeneous and independent of each other in the sense that, for any ν_0, there are no $\mathbb{P}(\nu_0)$-generic reals in a $(\prod_{\nu \neq \nu_0} \mathbb{P}(\nu))$-generic extensions of $\mathbf{L}$.
2. The property of a real x being $\mathbb{P}(\nu)$-generic over $\mathbf{L}$ is $\Pi^1_{\mathfrak{n}}$ as a binary relation, where $\mathfrak{n} \geq 2$ is a number chosen in Theorems 1–3.
3. A condition which makes $\mathbb{P}(\nu)$-generic reals for different values $\nu < \omega_1$ undistinguishable from each other below the $\Pi^1_{\mathfrak{n}}$ definability level (at which they are distinguishable by condition 2).

Each $\mathbb{P}(\nu)$ will be a forcing notion of almost-disjoint type, determined by a set $\mathbb{U}(\nu) \subseteq \omega^\omega$. To make the exposition self-contained, we review some basic details related to almost-disjoint forcing, finite-support products, and related generic extensions, taken mainly from [9], in Sections 2 and 3.

Having the construction of $\mathbb{P}(\nu)$, $\nu < \omega_1$, accomplished in Section 4, the proof of, e.g., Theorem 1 (Section 7.1) is performed as follows. Let $b \in \mathbf{L}$, $b \subseteq \omega$ be chosen as in Theorem 1 for a given $\mathfrak{n} \geq 2$. We consider a $\mathbb{P}$-generic extension $\mathbf{L}[G]$ of $\mathbf{L}$, where $\mathbb{P} = \prod_{i<\omega} \mathbb{P}(i)$. Let $a_i \subseteq \omega$ be the $\mathbb{P}(i)$-generic real generated by the ith projection G_i of G; these reals are nonconstructible and $\mathbf{L}[G] = \mathbf{L}[\{a_i\}_{i<\omega}]$. Let $z = \{0\} \cup \{2k : k \in b\} \cup \{2k+1 : k \notin b\}$ Consider the subextension $\mathbf{L}[\{a_i\}_{i\in z}]$. Then it is true in $\mathbf{L}[\{a_i\}_{i\in z}]$ by condition 1, that

$$\begin{aligned} b &= \{k < \omega : \text{there exist } \mathbb{P}(2k)\text{-generic reals}\} \\ &= \{k < \omega : \text{there are no } \mathbb{P}(2k+1)\text{-generic reals}\}, \end{aligned}$$

so using condition 2, we easily get $b \in \Delta^1_{\mathfrak{n}+1}$ in $\mathbf{L}[\{a_i\}_{i\in z}]$. A similar construction (but with b being generic over $\mathbf{L}$) was carried out in the early years of forcing in [9] for $\mathfrak{n} = 2$, which is the least possible value. In the case $\mathfrak{n} = 2$, the fact, that all Σ^1_2 sets $x \subseteq \omega$ in the extension belong to $\mathbf{L}$ and are Σ^1_2 in $\mathbf{L}$, is guaranteed by the Shoenfield absoluteness theorem.

If $\mathfrak{n} \geq 3$, then the Shoenfield absoluteness argument does not work, of course. Still we can argue that any lightface $\Sigma^1_{\mathfrak{n}}$ set $x \subseteq \omega$ in $\mathbf{L}[\{a_i\}_{i\in z}]$ belongs to $\mathbf{L}$ by the general forcing theory, because the product forcing $\mathbb{P}_z = \prod_{i\in z} \mathbb{P}(i) \in \mathbf{L}$ is homogeneous by condition 1. However this does not immediately imply the lightface definability of b in $\mathbf{L}$, as $\mathbb{P}_z$ is defined via z, hence via b. To solve this difficulty, we make use of condition 3 to prove another absoluteness property: $\Sigma^1_{\mathfrak{n}}$ formulas turn out to be absolute between $\mathbf{L}[\{a_i\}_{i\in z}]$ and the entire extension $\mathbf{L}[G] = \mathbf{L}[\{a_i\}_{i<\omega}]$, which is an $\mathbb{P}$-generic extension of $\mathbf{L}$. Here $\mathbb{P} = \prod_{i<\omega} \mathbb{P}(i)$ is a parameter-free definable forcing in $\mathbf{L}$, leading to the parameter-free definability of b in $\mathbf{L}$. There are two issues here that need to be explained.

First, how to secure condition 3 in a sufficiently effective form. To explain the main technical device, we recall that by [9] the system of forcing notions $\mathbb{P}(\nu)$ is the result of certain transfinite ω_1-long construction of assembling it from countable fragments in $\mathbf{L}$. The construction can be viewed as a maximal branch in a certain "mega-tree", say $\mathcal{T}$, whose nodes are such countable fragments, and each of them is chosen to be the Gödel-least appropriate one over the previous one. The complexity of this construction is Δ^1_2 in the codes, leading in [9] to the Π^1_2 definability of the property of being generic, as in condition 2, in case $\mathfrak{n} = 2$.

To adapt this construction for the case $\mathfrak{n} \geq 3$, our method requires us to define a maximal branch in $\mathcal{T}$ that intersects all dense sets in $\mathcal{T}$ of class $\Sigma^1_{\mathfrak{n}-1}$. Such a construction is carried out in Section 4. This genericity-like condition of meeting all dense $\Sigma^1_{\mathfrak{n}-1}$ sets, results in the $\Pi^1_{\mathfrak{n}}$ definability of the property of being generic in condition 2, and also yields condition 3, since the abundance of order automorphisms of the "mega-tree" $\mathcal{T}$ (including those related to index permutations) allows to establish some homogeneity properties of a certain auxiliary forcing-style relation.

This auxiliary forcing-style relation, defined and studied in Sections 5 and 6. The auxiliary relation approximates the truth in $\mathbb{P}'$-generic extensions, as $\mathbf{L}[\{a_i\}_{i\in z}]$ above, up to $\Sigma^1_{\mathfrak{n}}$ formulas, but, unlike the ordinary $\mathbb{P}'$-forcing relation, is sufficiently homogeneous. In particular, it helps to obtain the mentioned absoluteness property. This will allow us to accomplish the proof of the main results, Theorem 1 together with part (i) of Theorem 2 in Section 7, part (ii) of Theorem 2 in Section 8, Theorem 3 in Section 9. The flowchart can be seen in Figure 1.

The flowchart can be seen in Figure 1. And we added the index and contents as Supplementary Materials for easy reading.

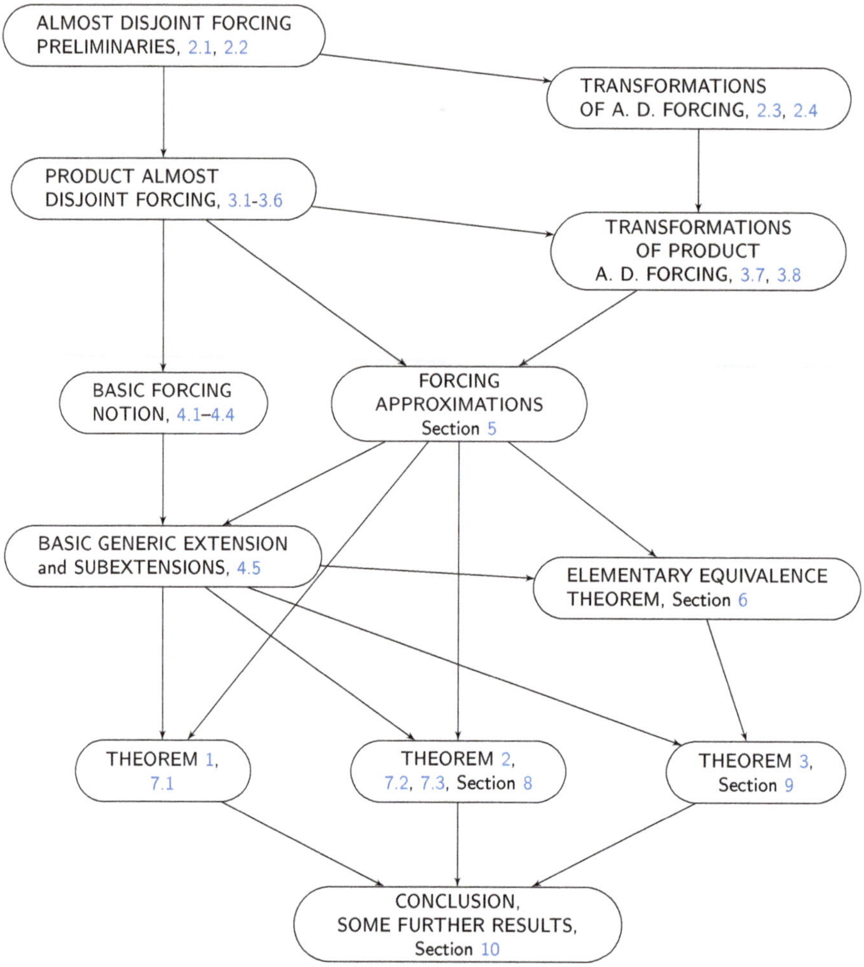

Figure 1. Flowchart.

General Set-Theoretic Notation Used in This Paper

- $\omega = \{0,1,2,\dots\}$: natural numbers; $\omega^2 = \omega \times \omega$.
- $X \subseteq Y$ iff $\forall x\,(x \in X \Longrightarrow x \in Y)$: the inclusion.
- $X \subsetneqq Y$ means that $X \subseteq Y$ but $Y \not\subseteq X$: strict inclusion.
- $\mathtt{card}\, X$ is the cardinality of a set X, equal to the number of elements of X in case X is finite.
- $\mathtt{dom}\, P = \{x : \exists y\,(\langle x,y\rangle \in P)\}$ and $\mathtt{ran}\, P = \{y : \exists x\,(\langle x,y\rangle \in P)\}$ — the domain and range of any set P that consists of pairs.
- In particular if $P = f$ is a function then $\mathtt{dom}\, f$ and $\mathtt{ran}\, f$ are the domain and the range of f.

- Functions are identified with their graphs: if $P = f$ is a function then $f = \{\langle x, f(x)\rangle : x \in \mathtt{dom}\, f\}$, so that $y = f(x)$ is equivalent to $\langle x, y\rangle \in f$.
- $f[X] = \{f(x) : x \in X \cap \mathtt{dom}\, f\}$, the *f-image* of X.
- $f^{-1}[Y] = \{x \in \mathtt{dom}\, f : f(x) \in Y\}$, the f-pre-image of a set Y.
- $f^{-1}(y) = \{x \in \mathtt{dom}\, f : f(x) = y\}$, the f- pre-image of an element y.
- Δ is the symmetric difference.
- $\{x_a\}_{a\in A}$ is the map f defined on A by $f(a) = x_a$, $\forall a$.
- $\mathscr{P}(X) = \{x : x \subseteq X\}$, the power set.
- $X^{<\omega}$ is the set of all strings (finite sequences) of elements of a set X.
- In particular $\omega^{<\omega}$ is the set of strings of natural numbers.
- $\mathtt{lh}\, s < \omega$ is the length of a string s.
- $s^\frown x$ is the string obtained by adjoining x as the rightmost term to a given string s.
- $s \subset t$ means that the string t is a proper extension of s.
- $\varnothing = \Lambda$ is resp. the empty set and the empty string.
- ω^ω is the Baire space.

2. Almost Disjoint Forcing

In this section, we review basic definitions and results related to almost disjoint forcing, as well as some rarely used results related, for instance, to symmetries of almost disjoint forcing notions.

Assumption 1. *In this paper, we assume that* $\mathbf{L}$ *is the ground universe. Thus all forcing notions are defined in* $\mathbf{L}$ *while all generic extensions are those of* $\mathbf{L}$*. (In fact many intermediate results remain true w.r.t. any ground universe.)*

2.1. Almost Disjoint Forcing

We present this forcing in a form based on the fact that the set **Fun** of all functions $f : \omega \to \omega$ is almost disjoint in the sense that if $f \neq g$ belong to **Fun** then the infinite sets $\{f \restriction m : m \in \omega\}$ and $\{g \restriction m : m \in \omega\}$ of finite strings have a finite intersection.

Definition 1. $\mathbf{Seq} = \omega^{<\omega} \smallsetminus \{\Lambda\}$ *= all finite non-empty strings of natural numbers. A recursive enumeration* $\omega^{<\omega} = \{s_k : k \in \omega\}$ *is fixed, such that* $s_0 = \Lambda$*, the empty string, and* $s_k \subseteq s_\ell \implies k \leqslant \ell$*. Thus* $\mathbf{Seq} = \omega^{<\omega} \smallsetminus \{\Lambda\} = \{s_k : k \geq 1\}$*. For any* $s = s_k$*, we let* $\mathtt{num}\, s = k$*; in particular* $\mathtt{num}\, \Lambda = 0$*.*

Fun $= \omega^\omega$ *= all infinite sequences of natural numbers. A set* $X \subseteq \mathbf{Fun}$ *is dense iff for any* $s \in \mathbf{Seq}$ *there is* $f \in X$ *such that* $s \subset f$*.*

Let $S \subseteq \mathbf{Seq}$*,* $f \in \mathbf{Fun}$*. If the set* $S/f = \{n : f \restriction n \subset S\}$ *is infinite then we say that* S *covers* f*, otherwise* S *does not cover* f*.*

We underline that Λ, the empty string, does not belong to **Seq**.

Given a set $u \subseteq \mathbf{Fun}$ in the ground universe, the general goal of almost disjoint forcing is to find a generic set $S \subseteq \mathbf{Seq}$ such that the equivalence

$$f \in u \iff S \text{ does not cover } f \tag{1}$$

holds for each $f \in \mathbf{Fun}$ in the ground universe. This goal will be achieved by a forcing $P[u]$ introduced in Definition 4. In fact $P[u]$ will be a part, determined by u, of a common reservoir P^*.

Definition 2. P^* *is the set of all pairs* $p = \langle S_p; F_p\rangle$ *of finite sets* $F_p \subseteq \mathbf{Fun}$*,* $S_p \subseteq \mathbf{Seq}$*. Elements of* P^* *will sometimes be called (forcing) conditions. If* $p \in P^*$ *then put* $F_p^\vee = \{f \restriction n : f \in F_p \wedge n \geq 1\}$*. The set* $F_p^\vee$ *is an infinite (or else* $F_p^\vee = F_p = \varnothing$*) tree in* $\mathbf{Seq}$*, without terminal nodes.*

Definition 3 (order)**.** *Let* $p, q \in P^*$*. We define* $q \leq p$ *(*q *is stronger) iff* $S_p \subseteq S_q$*,* $F_p \subseteq F_q$*, and the difference* $S_q \smallsetminus S_p$ *does not intersect* $F_p^\vee$*, that is,* $S_q \cap F_p^\vee = S_p \cap F_p^\vee$*.*

Thus any condition $p \in P^*$ is a pair that consists of a "finite" component S_p and an "infinite" component F_p. Either of the components is a finite set (possibly, empty), but S_p consists of finite strings of integers while F_p consists of infinite sequences of integers that will be called functions (from ω to ω). Both components of a stronger condition q, naturally, increase, but strings $t \in S_q \smallsetminus S_p$ must satisfy $t \notin F_p^{\vee}$—in other words, t is not a substring of any function (infinite sequence) $f \in F_p$.

If $p \in P^*$ then both $\langle \varnothing; F_p \rangle$ and $\langle S_p; \varnothing \rangle$ belong to P^* and $p \le \langle S_p; \varnothing \rangle$, but $p \le \langle \varnothing; F_p \rangle$ may fail. In fact $p \le \langle \varnothing; F_p \rangle$ iff $S_p \cap F_p^{\vee} = \varnothing$.

Lemma 1. *Conditions $p, q \in P^*$ are compatible in P^* iff 1) $S_q \smallsetminus S_p$ does not intersect $F_p^{\vee}$, and 2) $S_p \smallsetminus S_q$ does not intersect $F_q^{\vee}$. Therefore, any $p, q \in P^*$ are compatible in P^* iff $p \wedge q \le p$ and $p \wedge q \le q$.*

Proof. The pair $p \wedge q = \langle S_p \cup S_q; F_p \cup F_q \rangle$ is a condition in P^*. Moreover if 1) and 2) hold then we have $p \wedge q \le p$ and $p \wedge q \le q$, thus p, q are compatible. □

Now let us introduce a relativized version of P^*. The parameter of relativization will be an arbitrary set $u \subseteq \mathbf{Fun}$ served as a reservoir of functions allowed to occur in sets F_p.

Definition 4. *If $u \subseteq \mathbf{Fun}$ then put $P[u] = \{p \in P^* : F_p \subseteq u\}$.*

Note that if $p, q \in P[u]$ then $p \wedge q \in P[u]$. Thus in this case if conditions p, q are compatible in P^* then they are compatible in $P[u]$, too. Therefore, we will say that conditions $p, q \in P^*$ are compatible (or incompatible) without an indication which set $P[u]$ containing both conditions is considered.

Lemma 2. *If $u \subseteq \mathbf{Fun}$ then $P[u]$ is a ccc forcing.*

Proof. If $S_p = S_q$ then p and q are compatible by Lemma 1. However there are only countably many sets of the form S_p. □

2.2. *Almost-Disjoint Generic Extensions*

Fix, in $\mathbf{L}$, a set $u \subseteq \mathbf{Fun}$ and consider a $P[u]$-generic extension $\mathbf{L}[G]$ of the ground (constructible by Assumption 1) set universe $\mathbf{L}$, obtained by adjoining a $P[u]$-generic set $G \subseteq P[u]$. Put $S_G = \bigcup_{p \in G} S_p$; thus $S_G \subseteq \mathbf{Seq}$. The next lemma reflects the idea of almost-disjoint forcing: elements of u are distinguished by the property of S_G not covering f in the sense of Definition 1.

Lemma 3. *Suppose that $u \subseteq \mathbf{Fun}$ in the universe $\mathbf{L}$, and $G \subseteq P[u]$ is a set $P[u]$-generic over $\mathbf{L}$. Then*

(i) *G belongs to $\mathbf{L}[S_G]$;*
(ii) *if $f \in \mathbf{Fun} \cap \mathbf{L}$ then $f \in u$ iff S_G does not cover f;*
(iii) *if $p \in P[u]$ then $p \in G$ iff $s_p \subseteq S_G \wedge (S_G \smallsetminus s_p) \cap (F_p^{\vee} \cup S_p^{\vee}) = \varnothing$.*

Proof. (ii) Let $f \in u$. The set $D_f = \{p \in P[u] : f \in F_p\}$ is dense in $P[u]$. (Let $q \in P[u]$. Define $p \in P[u]$ so that $S_p = S_q$ and $F_p = F_q \cup \{f\}$. Then $p \in D_f$ and $p \leqslant q$.) Therefore $D_f \cap G \neq \varnothing$. Pick any $p \in D_f \cap G$. Then $f \in F_p$. Now every $r \in G$ is compatible with p, and hence $S_r/f \subseteq S_p/f$ by Lemma 1. Thus $S_G/f = S_p/f$ is finite.

Let $f \notin u$. The sets $D_{fl} = \{p \in P[u] : \sup(S_p/f) > l\}$ are dense in $P[u]$. (If $q \in P[u]$ then F_q is finite. As $f \notin u$, there is $m > l$ with $f \restriction m \notin F_q^{\vee}$. Define p so that $F_p = F_q$ and $S_p = S_q \cup \{f \restriction m\}$. Then $p \in D_{fl}$ and $p \leqslant q$.) Let $p \in D_{fl} \cap G$. Then $\sup(S_G/f) > l$. As l is arbitrary, S_G/f is infinite.

(iii) Consider any $p \in P[u]$. Suppose that $p \in G$. Then obviously $s_p \subseteq S_G$. If there exists $s \in (S_G \smallsetminus S_p) \cap F_p^{\vee}$ then by definition we have $s \in S_q$ for some $q \in G$. However, then p, q are incompatible by Lemma 1, a contradiction.

Now suppose that $p \notin G$. Then there exists $q \in G$ incompatible with p. By Lemma 1, there are two cases. First, there exists $s \in (S_q \smallsetminus S_p) \cap F_p^{\vee}$. Then $s \in S_G \smallsetminus S_p$, so p is not compatible with S_G.

Second, there exists $s \in (S_p \smallsetminus S_q) \cap F_q^{\vee}$. Then any condition $r \leq q$ satisfies $s \notin S_r$. Therefore $s \notin S_G$, so $S_p \not\subseteq S_G$, and p is not compatible with S_G.

(i) $G = \{p \in P[u] : s_p \subseteq S_G \wedge (S_G \smallsetminus s_p) \cap F_p^{\vee} = \varnothing\}$ by (iii). □

2.3. Lipschitz Transformations

Let **Lip** be the group of all $\subseteq$-automorphisms of **Seq**; these transformations may be called Lipschitz by obvious association. Any $\boldsymbol{\lambda} \in \mathbf{Lip}$ preserves the length $\mathtt{lh}$ of finite strings, that is, $\mathtt{lh}\, s = \mathtt{lh}\,(\boldsymbol{\lambda} \cdot s)$ for all $s \in \mathbf{Seq}$. Define the action of any transformation $\boldsymbol{\lambda} \in \mathbf{Lip}$ on:

- finite strings $s \in \mathbf{Seq}$ by: $\boldsymbol{\lambda} \cdot s = \boldsymbol{\lambda}(s)$;
- functions $f \in \mathbf{Fun}$: $\boldsymbol{\lambda} \cdot f \in \mathbf{Fun}$ is defined so that $(\boldsymbol{\lambda} \cdot f) \restriction m = \boldsymbol{\lambda} \cdot (f \restriction m)$;
- sets $S \subseteq \mathbf{Seq}$, $F \subseteq \mathbf{Fun}$ by: $\boldsymbol{\lambda} \cdot S = \{\boldsymbol{\lambda} \cdot s : s \in S\}$, $\boldsymbol{\lambda} \cdot F = \{\boldsymbol{\lambda} \cdot f : f \in F\}$;
- conditions $p \in P^*$, by: $\boldsymbol{\lambda} \cdot p = \langle \boldsymbol{\lambda} \cdot S_p ; \boldsymbol{\lambda} \cdot F_p \rangle$.

Lemma 4 (routine). *The action of any $\boldsymbol{\lambda} \in \mathbf{Lip}$ is an order-preserving automorphism of P^*. If $u \subseteq \mathbf{Fun}$ and $p \in P[u]$ then $\boldsymbol{\lambda} \cdot p \in P[\boldsymbol{\lambda} \cdot u]$.*

Lemma 5. *Suppose that $u, v \subseteq \mathbf{Fun}$ are countable sets topologically dense in $\mathbf{Fun}$, and $p \in P[u]$, $q \in P[v]$. Then there is $\boldsymbol{\lambda} \in \mathbf{Lip}$ and conditions $p' \in P[u]$, $p' \leq p$ and $q' \in P[v]$, $q' \leq q$, such that $\boldsymbol{\lambda} \cdot u = v$, and $\boldsymbol{\lambda} \cdot p' = q'$ — therefore conditions $\boldsymbol{\lambda} \cdot p$ and q are compatible in $P[v]$.*

Proof. Put $\mathtt{bas}\, r = \{s(0) : s \in S_r\} \cup \{f(0) : f \in F_r\}$ for any $r \in P^*$; $\mathtt{bas}\, r \subseteq \omega$ is finite. Let $M < \omega$ satisfy $\mathtt{bas}\, p \cup \mathtt{bas}\, q \subseteq M$. Because of density, for any $i < M$ there exist $f_i \in u$ and $g_i \in u'$ such that $f_i(0) = i$ and $g_i(0) = M + i$.

For any $f \neq g \in \mathbf{Fun}$, let $\mathrm{N}(f, g)$ be the largest n with $f \restriction n = g \restriction n$.

We will define enumerations $u = \{f_k : k < \omega\}$ and $u' = \{g_k : k < \omega\}$, without repetitions, which agree with the above definition for $k < M$ and satisfy $\mathrm{N}(f_k, f_l) = \mathrm{N}(g_k, g_l)$ for all k, l, and $g_k(0) = f_k(0)$ for all $k \geq M$. As soon as this is accomplished, define $\boldsymbol{\lambda} \in \mathbf{Lip}$ as follows. Consider any $s \in \mathbf{Seq}$ of length $m = \mathtt{lh}\, s$. As u is dense, $s = f_k \restriction m$ for some k. Put $\boldsymbol{\lambda}(s) = g_k \restriction m$. Clearly $\boldsymbol{\lambda} \cdot u = u'$, and in particular $\boldsymbol{\lambda} \cdot f_k = g_k$ for all k, and hence

$(*)$ if $k < M$ then $\boldsymbol{\lambda}(\langle k \rangle) = \langle M + k \rangle$ and $\boldsymbol{\lambda}(\langle M + k \rangle) = \langle k \rangle$, but if $k \geq 2M$ then $\boldsymbol{\lambda}(\langle k \rangle) = \langle k \rangle$.

Now to define q' put $r' = \boldsymbol{\lambda} \cdot p$. Then $r' \in P[v]$, and $\mathtt{bas}\, r' = \boldsymbol{\beta} \cdot \mathtt{bas}\, p \subseteq \omega \smallsetminus M$ by $(*)$, since $\mathtt{bas}\, p \subseteq M$. Therefore, $\mathtt{bas}\, r' \cap \mathtt{bas}\, q = \varnothing$ because $\mathtt{bas}\, q \subseteq M$ as well. It follows that conditions r' and q are compatible in $P[v]$, and hence condition $q' = r' \wedge q$ (that is, $S_{q'} = S_{r'} \cup S_q$ and $X_{q'} = X_{r'} \cup X_q$) belongs to $P[v]$, and obviously $q' \leq q$. Pretty similarly, to define q, we put $r = \boldsymbol{\lambda}^{-1} \cdot q \in P[u]$, thus $\mathtt{bas}\, r \subseteq \omega \smallsetminus M$, $\mathtt{bas}\, r \cap \mathtt{bas}\, p = \varnothing$, conditions r, p are compatible, condition $p' = p \wedge r$ (that is, $S_{p'} = S_p \cup S_r$ and $X_{p'} = X_p \cup X_r$) belongs to $P[u]$, and $p' \leq p$. Note that $q = \boldsymbol{\lambda} \cdot r$ and $r' = \boldsymbol{\lambda} \cdot p$ by construction. It follows that $q' = r' \wedge q = \boldsymbol{\lambda} \cdot (p \wedge r) = \boldsymbol{\lambda} \cdot p'$, as required.

To define f_k and g_k by induction, suppose that $k \geq M$, $f_0, \dots, f_{k-1}$ and $g_0, \dots, g_{k-1}$ are defined, and $\mathrm{N}(f_i, f_j) = \mathrm{N}(g_i, g_j)$ holds in this domain. Consider any next function $f \in u \smallsetminus \{f_0, \dots, f_{k-1}\}$, and let it be f_k. There are functions $g \in \mathbf{Fun}$ satisfying $\mathrm{N}(f_j, f_k) = \mathrm{N}(g_j, g)$ for all $j < k$. This property of g is determined by a certain finite part $g \restriction m$. By the density the set v contains a function g of this type. Let g_k be any of them. In the special case when $\mathrm{N}(f_j, f_k) = 0$ for all $j < k$ (then $k \geq 2M$), we take any $g_k \in v$ satisfying $\mathrm{N}(f_j, f_k) = 0$ for all $j < k$ and $g_k(0) = f_k(0)$. □

2.4. Substitution Transformations

The next lemma (Lemma 6) will help to prove that the forcing notions considered are sufficiently homogeneous. Assume that $p, q \in P^*$ satisfy the following requirement:

$$F_p = F_q \quad \text{and} \quad S_p \cup S_q \subseteq F_p^{\vee} = F_q^{\vee}. \tag{2}$$

We define a transformation H^p_q acting as follows. Let $p' \in P^*$, $p' \leq p$. Then by definition $S_p \subseteq S_{p'}$, $F_p \subseteq F_{p'}$, and $S_{p'} \cap F^\vee_p = S_p$ (by (2)). We put $H^p_q(p') = q' := \langle S_{q'}, F_{q'} \rangle$, where $F_{q'} = F_{p'}$ and $S_{q'} = (S_{p'} \smallsetminus S_p) \cup S_q$. Thus the difference between $S_{q'}$ and $S_{p'}$ lies entirely within the set $F^\vee_p = F^\vee_q$, and in particular $S_{q'}$ has S_q there while $S_{p'}$ has S_p there.

Lemma 6 (routine). *If $p, q \in P^*$, $F_p = F_q$, and $S_p \cup S_q \subseteq F^\vee_p = F^\vee_q$, then*

$$H^p_q \ : \ P = \{p' \in P^* : p' \leq p\} \stackrel{\text{onto}}{\longrightarrow} Q = \{q' \in P^* : q' \leq q\}$$

*is an order isomorphism, and $H^p_q = (H^q_p)^{-1}$. If moreover $u \subseteq$ **Fun** and $p, q \in P[u]$ then H^p_q maps the set $\{p' \in P[u] : p' \leq p\}$ onto $\{q' \in P[u] : q' \leq q\}$ order-preservingly.*

3. Almost Disjoint Product Forcing

Here we review the structure and basic properties of product almost-disjoint forcing over **L** and corresponding generic extensions of **L**. In order to support various applications, we make use of ω_1-many independent forcing notions.

3.1. Product Forcing, Systems, Restrictions

We begin with ω_1-products of P^* after which we consider more complicated forcing notions.

Definition 5. *Let $\mathcal{I} = \omega_1$. This is the index set for the forcing products considered below. Let $\mathbf{P}^*$ be the product of $\mathcal{I}$ copies of the set P^* (Definition 2), with finite support. That is, $\mathbf{P}^*$ consists of all functions $p : |p| \to P^*$ such that the set $|p| = \mathtt{dom}\, p \subseteq \mathcal{I}$ is finite.*

If $p \in \mathbf{P}^*$ then put $F_p(\nu) = F_{p(\nu)}$ and $S_p(\nu) = S_{p(\nu)}$ for all $\nu \in |p|$, so that $p(\nu) = \langle S_p(\nu); F_p(\nu) \rangle$. We order $\mathbf{P}^*$ componentwise: $p \leq q$ iff $|q| \subseteq |p|$ and $p(\nu) \leq q(\nu)$ for all $\nu \in |q|$. Put

$$F^\vee_p(\nu) = F^\vee_{p(\nu)} = \{f \restriction m : f \in F_p(\nu) \wedge m \geq 1\}.$$

If $p, q \in \mathbf{P}^*$ then define a condition $r = p \wedge q \in \mathbf{P}^*$ so that $|p \wedge q| = |p| \cup |q|$, $(p \wedge q)(\nu) = p(\nu) \wedge q(\nu)$ whenever $\nu \in |p| \cap |q|$, and if $\nu \in |p| \smallsetminus |q|$ or $\nu \in |q| \smallsetminus |p|$, then $(p \wedge q)(\nu) = p(\nu)$, resp., $(p \wedge q)(\nu) = q(\nu)$. Then Conditions p, q are compatible iff $p \wedge q \leq p$ and $p \wedge q \leq q$.

We consider certain subforcings of the total product almost disjoint forcing notion $\mathbf{P}^*$. This involves the following notion of a system.

Definition 6. *A system is any map $U : |U| \to \mathscr{P}(\mathbf{Fun})$ such that $|U| \subseteq \mathcal{I}$ and each set $U(\nu)$ $(\nu \in |U|)$ is topologically dense in* **Fun**. *A system U is:*

- *disjoint, if its components $U(\nu) \subseteq \mathbf{Fun}$ $(\nu \in \mathcal{I})$ are pairwise disjoint;*
- *countable, if the set $|U|$ and each $U(\nu)$ $(\nu \in |U|)$ are at most countable.*
- *If U, V are systems, $|U| \subseteq |V|$, and $U(\nu) \subseteq V(\nu)$ for all $\nu \in |U|$ then we write that V extends U, in symbol $U \preccurlyeq V$.*
- *If $\{U_\xi\}_{\xi<\lambda}$ is a sequence of systems then define a system $U = \bigvee_{\xi<\lambda} U_\xi$ by $|U| = \bigcup_{\xi<\lambda} |U_\xi|$ and $U(\nu) = \bigcup_{\xi<\lambda, \nu \in |U_\xi|} U_\xi(\nu)$ for all $\nu \in |U|$.*
- *If U is a system then let $\mathbf{P}[U]$ be the finite support product of sets $P[U(\nu)]$, $\nu \in |U|$, that is, $\mathbf{P}[U] = \{p \in \mathbf{P}^* : |p| \subseteq |U| \wedge \forall \nu\, (F_p(\nu) \subseteq U(\nu))\}$.*

Definition 7 (restrictions). *Suppose that $c \subseteq \mathcal{I}$.*

If $p \in \mathbf{P}^*$ then define $p' = p \restriction c \in \mathbf{P}^*$ so that $|p'| = c \cap |p|$ and $p'(\nu) = p(\nu)$ whenever $\nu \in |p'|$. Accordingly if U is a system then define a system $U \restriction c$ so that $|U \restriction c| = c \cap |U|$ and $(U \restriction c)(\nu) = U(\nu)$ for $\nu \in |U \restriction c|$. A special case: if $\nu \in \mathcal{I}$ then let $p \restriction_{\neq \nu} = p \restriction (|p| \smallsetminus \{\nu\})$ and $U \restriction_{\neq \nu} = U \restriction (|U| \smallsetminus \{\nu\})$.

Note that writing $p \restriction c$ or $U \restriction c$, it is not assumed that $c \subseteq |p|$, resp., $c \subseteq |U|$.

3.2. Regular Forcing Notions

Unfortunately, product forcing notions of the form $\mathbf{P}[U]$ (U being a system in $\mathbf{L}$) do not provide us with all the definability effects we need. We will make use of certain more complicated forcing notions $K \subseteq \mathbf{P}^*$ in $\mathbf{L}$. To explain the idea, let a system $U \in \mathbf{L}$ satyisfy $|U| = \omega$. Let $G \subseteq \mathbf{P}[U]$ be generic over $\mathbf{L}$. The sets $S_G(\nu) = S_{G(\nu)} = \bigcup_{p \in G} S_p(\nu) \subseteq \mathbf{Seq}$ then belong to $\mathbf{L}[G]$, and in fact $\mathbf{L}[G] = \mathbf{L}[\{S_G(\nu)\}_{\nu<\omega}]$. As $\mathbf{Seq} = \{\mathfrak{s}_k : k \geq 1\}$ (a fixed recursive enumeration, Definition 1), let $a_0[G] = \{k \geq 1 : \mathfrak{s}_k \in S_0[G]\}$ and $c = \{0\} \cup a_G(0)$. Consider the model $\mathbf{L}[\{S_G(\nu)\}_{\nu \in c}]$. The first idea is to make use of $U \restriction c$, but oops, clearly $c \notin \mathbf{L}$, and consequently $U \restriction c \notin \mathbf{L}$ and $\mathbf{P}[U \restriction c] \notin \mathbf{L}$, so that many typical product forcing results do not apply in this case. The next definition attempts to view the problem from another angle.

Definition 8 (in $\mathbf{L}$). *A set $K \subseteq \mathbf{P}^*$ is called a regular subforcing if:*

(1) *if conditions $p, q \in K$ are compatible then $p \wedge q \in K$;*
(2) *if $p, q \in K$ then $p \restriction |q| \in K$ — but it is not assumed that $p \in K$ necessarily implies $p \restriction c \in K$ for an arbitrary $c \subseteq |p|$;*
(3) *if $p, q \in \mathbf{P}^*$, $q \leq p$, and $|q| = |p|$ exactly, then $p \in K$ implies $q \in K$;*
(4) *for any condition $p \in \mathbf{P}^*$, there exist: a condition $p^* \in \mathbf{P}^*$ and a set $d \subseteq |p^*|$ such that $p^* \leq p$, $F_{p^*}(\nu) = F_p(\nu)$ for all $\nu \in |p|$, $F_{p^*}(\nu) = \varnothing$ for all $\nu \in |p^*| \smallsetminus |p|$, $p^* \restriction d \in K$, and every condition $q \in K$, $q \leq p^* \restriction d$, satisfies $|q| \cap |p^*| = d$, and hence q is compatible with p^* and with p.*

In this case, if U is a system then define $K[U] = K \cap \mathbf{P}[U]$. In particular, if simply $K = \mathbf{P}^*$ then $\mathbf{P}^*[U] = \mathbf{P}^* \cap \mathbf{P}[U] = \mathbf{P}[U]$.

Example 1 (trivial). *If $c \subseteq \mathcal{I}$ in the ground universe $\mathbf{L}$, then $\mathbf{P}^* \restriction c$ is a regular forcing. To prove (4) of Definition 8 let $p^* = p$ and $d = |p| \cap c$.*

Example 2 (less trivial). *Consider the set K of all conditions $p \in \mathbf{P}^*$ such that $|p| \subseteq \omega$ and if $\nu \in |p|$, $\nu \geq 1$, then $\mathfrak{s}_\nu \in S_p(0)$. We claim that K is a regular subforcing.*

To verify 8(2), note that if $q \in K$ then either $0 \in |q|$ or $|q| = \varnothing$.

To verify 8(4), let $p \in \mathbf{P}^*$. If $|p| \subseteq \{0\}$, then setting $p^* = p$ and $d = |p|$ works, so we assume that $|p| \nsubseteq \{0\}$. Define $p^* \in \mathbf{P}^*$ so that $p^*(\nu) = p(\nu)$ for all $\nu \geq 1$, $F_{p^*}(0) = F_p(0)$, and $S_{p^*}(0) = S_p(0) \cup \{\mathfrak{s}_\nu : \nu \in I'\}$, where I' consists of all $\nu \in |p|$, $\nu \geq 1$, such that $\mathfrak{s}_\nu \notin S_p(0) \cup F_p^\vee(0)$. Then $|p^*| = |p| \cup |0|$, $p^* \leq p$, and we have $\mathfrak{s}_\nu \in S_{p^*}(0) \cup F_{p^*}^\vee(0)$ (not necessarily $\mathfrak{s}_\nu \in S_{p^*}(0)$) for all $\nu \in |p|$, $\nu \geq 1$. Let $d \subseteq |p^*|$ contain 0 and all $\nu \in |p|$, $\nu \geq 1$ with $\mathfrak{s}_\nu \in S_{p^*}(0)$; easily $p^* \restriction d \in K$.

Now let $q \in K$, $q \leq r = p^* \restriction d$. Consider any index $\nu \in |p^*| \smallsetminus d$. Then $\mathfrak{s}_\nu \notin S_{p^*}(0) = S_r(0)$, hence $\mathfrak{s}_\nu \in F_{p^*}^\vee(0) = F_r^\vee(0)$. We claim that $\nu \notin |q|$. Indeed otherwise $\mathfrak{s}_\nu \in S_q(0)$ as $q \in K$. However $\mathfrak{s}_\nu \in F_r^\vee(0) \smallsetminus S_r(0)$ (see above). However, this contradicts $\mathfrak{s}_\nu \in S_q(0)$, because $q \leq r$.

Theorem 4 (in $\mathbf{L}$). *The partially ordered set $\mathbf{P}^*$, and hence each $\mathbf{P}[U]$, and generally each regular subforcing of $\mathbf{P}[U]$ (for any system U) satisfies* CCC (countable antichain condition).

Proof. Suppose towards the contrary that $A \subseteq \mathbf{P}^*$ is an uncountable antichain. We may assume that there is $m \in \omega$ such that $|p| = m$ for all $p \in A$. Applying the Δ-lemma argument, we obtain an uncountable set $A' \subseteq A$ and a finite set $w \subseteq \mathcal{I}$ with $\mathtt{card}\, w < m$ strictly, such that $|p| \cap |q| = w$ for all $p \neq q$ in A'. Then $A'' = \{p \restriction w : p \in A'\}$ is still an uncountable antichain, with $|p| = w$ for all $p \in A'$, easily leading to a contradiction (see the proof of Lemma 2). □

Lemma 7 (in $\mathbf{L}$). *If $K \subseteq \mathbf{P}^*$ is a regular forcing and U is a system then $K[U] = K \cap \mathbf{P}[U]$ is a regular subforcing of $\mathbf{P}[U]$.*

To show how (4) of Definition 8 works, we prove

Lemma 8 (in $\mathbf{L}$). *If U is a system and $K \subseteq \mathbf{P}[U]$ is a regular subforcing of $\mathbf{P}[U]$ then any set $D \subseteq K$ pre-dense in K remains pre-dense in $\mathbf{P}[U]$.*

Proof. Consider any $p \in \mathbf{P}[U]$. Let $p^* \in \mathbf{P}[U]$ and $d \subseteq |p^*|$ satisfy (4) of Definition 8. In particular, $p^* \leq p$ and $p^* \restriction d \in K$. By the pre-density, there is a condition $q \in D$ compatible with $p^* \restriction d$. Then by (1) of Definition 8 there is a condition $r = q \wedge (p^* \restriction d) \in K$ such that $r \leq q$ and $r \leq p^* \restriction d$. Then r is compatible with p by the choice of p^* and d. $\square$

3.3. Outline of Product and Regular Extensions

We consider sets of the form $\mathbf{P}[U]$, U being a system in $\mathbf{L}$, as well as regular subforcings $K \subseteq \mathbf{P}[U]$, as forcing notions over $\mathbf{L}$. Accordingly, we will study $\mathbf{P}[U]$-generic and K-generic extensions $\mathbf{L}[G]$ of the ground universe $\mathbf{L}$. Define some elements of these extensions.

Definition 9. *Suppose that $G \subseteq \mathbf{P}^*$. Put $|G| = \bigcup_{p \in G} |p|$; $|G| \subseteq \mathcal{I}$. Let*

$$S_G(\nu) = S_{G(\nu)} = \bigcup_{p \in G} S_p(\nu) \quad \text{and} \quad a_{G(\nu)} = a_G(\nu) = \{k \geq 1 : s_k \in S_G(\nu)\},$$

for any $\nu \in \mathcal{I}$, where $G(\nu) = \{p(\nu) : p \in G\} \subseteq P^*$, and $\mathbf{Seq} = \{s_k : k \geq 1\}$ is a fixed recursive enumeration (see Definition 1).

Thus $S_G(\nu) \subseteq \mathbf{Seq}$, $a_G(\nu) \subseteq \omega \smallsetminus \{0\}$, and $S_G(\nu) = a_G(\nu) = \varnothing$ for any $\nu \notin |G|$.

By the way, this defines a sequence $\vec{S}_G = \{S_G(\nu)\}_{\nu \in \mathcal{I}}$ of subsets of $\mathbf{Seq}$.

If $c \subseteq \mathcal{I}$ then let $G \restriction c = \{p \in G : |p| \subseteq c\}$. It will typically happen that $G \restriction c = \{p \restriction c : p \in G\}$. Put $G \restriction_{\neq \nu} = \{p \in G : \nu \notin |p|\} = G \restriction (\mathcal{I} \smallsetminus \{\nu\})$.

If U is a system in $\mathbf{L}$, the ground universe, then any $\mathbf{P}[U]$-generic set $G \subseteq \mathbf{P}[U]$ splits into the family of sets $G(\nu)$, $\nu \in \mathcal{I}$, and each $G(\nu)$ is $P[U(\nu)]$-generic.

Lemma 9. *Let U be a system and $K \subseteq \mathbf{P}[U]$ be a regular subforcing in the ground universe $\mathbf{L}$. Let $G \subseteq \mathbf{P}[U]$ be a set $\mathbf{P}[U]$-generic over $\mathbf{L}$. Then:*

(i) $G \in \mathbf{L}[S_G]$;
(ii) *the set $G \cap K$ is K-generic over $\mathbf{L}$*;
(iii) $\mathbf{L}[G \cap K] = \mathbf{L}[G \restriction c]$, *where $c = |G \cap K|$* (it is not necessary that $c \in \mathbf{L}$!);
(iv) *if $\nu \notin |G \cap K|$ then $\mathbf{L}[G \cap K] \subseteq \mathbf{L}[G \restriction_{\neq \nu}]$*;
(v) *if $\nu \in \mathcal{I}$ then $S_G(\nu) \notin \mathbf{L}[G \restriction_{\neq \nu}]$*;
(vi) *if $\nu \in |G|$ then the set $G(\nu) = \{p(\nu) : p \in G\} \in \mathbf{L}[G]$ is $P[U(\nu)]$-generic over $\mathbf{L}$, hence if $f \in \mathbf{Fun} \cap \mathbf{L}$ then $f \in U(\nu) \iff S_G(\nu)/f$ is finite.*

Proof. (ii) This follows from Lemma 8.

(iii) Let us show that $G \restriction c = \{q \in \mathbf{P}^* : \exists p \in G \cap K\, (p \leq q)\}$; this proves $G \restriction c \in \mathbf{L}[G \cap K]$. Suppose that $q \in G \restriction c$, so that $q \in G$ and $|q| \subseteq c$, in other words, $|q| \subseteq |p_1| \cup \cdots \cup |p_n|$ for a finite set of conditions $p_1, \ldots, p_n \in G \cap K$. Note that $p = p_1 \wedge \cdots \wedge p_n \in K$ by Definition 8(1). Thus $p \in G \cap K$, and $|q| \subseteq |p|$. Yet $q \in G$ as well, therefore, $p' = p \wedge q \in G$, and $|p'| = |p|$. It follows that $p' \in K$, by Definition 8(3), so that $p' \in G \cap K$. Finally $p' \leq q$.

Now suppose that $p \in G \cap K$ and $p \leq q \in \mathbf{P}^*$. Then obviously q belongs to $\mathbf{P}[U]$ (since so does p), hence $q \in G$ (since G is generic). Finally $|q| \subseteq |p| \subseteq c$.

Let us show that $G \cap K = (G \restriction c) \cap K$; this proves $G \cap K \in \mathbf{L}[G \restriction c]$. Indeed if $p \in G \cap K$ then by definition $|p| \subseteq c = |G \cap K|$, therefore $p \in G \restriction c$, as required.

(iv) This is clear since we have $G \cap K = G \restriction_{\neq \nu} \cap K$ in the case considered.

(v) The set $\mathbf{P}[U]$ can be identified with the product $\mathbf{P}[U]\restriction_{\neq\nu} \times P[U(\nu)]$. Thus $G(\nu)$ and $S_G(\nu)$ are $P[U(\nu)]$-generic over $\mathbf{L}[\mathbf{P}[U]\restriction_{\neq\nu}]$.

(vi) The genericity easily follows from Definition 8(3). Then use Lemma 3.

(i) First of all, $G = \prod_\nu G(\nu)$ by the product-forcing theorem. Then, each $G(\nu)$ is recovered from the associated $S_G(\nu)$ by means of a simple uniform formula, see the proof of Lemma 3(i). □

3.4. Names for Sets in Product and Regular Extensions

For any set X we let $\mathbf{N}_X$ be the set of all $\mathbf{P}^*$-names for subsets of X. Thus $\mathbf{N}_X$ consists of all sets $\tau \subseteq \mathbf{P}^* \times X$. Let $\mathbf{SN}_X$ (small names) consist of all at most countable names $\tau \in \mathbf{N}_X$.

We define $\mathtt{dom}\,\tau = \{p : \exists\, x\, (\langle p, x\rangle \in \tau)\}$, $|\tau| = \bigcup\{|p| : p \in \mathtt{dom}\,\tau\}$ for any name τ.

Say that a name τ is below a given $p \in \mathbf{P}^*$ if all $p' \in \mathtt{dom}\,\tau$ satisfy $p' \leq p$.

For any set $K \subseteq \mathbf{P}^*$, we let $\mathbf{N}_X(K)$ be the set of all names $\tau \in \mathbf{N}_X$ such that $\mathtt{dom}\,\tau \subseteq K$, and accordingly $\mathbf{SN}_X(K) = \mathbf{N}_X(K) \cap \mathbf{SN}_X$ (small names). In particular, we'll consider such sets of names as $\mathbf{SN}_X(\mathbf{P}[U])$ and $\mathbf{SN}_X(\mathbf{P}[U]\restriction c)$. Names in $\mathbf{N}_X(K)$ for different sets X will be called K-names. Accordingly, names in $\mathbf{SN}_X(K)$ for different sets X will be called small K-names.

Definition 10 (valuations). *If $\tau \in \mathbf{N}_X$ and $G \subseteq \mathbf{P}^*$ then define $\tau[G] = \{x : \exists\, p \in G\, (\langle p, x\rangle \in \tau)\}$, the G-valuation of τ; $\tau[G]$ is a subset of X.*

Example 3 (some names). *Let $\odot \in \mathbf{P}^*$ be the empty condition, that is, $|\odot| = \varnothing$. This is the weakest condition in any $\mathbf{P}[U]$. If X is a set in the ground universe then $\breve{X} = \{\langle \odot, x\rangle : x \in X\}$ is a K-name for any regular forcing $K \subseteq \mathbf{P}^*$, and $\breve{X}[G] = X$ for any set G containing $\odot$.*

We will typically use breve-names like $\breve{X}$ for sets in the ground universe, and dot-names (like $\dot{x}$) for sets in generic extensions.

Suppose that $K \subseteq \mathbf{P}^*$. Let $\underline{G} = \{\langle p, p\rangle : p \in K\}$. (In principle, $\underline{G}$ depends on K but this dependence will usually be suppressed.) Clearly $\underline{G} \in \mathbf{N}_K(K)$ (but $\underline{G} \notin \mathbf{SN}_K(K)$ unless K is countable), and in addition $\underline{G}[G] = G$ for any $\varnothing \neq G \subseteq K$. Thus $\underline{G}$ is a name for the generic set $G \subseteq K$.

Similarly, $\underline{G}\restriction c = \{\langle p, p\rangle : p \in K\restriction c\}$ ($c \subseteq \mathcal{I}$) is a name for $G\restriction c$ (see Definition 9).

3.5. Names for Functions

For any sets X, Y let $\mathbf{N}^X_Y$ be the set of all $\mathbf{P}^*$-names for functions $X \to Y$; it consists of all $\tau \subseteq \mathbf{P}^* \times (X \times Y)$ such that the sets $\tau''\langle x, y\rangle = \{p : \langle p, \langle x, y\rangle\rangle \in \tau\}$ satisfy the following requirement:

$$\text{if } y \neq y',\ p \in \tau''\langle x, y\rangle,\ p' \in \tau''\langle x, y'\rangle,\ \text{then } p, p' \text{ are incompatible.}$$

Let $\mathtt{dom}\,\tau = \bigcup_{x,y} \tau''\langle x, y\rangle$ and $|\tau| = \bigcup\{|p| : p \in \mathtt{dom}\,\tau\}$.

As above, $\mathbf{SN}^X_Y$ consists of all at most countable names $\tau \in \mathbf{N}^X_Y$.

For any set $K \subseteq \mathbf{P}^*$, we let $\mathbf{N}^X_Y(K)$ be the set of all names $\tau \in \mathbf{N}^X_Y$ such that $\mathtt{dom}\,\tau \subseteq K$, and accordingly $\mathbf{SN}^X_Y(K) = \mathbf{N}^X_Y(K) \cap \mathbf{SN}^X_Y$ (small names).

A name $\tau \in \mathbf{N}^X_Y(K)$ is K-full iff the union $\tau''x = \bigcup_y \tau''\langle x, y\rangle$ is pre-dense in K for any $x \in X$. A name $\tau \in \mathbf{N}^X_Y(K)$ is K-full below some $p_0 \in K$, iff all sets $\tau''x$ are pre-dense in K below p_0, that is, any condition $q \in K$, $q \leq p_0$, is compatible with some $r \in \tau_x$ (and this holds for all $x \in X$).

Note that $\mathbf{N}^X_Y(K) \subseteq \mathbf{N}_{X\times Y}(K)$, and accordingly $\mathbf{SN}^X_Y(K) \subseteq \mathbf{SN}_{X\times Y}(K)$. Thus all names in $\mathbf{N}^X_Y(K)$ and in $\mathbf{SN}^X_Y(K)$ are still K-names in the sense above.

Corollary 1 (of Lemma 8, in $\mathbf{L}$). *If U is a system, $K \subseteq \mathbf{P}[U]$ is a regular subforcing, X, Y any sets, and τ is a name in $\mathbf{N}^X_Y(K)$, then τ is K-full (resp., K-full below $p \in K$) iff τ is $\mathbf{P}[U]$-full (resp., $\mathbf{P}[U]$-full below p).*

Suppose that $\tau \in \mathbf{N}^X_Y$. Call a set $G \subseteq \mathbf{P}^*$ minimally τ-generic iff it is compatible in itself (if $p, q \in G$ then there is $r \in G$ with $r \leq p$, $r \leq q$), and intersects each set of the form $\tau''x$, $x \in X$. In this case put

$$\tau[G] = \{\langle x, y\rangle \in X \times Y : (\tau''\langle x, y\rangle) \cap G \neq \varnothing\},$$

so that $\tau[G] \in Y^X$ and $\tau[G](x) = y \iff \tau''\langle x, y\rangle \cap G \neq \varnothing$. If φ is a formula in which some names $\tau \in \mathbf{N}_Y^X$ occur (for various sets X, Y), and a set $G \subseteq \mathbf{P}^*$ is minimally τ-generic for any name τ in φ, then accordingly $\varphi[G]$ is the result of substitution of $\tau[G]$ for each name τ in φ.

Claim 1 (obvious). *Suppose that, in* $\mathbf{L}$, X, Y *are any sets,* $p \in K \subseteq \mathbf{P}^*$ *and* $\tau \in \mathbf{N}_Y^X(K)$ *is K-full (resp., K-full below* p*). Then, any set* $G \subseteq K$*, K-generic over* $\mathbf{L}$ *(resp., K-generic over* $\mathbf{L}$ *and containing* p*), is minimally* τ*-generic.*

Definition 11 (equivalent names). *Names* $\tau, \mu \in \mathbf{SN}_\omega^\omega(\mathbf{P}^*)$ *are called equivalent iff conditions* q, r *are incompatible whenever* $q \in \tau''\langle m, j\rangle$ *and* $r \in \mu''\langle m, k\rangle$ *for some* m *and* $j \neq k$*. (Recall that* $\tau''\langle m, k\rangle = \{p : \langle p, \langle m, k\rangle\rangle \in \tau\}$*.) Similarly, names* τ, μ *are equivalent below some* $p \in \mathbf{P}^*$ *iff the triple of conditions* p, q, r *is incompatible (that is,* $p \wedge q \wedge r$ *is not* $\leq$ *than at least one of* p, q, r*) whenever* $q \in \tau''\langle m, j\rangle$ *and* $r \in \mu''\langle m, k\rangle$ *for some* m *and* $j \neq k$*.*

Claim 2 (obvious). *Suppose that, in* $\mathbf{L}$, $p \in K \subseteq \mathbf{P}^*$*, and names* $\mu, \tau \in \mathbf{SN}_\omega^\omega(K)$ *are equivalent (resp., equivalent below* p*). Then, for any* $G \subseteq K$ *both minimally* μ*-generic and minimally* τ*-generic (resp., and containing* p*),* $\mu[G] = \tau[G]$*.*

Lemma 10. *Suppose that, in* $\mathbf{L}$, U *is a system,* $K \subseteq \mathbf{P}[U]$ *is a regular subforcing,* $p_0 \in K$, $A \subseteq P = \{p \in K : p \leq p_0\}$ *is a countable antichain, and, for any* $p \in A$, $\tau_p \in \mathbf{SN}_\omega^\omega(K)$ *is a name K-full below* p_0*. Then there is a K-full name* $\tau \in \mathbf{SN}_\omega^\omega(K)$*, equivalent to* τ_p *below* p *for any* $p \in A$*.*

Proof. Let B be a maximal (countable) antichain in the set of all conditions $q \in K$ incompatible with p_0. Then $A \cup B$ is a countable maximal antichain in K. We let τ consist of: 1) all triples $\langle r \wedge q, \langle k, m\rangle\rangle$, such that $q \in A$ and $\langle r, \langle k, m\rangle\rangle \in \tau_q$, and 2) all triples $\langle q, \langle k, 0\rangle\rangle$, such that $q \in B$ and $m \in \omega$. □

3.6. Names and Sets in Generic Extensions

For any forcing P, let $\Vdash_P$ denote the P-forcing relation over $\mathbf{L}$ as the ground model.

Theorem 5. *Suppose that* U *is a system and* $K \subseteq \mathbf{P}[U]$ *a regular subforcing in* $\mathbf{L}$*. Let* $G \subseteq K$ *be a set K-generic over* $\mathbf{L}$*. Then:*

(i) *if* $p \in K$ *and* φ *is a closed formula with K-names as parameters, then*

$$p \Vdash_K \varphi \quad \textit{iff} \quad p \Vdash_{\mathbf{P}[U]} \text{``}\mathbf{L}[\underline{G} \cap \check{K}] \models \varphi[\underline{G}]\text{''};$$

(ii) *if* X, Y *are countable sets in* $\mathbf{L}$*, and* $f \in \mathbf{L}[G]$, $f : X \to Y$*, then there is a K-full name* $\tau \in \mathbf{SN}_Y^X(K)$ *in* $\mathbf{L}$ *such that* $f = \tau[G]$*.*

(iii) *if* $X \in \mathbf{L}$, $y \in \mathbf{L}[G]$, $y \subseteq X$*, then there is a name* $\tau \in \mathbf{N}_X(K)$ *in* $\mathbf{L}$ *such that* $y = \tau[G]$*, and in addition if* X *is countable in* $\mathbf{L}$ *then* $\tau \in \mathbf{SN}_X(K)$*.*

(iv) *if* X, Y *are countable sets in* $\mathbf{L}$, $p \in K$, $\varphi(f)$ *is a formula with K-names as parameters, and* $p \Vdash_K \exists f \in Y^X\, \varphi(f)$*, then there is a K-full name* $\tau \in \mathbf{SN}_Y^X(K)$ *in* $\mathbf{L}$ *such that* $p \Vdash_K \varphi(\tau)$*.*

Proof. (i) Suppose $p \Vdash_K \varphi$. To prove $p \Vdash_{\mathbf{P}[U]}$ "$\mathbf{L}[\underline{G} \cap \check{K}] \models \varphi[\underline{G}]$", consider a set $G \subseteq \mathbf{P}[U]$, $\mathbf{P}[U]$-generic over $\mathbf{L}$. Then $G \cap K$ is K-generic over $\mathbf{L}$ by Lemma 8, hence $\varphi[G]$ is true in $\mathbf{L}[G \cap K]$, as required. Conversely assume $\neg\, p \Vdash_K \varphi$. There is a condition $q \in K$, $q \leq p$, $q \Vdash_k \neg\, \varphi$. Then $q \Vdash_{\mathbf{P}[U]}$ "$\mathbf{L}[\underline{G} \cap \check{K}] \models \neg\, \varphi[\underline{G}]$" by the above, thus $p \Vdash_{\mathbf{P}[U]}$ "$\mathbf{L}[\underline{G} \cap K] \models \varphi[\underline{G}]$" fails.

(ii) It follows from general forcing theory that there is a K-full name $\sigma \in \mathbf{N}_Y^X(K)$, not necessarily countable, such that $f = \sigma[G]$. Then all sets $Q_x = \sigma''x$, $x \in X$, are pre-dense in K. Put $\tau = \{\langle p, \langle x, y\rangle\rangle \in \sigma : x \in X \wedge y \in Y \wedge p \in A_x\}$, where $A_x \subseteq Q_x$ is a maximal (countable, by Theorem 4) antichain for any x.

(iv) We conclude from (ii) that the set Q of all conditions $q \in K, q \leq p$, such that $q \Vdash_K \varphi(\tau)$ for some name $\tau = \tau_q \in \mathbf{SN}_Y^X(K)$, is dense in K below p. Let $A \subseteq Q$ be a maximal antichain in Q; A is countable and pre-dense in K below p. Apply Lemma 10 to get a name τ as required. $\square$

Example 4. *Consider the regular forcing $K = \mathbf{P}[U \restriction c]$, where U is a system and $c \subseteq \mathcal{I}$ in $\mathbf{L}$. If $G \subseteq \mathbf{P}[U]$ is $\mathbf{P}[U]$-generic over $\mathbf{L}$ then the restricted set $G \restriction c = G \cap (\mathbf{P}[U \restriction c])$ is $\mathbf{P}[U \restriction c]$-generic over $\mathbf{L}$, by Lemma 9 (with $K = \mathbf{P}[U \restriction c]$). Furthermore, it follows from Lemma 9 and Theorem 5 that if $\nu \in \mathcal{I}$ then $S_G(\nu) \in \mathbf{L}[G \restriction c]$ iff $\nu \in c$, so that $\mathbf{L}[G \restriction c] = \mathbf{L}[\{S_G(\nu)\}_{\nu \in c}]$.*

Example 5. *Consider the regular forcing K defined in Example 2 in Section 3.2. Suppose that U is a system in $\mathbf{L}$ and $G \subseteq \mathbf{P}[U]$ is a set $\mathbf{P}[U]$-generic over $\mathbf{L}$. Then $K[U] = K \cap \mathbf{P}[U]$ is a regular subforcing of $\mathbf{P}[U]$ by Lemma 7. We conclude that $G' = G \cap K$ is a set $K[U]$-generic over $\mathbf{L}$, by Lemma 9.*

It follows by the definition of K that the set $|G'| = \bigcup_{p \in G'} |p|$ satisfies $|G'| \subseteq \omega$, contains 0, and if $\nu \geq 1$ then $\nu \in |G'|$ iff $\mathbb{s}_\nu \in S_G(0)$. Therefore, by Lemma 9 and Theorem 5, the sets $G(0)$ and $S_G(0)$ belong to $\mathbf{L}[G']$, and if $1 \leq \nu < \omega$ then $S_G(\nu) \in \mathbf{L}[G']$ iff $\mathbb{s}_\nu \in S_G(0)$. Thus

$$\mathbf{L}[G'] = \mathbf{L}[S_G(0), \{S_G(\nu)\}_{\mathbb{s}_\nu \in S_G(0)}] = \mathbf{L}[G'] = \mathbf{L}[G \restriction c],$$

where $c = |G'| = \{0\} \cup \{\nu < \omega : \mathbb{s}_\nu \in S_G(0)\} \notin \mathbf{L}$.

3.7. Transformations Related to Product Forcing

There are three important families of transformations of the whole system of objects related to product forcing. Two of them are considered in this Subsection.

Family 1: permutations. If $c, c' \subseteq \mathcal{I}$ are sets of equal cardinality then let $\mathrm{BIJ}_{c'}^{c}$ be the set of all bijections $\pi : c \xrightarrow{\text{onto}} c'$. Let $|\pi| = \{\nu \in c : \pi(\nu) \neq \nu\} \cup \{\nu \in c' : \pi^{-1}(\nu) \neq \nu\}$, so that π is essentially a bijection $c \cap |\pi| \xrightarrow{\text{onto}} c' \cap |\pi|$, equal to the identity on $c \smallsetminus |\pi| = c' \smallsetminus |\pi|$. Define the action of any $\pi \in \mathrm{BIJ}_{c'}^{c}$ onto:

- sets $e \subseteq c$: $\pi \cdot e := \{\pi(\nu) : \nu \in e\}$ — then $\pi \cdot e \subseteq c'$ and $\pi \cdot c = c'$;
- systems U with $|U| \subseteq c$: $(\pi \cdot U)(\pi(\nu)) := U(\nu)$ for all $\nu \in |U|$ — then $|\pi \cdot U| = \pi \cdot |U| \subseteq c'$;
- conditions $p \in \mathbf{P}^*$ with $|p| \subseteq c$: $(\pi \cdot p)(\pi(\nu)) := p(\nu)$ for all $\nu \in |p|$;
- sets $G \subseteq \mathbf{P}^* \restriction c$: $\pi \cdot G := \{\pi \cdot p : p \in G\}$ — then $\pi \cdot G \subseteq \mathbf{P}^* \restriction c'$,
  in particular, $\pi \cdot K = \{\pi \cdot p : p \in K\} \subseteq \mathbf{P}^* \restriction c'$ for any regular subforcing $K \subseteq \mathbf{P}^* \restriction c$;
- names $\tau \in \mathbf{N}_Y^X(\mathbf{P}^* \restriction c)$: $\pi \cdot \tau := \{\langle \pi \cdot p, \langle \ell, k \rangle \rangle : \langle p, \langle \ell, k \rangle \rangle \in \tau\}$ — then $\pi \cdot \tau \in \mathbf{N}_Y^X(\mathbf{P}^* \restriction c')$;

Lemma 11. *If $c, c' \subseteq \mathcal{I}$ are sets of equal cardinality and $\pi \in \mathrm{BIJ}_{c'}^{c}$ then $p \longmapsto \pi \cdot p$ is an order preserving bijection of $\mathbf{P}^* \restriction c$ onto $\mathbf{P}^* \restriction c'$, and if U is a system and $|U| \subseteq c$ then $|\pi \cdot U| \subseteq c'$, and we have $p \in \mathbf{P}[U] \iff \pi \cdot p \in \mathbf{P}[\pi \cdot U]$.*

Family 2: Lipschitz transformations. Let $\mathbf{Lip}^{\mathcal{I}}$ be the $\mathcal{I}$-product of the group $\mathbf{Lip}$ (see Section 2.3), with countable support; this will be our second family of transformations. Thus a typical element $\boldsymbol{\alpha} \in \mathbf{Lip}^{\mathcal{I}}$ is $\boldsymbol{\alpha} = \{\alpha_\nu\}_{\nu \in |\boldsymbol{\alpha}|}$, where $|\boldsymbol{\alpha}| = \mathtt{dom}\,\boldsymbol{\alpha} \subseteq \mathcal{I}$ is at most countable, and $\alpha_\nu \in \mathbf{Lip}$, $\forall \nu$. We will routinely identify each $\boldsymbol{\alpha} \in \mathbf{Lip}^{\mathcal{I}}$ with its extension on $\mathcal{I}$ defined so that α_ν is the identity map (on $\mathbf{Seq}$) for all $\nu \in \mathcal{I} \smallsetminus |\boldsymbol{\alpha}|$. Keeping this identification in mind, define the action of any $\boldsymbol{\alpha} \in \mathbf{Lip}^{\mathcal{I}}$ on:

- systems U: $|\boldsymbol{\alpha} \cdot U| := |U|$ and $(\boldsymbol{\alpha} \cdot U)(\nu) := \alpha_\nu \cdot U(\nu)$;
- conditions $p \in \mathbf{P}^*$, by $|\boldsymbol{\alpha} \cdot p| = |p|$ and $(\boldsymbol{\alpha} \cdot p)(\nu) = \alpha_\nu \cdot p(\nu)$;
- sets $G \subseteq \mathbf{P}^*$: $\boldsymbol{\alpha} \cdot G := \{\boldsymbol{\alpha} \cdot p : p \in G\}$,
  in particular, $\boldsymbol{\alpha} \cdot K = \{\boldsymbol{\alpha} \cdot p : p \in K\}$ for any regular subforcing $K \subseteq \mathbf{P}^*$;
- names $\tau \in \mathbf{N}_Y^X$: $\boldsymbol{\alpha} \cdot \tau := \{\langle \boldsymbol{\alpha} \cdot p, \langle n, k \rangle \rangle : \langle p, \langle n, k \rangle \rangle \in \tau\}$;

In the first two lines, we refer to the action of $\alpha_v \in \mathbf{Lip}$ on sets $u \subseteq \mathbf{Fun}$ and on forcing conditions, as defined in Section 2.3.

Lemma 12. *If $\alpha \in \mathbf{Lip}^{\mathcal{I}}$ then $p \longmapsto \pi \cdot p$ is an order preserving bijection of $\mathbf{P}^*$ onto $\mathbf{P}^*$, and if U is a system then we have $p \in \mathbf{P}[U] \iff \alpha \cdot p \in \mathbf{P}[\alpha \cdot U]$.*

Corollary 2 (of Lemma 5). *Suppose that U, V are countable systems, $|U| = |V|$, and $p \in \mathbf{P}[U]$, $q \in \mathbf{P}[V]$. Then there is a transformation $\alpha \in \mathbf{Lip}^{\mathcal{I}}$ such that*

(i) $|\alpha| = |U| = |V|$, $\alpha \cdot U = V$, *and*
(ii) *there are conditions $p' \in \mathbf{P}[U]$, $p' \leq p$ and $q' \in \mathbf{P}[V]$, $q' \leq q$ such that $\alpha \cdot p' = q'$—in particular, conditions $\alpha \cdot p$ and q are compatible in $\mathbf{P}[V]$.*

Proof. Apply Lemma 5 componentwise for every $v \in |U| = |U'|$. □

3.8. Substitutions and Homogeneous Extensions

Assume that conditions $p, q \in \mathbf{P}^*$ satisfy (2) of Section 2.4 for all v, that is:

$$|p| = |q|, \text{ and } S_p(v) \cup S_q(v) \subseteq F_p^{\vee}(v) = F_q^{\vee}(v) \text{ for all } v \in |p| = |q|. \tag{3}$$

Definition 12. *If (3) holds and $p' \in \mathbf{P}^*$, $p' \leq p$, then define $q' = H_q^p(p')$ so that $|q'| = |p'|$, $q'(v) = p'(v)$ whenever $v \in |p'| \smallsetminus |p|$, but $q'(v) = H_{q(v)}^{p(v)}(p'(v))$ for all $v \in |p|$, where $H_{q(v)}^{p(v)}$ is defined as in Section 2.4. This is Family 3 of transformations, called substitutions.*

Theorem 6. *If U is a system, and conditions $p, q \in \mathbf{P}[U]$ satisfy (3) above, then*

$$H_q^p \,:\, P = \{p' \in \mathbf{P}[U] : p' \leq p\} \overset{\text{onto}}{\longrightarrow} Q = \{q' \in \mathbf{P}[U] : q' \leq q\}$$

is an order isomorphism.

Proof. Apply Lemma 6 componentwise. □

Suppose that U, $p, q \in \mathbf{P}[U]$, H_q^p are as in Theorem 6. Extend the action of H_q^p onto names and formulas. Recall that a name $\tau \in \mathbf{N}_Y^X$ is below p iff $p' \leq p$ holds for any triple $\langle p', \langle n,k\rangle\rangle \in \tau$.

- If X, Y are any sets and $\tau \in \mathbf{N}_Y^X$ is a name below p then put $H_q^p(\tau) = \{\langle H_q^p(p'), \langle n,k\rangle\rangle : \langle p', \langle n,k\rangle\rangle \in \tau\}$, so $H_q^p(\tau) \in \mathbf{N}_Y^X$ is a name below q.
- If φ is a formula with names below p as parameters then $H_q^p(\varphi)$ denotes the result of substitution of $H_q^p(\tau)$ for any name τ in φ.

Forcing notions of the form $\mathbf{P}[U]$ are quite homogeneous by Theorem 6. The next result is a usual product forcing application of such a homogeneity.

Theorem 7. *Suppose that, in $\mathbf{L}$, U is a system, $d \subseteq c \subseteq \mathcal{I}$, K is a regular subforcing of $\mathbf{P}[U \restriction d]$, and $Q = \{p \in \mathbf{P}[U \restriction c] : p \restriction d \in K\} = K \times \mathbf{P}[U \restriction (c \smallsetminus d)]$. Let φ be a formula which contains as parameters: $(*)$ K-names, and $(\dagger)$ names of the form $\underline{G} \restriction e$, where $e \in \mathbf{L}$, $e \subseteq c$, and $\underline{G} \restriction e$ enters φ only via $\mathbf{L}[\underline{G} \restriction e]$. Then:*

(i) *if $p \in Q$ and $p \Vdash_Q \varphi$ then $p \restriction d \Vdash_Q \varphi$;*
(ii) *in particular, for $d = \varnothing$ (and $Q = \mathbf{P}[U \restriction c]$), Q decides any formula Φ which contains only names for sets in $\mathbf{L}$ and names $\underline{G} \restriction e$ via $\mathbf{L}[\underline{G} \restriction e]$ of the form $(\dagger)$ with $e \subseteq c$, as parameters;*
(iii) *if $p \in Q$ and $p \Vdash_Q \exists x \in \mathbf{L}[\underline{G} \restriction c]\, \varphi(x)$ then $p \restriction d \Vdash_Q \exists x \in \mathbf{L}[\underline{G} \restriction c]\, \varphi(x)$.*

Proof. (i) Otherwise there are conditions $p, q \in Q$ with $p \restriction d = q \restriction d$, $p \Vdash_Q \varphi$, but $q \Vdash_Q \neg \varphi$. We can w.l.o.g. assume that p, q satisfy (3) above (otherwise extend p, q appropriately). Define P, Q, H_q^p as in Definition 12 and Theorem 6.

Let $G \subseteq Q$ be a generic set containing p. Assuming w.l.o.g. that $G \subseteq P$, the set $H = \{H^p_q(p') : p' \in G\} \subseteq Q$ will be generic as well by Theorem 6, and $q \in H$. Therefore $\varphi[G]$ is true in $\mathbf{L}[G]$ but $\varphi[H]$ is false in $\mathbf{L}[H]$. Yet $\mathbf{L}[G] = \mathbf{L}[H]$ since $H^p_q \in \mathbf{L}$. Moreover $\varphi[G]$ coincides with $\varphi[H]$ since 1) H^p_q is the identity on d (indeed $p \restriction d = q \restriction d$), and 2) if $e \in \mathbf{L}$, $e \subseteq c$, then $\mathbf{L}[G \restriction e] = \mathbf{L}[H \restriction e]$ since $G \restriction e$, $H \restriction e$ can be obtained from each other via maps coded in $\mathbf{L}$. This is a contradiction.

(iii) This is a particular case. □

Corollary 3. *Under the assumptions of Theorem 7, suppose that X, Y are arbitrary sets in $\mathbf{L}$, $p \in Q$, and $p \Vdash_Q \exists f \in \mathbf{L}[\underline{G} \cap K] (f \in Y^X \wedge \varphi(f))$. Then there is a K-full name $\tau \in \mathbf{SN}^X_Y(K)$ such that $p \restriction d \Vdash_Q \varphi(\tau)$.*

Proof. We can assume that $|p| \subseteq d$ by Theorem 7(iii), thus $p = p \restriction d \in K$. It follows from Theorems 5(ii) and 7(i) that there exist: a (countable) antichain $A \subseteq K$ maximal below p, and, for any $q \in A$, a K-full name $\tau_q \in \mathbf{SN}^X_Y(K)$ such that $q \Vdash_Q \varphi(\tau_q)$. Now compose a K-full name $\tau \in \mathbf{SN}^X_Y(K)$, such that every $q \in A$ forces $\tau = \tau_q$, as in the proof of Theorem 5(iv). □

4. Basic Forcing Notion and Basic Generic Extension

The proofs of Theorems 1–3, that follow in Sections 7–9, will have something in common. Namely the generic extensions we employ to get the results required will be parts of a basic extension, introduced and studied in this section. To define the extension, we'll define (in $\mathbf{L}$ as the ground universe) an increasing sequence $\{\langle M_\xi, U_\xi\rangle\}_{\xi<\omega_1}$ of pairs of certain type—a Jensen–Solovay sequence, since this construction goes back to [9]—and make use of a forcing notion of the form $\mathbf{P}[U]$, where $U = \bigvee_{\xi<\omega_1} U_\xi$. It turns out that if such a sequence is n-complete, in sense that it meets all sets of n-complexity within the whole tree of possible constructions, then the truth of analytic formulas up to level n in corresponding generic extensions has a remarkable connection with the forcing approximations studied in Section 5. This will allow us to convert the homogeneity of the construction of Jensen–Solovay sequences into a uniformity of the corresponding generic extensions, expressed by Theorem 13.

Recall that $\mathbf{V} = \mathbf{L}$ assumed in the ground universe by Assumption 1.

4.1. Jensen–Solovay Sequences

If $U \preccurlyeq V$ are systems then by definition $\mathbf{P}[U] \subseteq \mathbf{P}[V]$ holds. However this is not necessarily a suitably good notion. For instance a dense set $X \subseteq \mathbf{P}[U]$ may not be pre-dense in $\mathbf{P}[V]$, thus if $G \subseteq \mathbf{P}[V]$ is a generic set then the "projection" $G \cap \mathbf{P}[U]$ is not necessarily $\mathbf{P}[U]$-generic. Yet there is a special type of extension of systems, introduced by Jensen and Solovay [9], which preserves the density. This method is based on the requirement that the functions in **Fun** that occur in V but not in U must be generic over a certain model that contains U.

Recall that $\mathbf{ZFC}^-$ is **ZFC** minus the Power Set axiom, see Section 5.1 below. Let $\mathbf{ZFC}^-_1$ be $\mathbf{ZFC}^-$ plus the axioms $\mathbf{V} = \mathbf{L}$ and "every set is at most countable".

Definition 13. *Let U, U' be a pair of systems. Suppose that M is any transitive model of $\mathbf{ZFC}^-$. Define $U \preccurlyeq_M U'$ iff $U \preccurlyeq U'$ and we have:*

(a) the set $\Delta(U, U') = \bigcup_{\nu\in|U|}(U'(\nu) \smallsetminus U(\nu))$ (note the union over $|U|$ rather than $|U'|$!) *is multiply Cohen generic over M, in the sense that every string $\langle f_1, \dots f_m\rangle$ of pairwise different functions $f_\ell \in \Delta(U, U')$ is Cohen generic over M, and*

(b) if $\nu \in |U|$ and $U'(\nu) \smallsetminus U(\nu) \neq \varnothing$ then $U'(\nu) \smallsetminus U(\nu)$ is dense in $\mathbf{Fun} = \omega^\omega$.

*Let **JS**, Jensen–Solovay pairs, be the set of all pairs $\langle M, U\rangle$ of a transitive model $M \models \mathbf{ZFC}^-$ and a disjoint ($\nu \neq \nu' \implies U(\nu) \cap U(\nu') = \varnothing$) system $U \in M$. Let **sJS**, small pairs, consist of all $\langle M, U\rangle \in \mathbf{JS}$ such that $M \models \mathbf{ZFC}^-_1$ and M (then U as well) is countable. Define the extension relations:*

$\langle M, U\rangle \preccurlyeq \langle M', U'\rangle$ iff $M \subseteq M'$ and $U \preccurlyeq_M U'$;

$$\langle M, U\rangle \prec \langle M', U'\rangle \quad \textit{iff} \;\; \langle M, U\rangle \preccurlyeq \langle M', U'\rangle \textit{ and } \forall\, \nu \in |U|\, (U(\nu) \subsetneqq U'(\nu)).$$

It would be a vital simplification to get rid of M as an explicit element of the construction, e.g., by setting $U \preccurlyeq^* U'$ iff $U \preccurlyeq U'$ and there is a CTM M containing U and such that $U \preccurlyeq_M U'$.

Lemma 13. *Suppose that pairs $\langle M, U\rangle \preccurlyeq \langle M', U'\rangle \preccurlyeq \langle M'', U''\rangle$ belong to* $\mathbf{JS}$. *Then $\langle M, U\rangle \preccurlyeq \langle M'', U''\rangle$. Thus $\preccurlyeq$ is a partial order on* $\mathbf{JS}$.

Proof. Prove that the set $F = \bigcup_{\nu\in|U|}(U''(\nu) \smallsetminus U(\nu))$ is multiply Cohen generic over M. Consider a simple case when $f \in U'(\nu) \smallsetminus U(\nu)$ and $g \in U''(\mu) \smallsetminus U'(\mu)$, where $\nu, \mu \in |U|$, and prove that $\langle f, g\rangle$ is Cohen generic over M. (The general case does not differ much.) By definition, f is Cohen generic over M and g is Cohen generic over M'. Therefore, g is Cohen generic over $M[f]$, which satisfies $M[f] \subseteq M'$ since $f \in M'$. It remains to apply the product forcing theorem. □

Remark 1. We routinely have $\langle M, U\rangle \preccurlyeq \langle M', U\rangle$ (the same U) provided $M \subseteq M'$. On the other hand, $\langle M, U\rangle \preccurlyeq \langle M, U'\rangle$ (with the same M) is possible only in the case when $\Delta(U, U') = \varnothing$, that is, $U(\nu) = U'(\nu)$ for all $\nu \in |U|$. In particular, if $\langle M, U\rangle \in \mathbf{JS}$, $c \in M$, $c \subseteq |U|$, then $\langle M, U \restriction c\rangle \preccurlyeq \langle M, U\rangle$.

Lemma 14 (extension). *If $\langle M, U\rangle \in \mathbf{sJS}$ and $z \subseteq \mathcal{I}$ is countable, then there is a pair $\langle M', U'\rangle \in \mathbf{sJS}$ such that $\langle M, U\rangle \prec \langle M', U'\rangle$ and $z \subseteq |U'|$.*

Proof. Let $d = |U| \cup z$, and let $\vec{f} = \{f_{\nu k}\}_{\nu\in d, k<\omega} \in (\mathbf{Fun})^{d\times\omega}$ be Cohen generic over M. Now define $U'(\nu) = U(\nu) \cup \{f_{\nu k} : k \in \omega\}$ for each $\nu \in d$, and let $M' \models \mathbf{ZFC}^-_1$ be any CTM satisfying $M \subseteq M'$ and containing U'. □

Definition 14. *A Jensen–Solovay sequence of length $\lambda \le \omega_1$ is any strictly $\prec$-increasing λ-sequence $\{\langle M_\xi, U_\xi\rangle\}_{\xi<\lambda}$ of pairs $\langle M_\xi, U_\xi\rangle \in \mathbf{sJS}$, which satisfies $U_\eta = \bigvee_{\xi<\eta} U_\xi$ on limit steps. Let $\overrightarrow{\mathbf{JS}}_\lambda$ be the set of all such sequences.*

Lemma 15. *Suppose that $\lambda \le \omega_1$ is a limit ordinal, and $\{\langle M_\xi, U_\xi\rangle\}_{\xi<\lambda}$ belongs to $\overrightarrow{\mathbf{JS}}_\lambda$. Put $U = \bigvee_{\xi<\lambda} U_\xi$, that is, $U(\nu) = \bigcup_{\xi<\lambda} U_\xi(\nu)$ for all $\nu \in \mathcal{I}$.*

Then $U_\xi \preccurlyeq_{M_\xi} U$ for every ξ.

If, moreover, $\lambda < \omega_1$ and M is a CTM of $\mathbf{ZFC}^-_1$ containing $\{\langle M_\xi, U_\xi\rangle\}_{\xi<\lambda}$ then $\langle M, U\rangle \in \mathbf{sJS}$ and $\langle M_\xi, U_\xi\rangle \prec \langle M, U\rangle$ for every ξ.

Proof. The same idea as in the proof of Lemma 13. □

4.2. Stability of Dense Sets

Assume that $\langle M, U\rangle \in \mathbf{sJS}$ and D is a pre-dense subset of $\mathbf{P}[U]$ (say, a maximal antichain). If U' is another system satisfying $U \preccurlyeq U'$, then it may well happen that D is not maximal in $\mathbf{P}[U']$. The role of the multiple genericity requirement (a) in Definition 13, first discovered in [9], is to somehow seal the property of pre-density of sets already in M for any further extensions. This is the content of the following key theorem. The product forcing arguments allow us to extend the stability result to pre-dense sets not necessarily in M, as in items (ii), (iii) of the following theorem.

Theorem 8. *Assume that, in* $\mathbf{L}$, *$\langle M, U\rangle \in \mathbf{sJS}$, U' is a disjoint system, and $U \preccurlyeq_M U'$. If D is a pre-dense subset of $\mathbf{P}[U]$ (resp., pre-dense below some $p \in \mathbf{P}[U]$) then D remains pre-dense in $\mathbf{P}[U']$ (resp., pre-dense in $\mathbf{P}[U']$ below p) in each of the following three cases:*

(i) *$D \in M$;*
(ii) *$D \in M[G]$, where $G \subseteq Q$ is Q-generic over* $\mathbf{L}$ *and $Q \in M$ is a PO set;*
(iii) *$D \in M[H]$, where $H \subseteq U'(\nu_0)$ is finite, $\nu_0 \in |U|$ is fixed, and $D \subseteq \mathbf{P}[U \restriction_{\neq \nu_0}]$.*

Proof. We consider only the case of sets D pre-dense in $\mathbf{P}[U]$ itself; the case of pre-density below some $p \in \mathbf{P}[U]$ is treated similarly.

(i) Suppose, towards the contrary, that a condition $p \in \mathbf{P}[U']$ is incompatible with each $q \in D$. As $D \subseteq \mathbf{P}[U]$, we can w.l.o.g. assume that $|p| \subseteq |U|$.

Our plan is to define a condition $p' \in \mathbf{P}[U]$, also incompatible with each $q \in D$, contrary to the pre-density. To maintain such a construction, consider the finite string $\vec{f} = \langle f_1, \dots, f_m \rangle$ of all elements $f \in \mathbf{Fun}$ occurring in $\bigcup_{\nu \in |p|} F_p(\nu)$ but not in U. It follows from $U \preccurlyeq_M U'$ that $\vec{f}$ is Cohen-generic over M. Further analysis shows that p being incompatible with D is implied by the fact that $\vec{f}$ meets a certain M-countable family of Cohen-dense sets. Therefore, we can simulate this in M, getting a string $\vec{g} \in M$ which meets the same Cohen-dense sets, and hence yields a condition $p' \in \mathbf{P}[U]$, also incompatible with each $q \in D$.

This argument was first carried out in [9] in full generality, where we address the reader. However, to present the key idea in sufficient detail in a somewhat simplified subcase, we assume that (1) $|p| = \{\nu\}$ is a singleton; $\nu \in |U|$. Then $p(\nu) = \langle S_p(\nu); F_p(\nu) \rangle \in P[U'(\nu)]$, where $S_p(\nu) \subseteq \mathbf{Seq}$ and $F_p(\nu) \subseteq U'(\nu)$ are finite sets. The (finite) set $X = F_p(\nu) \smallsetminus U(\nu)$ is multiply Cohen generic over M since $U \preccurlyeq_M U'$. To make the argument even more transparent, we suppose that (2) $X = \{f, g\}$, where $f \neq g$ and the pair $\langle f, g \rangle$ is Cohen generic over M. (The general case follows the same idea and can be found in [9]; we leave it to the reader.)

Thus $F_p(\nu) = F \cup \{f, g\}$, where $F = F_p(\nu) \cap U(\nu) \in M$ is by definition a finite set.

The plan is to replace the functions f, g by some functions $f', g' \in U(\nu)$ so that the incompatibility of p with conditions in D will be preserved.

It holds by the choice of p and Lemma 1 that $D = D_1(f, g) \cup D_2$, where

$$\begin{aligned} D_1(f,g) &= \{q \in D : A_q \cap F_p^{\vee}(\nu) \neq \varnothing\}, \text{ where } A_q = S_q(\nu) \smallsetminus S_p(\nu) \subseteq \mathbf{Seq}; \\ D_2 &= \{q \in D : (S_p(\nu) \smallsetminus S_q(\nu)) \cap F_q^{\vee}(\nu) \neq \varnothing\} \in M; \end{aligned}$$

and D_1 depends on f, g via $F_p(\nu)$. (See Section 3.1 on notation.) The equality $D = D_1(f, g) \cup D_2 \cup D_3$ can be rewritten as $\Delta \subseteq D_1(f, g)$, where $\Delta = D \smallsetminus D_2 \in M$. Further, $\Delta \subseteq D_1(f, g)$ is equivalent to

$(*)\quad \forall A \in \mathscr{A}\, (A \cap F_p^{\vee}(\nu) \neq \varnothing)$, where $\mathscr{A} = \{A_q : q \in D\} \in M$,

and each $A_q = S_q(\nu) \smallsetminus S_p(\nu) \subseteq \mathbf{Seq}$ is finite. Recall that $F_p(\nu) = F \cup \{f, g\}$, therefore $F_p^{\vee}(\nu) = Z \cup S(f, g)$, where $Z = \{h \restriction m : m \geq 1 \wedge h \in F\} \in M$ and $S(f, g) = \bigcup_{m \geq 1} \{f \restriction m, g \restriction m\}$. Thus $(*)$ is equivalent to

$(\dagger)\quad \forall A' \in \mathscr{A}'\, (A' \cap S(f, g) \neq \varnothing)$, where $\mathscr{A}' = \{A_q \smallsetminus Z : q \in D\} \in M$.

Note that each $A' \in \mathscr{A}'$ is a finite subset of $\mathbf{Seq}$, so we can reenumerate $\mathscr{A}' = \{A'_k : k < \omega\}$ in M and rewrite $(\dagger)$ as follows:

$(\ddagger)\quad \forall k\, (A'_k \cap S(f, g) \neq \varnothing)$, where each $A'_k \subseteq \mathbf{Seq}$ is finite.

As the pair $\langle f, g \rangle$ is Cohen-generic, there is a number m_0 such that $(\ddagger)$ is forced over M by $\langle \sigma_0, \tau_0 \rangle$, where $\sigma_0 = f \restriction m_0$ and $\tau_0 = g \restriction m_0$. In other words, $A'_k \cap S(f', g') \neq \varnothing$ holds for all k whenever $\langle f', g' \rangle$ is Cohen-generic over M and $\sigma_0 \subset f'$, $\tau_0 \subset g'$. It follows that for any k and strings $\sigma, \tau \in \mathbf{Seq}$ extending resp. σ_0, τ_0 there are strings $\sigma', \tau' \in \mathbf{Seq}$ extending resp. σ, τ, at least one of which extends one of $w \in A'_k$. This allows us to define, in M, a pair of $f', g' \in \mathbf{Fun}$ such that $\sigma_0 \subset f'$, $\tau_0 \subset g'$, and for any k at least one of f', g' extends one of $w \in A'_k$. In other words, we have

$$\forall k\, (A'_k \cap S(f', g') \neq \varnothing) \quad \text{and} \quad \forall A' \in \mathscr{A}'\, (A' \cap S(f', g') \neq \varnothing).$$

It follows that the condition p' defined by $|p'| = \{\nu\}$, $S_{p'}(\nu) = S_p(\nu)$, $F_{p'}(\nu) = F \cup \{f', g'\}$, still satisfies $\forall A \in \mathscr{A}\, (A \cap F_{p'}^{\vee}(\nu) \neq \varnothing)$ (compare with $(*)$), and further $D = D_1(f', g') \cup D_2 \cup D_3$, therefore, p' is incompatible with each $q \in D$. Yet $p' \in M$ since $f', g' \in M$, which contradicts the pre-density of D.

(ii) The above proof works with $M[G]$ instead of M since the set X as in the proof is multiple Cohen generic over $M[G]$ by the product forcing theorem.

(iii) Assuming w.l.o.g. that $H \subseteq U'(\nu_0) \smallsetminus U(\nu_0)$, we conclude that $M[H]$ is a Cohen generic extension of M. Following the above, let $\nu \in |U|$, $\nu \neq \nu_0$. By the definition of $\preccurlyeq$ the set $F = F_p(\nu) \smallsetminus U(\nu)$ is multiply Cohen generic not only over M but also over $M[H]$. This allows to carry out the same argument as above. □

Corollary 4. (i) *Assume that, in* $\mathbf{L}$, $\langle M, U\rangle \in \mathbf{sJS}$, *and* $\langle M, U\rangle \preccurlyeq \langle M', U'\rangle \in \mathbf{JS}$. *Let a set* $G \subseteq \mathbf{P}[U']$ *be* $\mathbf{P}[U']$*-generic over* M'. *Then* $G \cap \mathbf{P}[U]$ *is* $\mathbf{P}[U]$*-generic over* M.
(ii) *If moreover,* $K \in M$, $K \subseteq \mathbf{P}[U]$ *is a regular subforcing, then* $G \cap K$ *is* K*-generic over* M.

Proof. To prove (i), note that if a set $D \in M$, $D \subseteq \mathbf{Q}(U)$, is pre-dense in $\mathbf{Q}(U)$, then it is pre-dense in $\mathbf{Q}(U')$ by Theorem 8, and hence $G \cap D \neq \varnothing$ by the genericity. To prove (ii), apply Lemma 8. □

The next corollary returns us to names, the material of Sections 3.4 and 3.5.

Corollary 5 (of Theorem 8(i)). *In* $\mathbf{L}$, *suppose that* $\langle M, U\rangle \in \mathbf{sJS}$, $\langle M, U\rangle \preccurlyeq \langle M', U'\rangle \in \mathbf{JS}$, *and* X, Y *belong to* M. *Assume that* $\tau \in M \cap \mathbf{SN}^X_Y(\mathbf{P}[U])$ *is a* $\mathbf{P}[U]$*-full name. Then* τ *remains* $\mathbf{P}[U']$*-full. If moreover* $p \in \mathbf{P}[U]$ *and* τ *is* $\mathbf{P}[U]$*-full below* p, *then* τ *remains* $\mathbf{P}[U']$*-full below* p.

4.3. Digression: Definability in HC

The next subsection will contain a transfinite construction of a key forcing notion in $\mathbf{L}$ relativized to HC. Recall that HC is the collection of all *hereditarily countable* sets. In particular, $\mathrm{HC} = \mathbf{L}_{\omega_1}$ in $\mathbf{L}$. In matters of related definability classes, we refer to e.g., Part B, Chapter 5, Section 4 in [20], or Chapter 13 in [21], on the Lévy hierarchy of $\in$-formulas and definability classes Σ^X_n, Π^X_n, Δ^X_n for any set X, and especially on Σ^{HC}_n, Π^{HC}_n, Δ^{HC}_n for $X = \mathrm{HC}$ in Sections 8 and 9 in [22], or elsewhere. In particular,

Σ^{HC}_n = all sets $X \subseteq \mathrm{HC}$, definable in HC by a parameter-free Σ_n formula.
$\mathbf{\Sigma}^{\mathrm{HC}}_n$ = all sets $X \subseteq \mathrm{HC}$ definable in HC by a Σ_n formula with sets in HC as parameters.

Something like $\Sigma^{\mathrm{HC}}_n(x)$, $x \in \mathrm{HC}$, means that only x is admitted as a parameter, while $\Sigma^{\mathrm{HC}}_n(M)$, where $M \subseteq \mathrm{HC}$ is a transitive model, means that all $x \in M$ are admitted as parameters. Collections like Π^{HC}_n, $\Pi^{\mathrm{HC}}_n(x)$, $\Pi^{\mathrm{HC}}_n(M)$ are defined similarly, and $\Delta^{\mathrm{HC}}_n = \Sigma^{\mathrm{HC}}_n \cap \Pi^{\mathrm{HC}}_n$, *etc.*. The boldface classes are defined as follows: $\mathbf{\Sigma}^{\mathrm{HC}}_n = \Sigma^{\mathrm{HC}}_n(\mathrm{HC})$, $\mathbf{\Pi}^{\mathrm{HC}}_n = \Pi^{\mathrm{HC}}_n(\mathrm{HC})$, $\mathbf{\Delta}^{\mathrm{HC}}_n = \Delta^{\mathrm{HC}}_n(\mathrm{HC})$.

Remark 2. It is known that the classes Σ^{HC}_n, Π^{HC}_n, Δ^{HC}_n are equal to resp. Σ^1_{n+1}, Π^1_{n+1}, Δ^1_{n+1} for sets of reals, and the same for parameters and boldface classes. This well-known result was explicitly mentioned in [23] (Lemma on p. 281), a detailed proof see Lemma 25.25 in [21], or Theorem 9.1 in [22].

Remark 3. Recall that $<_{\mathbf{L}}$ is the Gödel wellordering of $\mathbf{L}$, the constructible universe.
It is known that the restriction $<_{\mathbf{L}} \restriction \mathrm{HC}$ is a Δ^{HC}_1 relation, and if $n \geq 1$, $p \in \omega^\omega$ is any parameter, and $R(x, y, z, \dots)$ is a finitary $\Delta^{\mathrm{HC}}_n(p)$ relation on HC then the relations $\exists x <_{\mathbf{L}} y\, R(x, y, z, \dots)$ and $\forall x <_{\mathbf{L}} y\, R(x, y, z, \dots)$ (with arguments $y, z, \dots$) are $\Delta^{\mathrm{HC}}_n(p)$ as well.

4.4. Complete Sequences and the Basic Notion of Forcing

Say that a pair $\langle M, U\rangle \in \mathbf{sJS}$ solves a set $D \subseteq \mathbf{sJS}$ iff either $\langle M, U\rangle \in D$, or there is no pair $\langle M', U'\rangle \in D$ extending $\langle M, U\rangle$. Let $D^{\mathtt{solv}}$ be the set of all pairs $\langle M, U\rangle \in \mathbf{sJS}$ which solve D.

Definition 15. *Let* $n \geq 3$. *A sequence* $\{\langle M_\xi, U_\xi\rangle\}_{\xi<\omega_1} \in \overrightarrow{\mathbf{JS}}_{\omega_1}$ *is* n*- complete iff it intersects every set of the form* $D^{\mathtt{solv}}$, *where* $D \subseteq \mathbf{sJS}$ *is* $\mathbf{\Sigma}^{\mathrm{HC}}_{n-2}$. *(See Section 4.3 on definability classes in* HC.*)*

Let us prove the existence of complete sequences.

Theorem 9 (in **L**). *Let $n \geq 2$. There is a sequence $\{\langle M_\xi, U_\xi\rangle\}_{\xi<\omega_1} \in \overrightarrow{\mathbf{JS}}_{\omega_1}$ of class $\Delta^{\mathrm{HC}}_{n-1}$, n-complete in case $n \geq 3$, and such that $\xi \in |U_{\xi+1}|$ for all ξ — hence the limit system $U = \bigvee_{\xi<\omega_1} U_\xi$ satisfies $|U| = \mathcal{I}$.*

Proof. Define pairs $\langle M_\xi, U_\xi\rangle$, $\xi < \omega_1$, by induction. Let U_0 be the null system with $|U_0| = \varnothing$, and M_0 be the least CTM of $\mathbf{ZFC}^-_1$. If $\lambda < \omega_1$ is limit then put $U_\lambda = \bigvee_{\xi<\lambda} U_\xi$ and let M_λ be the least CTM of $\mathbf{ZFC}^-_1$ containing the sequence $\{\langle M_\xi, U_\xi\rangle\}_{\xi<\lambda}$. If $\langle M_\xi, U_\xi\rangle \in \mathbf{sJS}$ is defined then by Lemma 14 there is a pair $\langle M', U'\rangle \in \mathbf{sJS}$ with $\langle M_\xi, U_\xi\rangle \prec \langle M', U'\rangle$ and $\xi \in |U'|$. Let $\Theta \subseteq \omega_1 \times \mathrm{HC}$ be a universal $\Sigma^{\mathrm{HC}}_{n-2}$ set, and $D_\xi = \{z \in \mathbf{sJS} : \langle \xi, z\rangle \in \Theta\}$. Let $\langle M_{\xi+1}, U_{\xi+1}\rangle$ be the $<_{\mathbf{L}}$-least pair $\langle M, U\rangle \in D_\xi{}^{\mathtt{solv}}$ satisfying $\langle M', U'\rangle \preccurlyeq \langle M, U\rangle$. To check the definability property use the fact mentioned by Remark 3 in Section 4.3. □

Now define the basic forcing notion.

Definition 16 (in **L**). *Fix a number $\mathfrak{n} \geq 2$. Let $\{\langle \mathbb{M}_\xi, \mathbb{U}_\xi\rangle\}_{\xi<\omega_1} \in \overrightarrow{\mathbf{JS}}_{\omega_1}$ be any $\mathfrak{n}$-complete Jensen–Solovay sequence of class $\Delta^{\mathrm{HC}}_{\mathfrak{n}-1}$ as in Theorem 9—in case $\mathfrak{n} \geq 3$, or just any Jensen–Solovay sequence of class Δ^{HC}_1—in case $\mathfrak{n} = 2$, and in both cases $\xi \in |\mathbb{U}_{\xi+1}|$ for every $\xi < \omega_1$, as in Theorem 9. Put $\mathbb{U} = \bigvee_{\xi<\omega_1} \mathbb{U}_\xi$, so $\mathbb{U}$ is a system, $|\mathbb{U}| = \mathcal{I} = \omega_1$, $\mathbb{U}(\nu) = \bigcup_{\xi<\omega_1, \nu\in|\mathbb{U}_\xi|} \mathbb{U}_\xi(\nu)$ for all $\nu \in \mathcal{I}$. We finally define $\mathbb{P} = \mathbf{P}[\mathbb{U}]$ and $\mathbb{P}_\gamma = \mathbf{P}[\mathbb{U}_\gamma]$ for $\gamma < \omega_1$.*

Thus $\mathbb{P}$ is the product of sets $\mathbb{P}(\nu) = P[\mathbb{U}(\nu)]$, $\nu \in \mathcal{I}$, with finite support.
We proceed with a couple of simple lemmas.

Corollary 6. *Suppose that, in **L**, M is a transitive model of $\mathbf{ZFC}^-$ containing the sequence $\{\langle \mathbb{M}_\xi, \mathbb{U}_\xi\rangle\}_{\xi<\omega_1} \in \overrightarrow{\mathbf{JS}}_{\omega_1}$ of Definition 16. Then, for every $\xi < \omega_1$:*

(i) *$\langle M, \mathbb{U}\rangle \in \mathbf{JS}$ and $\langle \mathbb{M}_\xi, \mathbb{U}_\xi\rangle \prec \langle M, \mathbb{U}\rangle$;*
(ii) *if $\nu \in \mathcal{I}$ then $\mathbb{U}(\nu)$ is uncountable and topologically dense in ω^ω, and if $\nu \neq \mu$ belong to $\mathcal{I}$ then $\mathbb{U}(\nu) \cap \mathbb{U}(\mu)$ is empty;*
(iii) *any set $D \in \mathbb{M}_\xi$, $D \subseteq \mathbb{P}_\xi$, pre-dense in $\mathbb{P}_\xi$ (resp., pre-dense in $\mathbb{P}_\xi$ below some $p \in \mathbb{P}_\xi$), is pre-dense in $\mathbb{P}$ (resp., pre-dense in $\mathbb{P}$ below p);*
(iv) *any name $\tau \in \mathbb{M}_\xi \cap \mathbf{SN}^\omega_\omega(\mathbb{P}_\xi)$, $\mathbb{P}_\xi$-full (resp., $\mathbb{P}_\xi$-full below some $p \in \mathbb{P}_\xi$), is $\mathbb{P}$-full (resp., $\mathbb{P}$-full below p);*
(v) *if $G \subseteq \mathbb{P}$ is a set $\mathbb{P}$-generic over the ground universe **L** then the set $G^\xi = G \cap \mathbb{P}_\xi$ is $\mathbb{P}_\xi$-generic over $\mathbb{M}_\xi$.*

Proof. To prove (i) use Lemma 15. Both claims of (ii) hold by Definition 13. To prove (iii) and (iv) use Corollary 5. Finally, (v) follows from (iii). □

Now let us address definability issues.

Lemma 16 (in **L**). *The binary relation $f \in \mathbb{U}(\nu)$ is $\Delta^{\mathrm{HC}}_{\mathfrak{n}-1}$.*
The sets $\mathbb{P}$ and $\mathbf{SN}^\omega_\omega(\mathbb{P})$ ($\mathbb{P}$-names for functions in **Fun**) *are $\Delta^{\mathrm{HC}}_{\mathfrak{n}-1}$.*
The set of all $\mathbb{P}$-full names in $\mathbf{SN}^\omega_\omega(\mathbb{P})$ is $\Delta^{\mathrm{HC}}_{\mathfrak{n}-1}$.

Proof. The sequence $\{\langle \mathbb{M}_\xi, \mathbb{U}_\xi\rangle\}_{\xi<\omega_1}$ is $\Delta^{\mathrm{HC}}_{\mathfrak{n}-1}$ by definition, hence the relation $f \in \mathbb{U}(\nu)$ is $\Sigma^{\mathrm{HC}}_{\mathfrak{n}-1}$. On the other hand, if $f \in \mathbf{Fun}$ belongs to some $\mathbb{M}_\xi$ then $f \in \mathbb{U}(\nu)$ obviously implies $f \in \mathbb{U}_\xi(\nu)$, leading to a $\Pi^{\mathrm{HC}}_{\mathfrak{n}-1}$ definition of the relation $f \in \mathbb{U}(\nu)$. To prove the last claim, note that by Corollary 5 if a name $\tau \in \mathbf{SN}^\omega_\omega(\mathbb{P}_\xi) \cap \mathbb{M}_\xi$ is $\mathbb{P}_\xi$-full then it remains $\mathbb{P}$-full. □

4.5. Basic Generic Extension and Regular Subextensions

Recall that an integer $\mathfrak{n} \geq 2$ and sets $\mathbb{U}_\xi$, $\mathbb{M}_\xi$, $\mathbb{U}$, $\mathbb{P}_\xi$, $\mathbb{P}$ are fixed in **L** by Definition 16. These sets are fixed for the remainder.

Suppose that, in $\mathbf{L}$, $K \subseteq \mathbb{P}$ is a regular subforcing. If $G \subseteq \mathbb{P}$ is a set $\mathbb{P}$-generic over $\mathbf{L}$ then $G \cap K$ is K-generic over $\mathbf{L}$ by Lemma 9(vi), and hence $\mathbf{L}[G \cap K]$ is a K-generic extension of $\mathbf{L}$. The following formulas Γ_i $(i \in \mathcal{I})$ will give us a useful coding tool in extensions of this form:

$$\Gamma_\nu(S) :=_{\text{def}} \nu \in \mathcal{I} \wedge S \subseteq \mathbf{Seq} \wedge \forall f \in \mathbf{Fun} \cap \mathbf{L}\, (f \in \mathbb{U}(\nu) \iff \mathtt{max}\,(S/f) < \omega).$$

This is based on the next two results. Recall that $|G \cap K| = \bigcup_{p \in G \cap K} |p|$.

Lemma 17. *$\Gamma_\nu(S)$ as a binary relation belongs to $\Pi^{\mathrm{HC}}_{\mathfrak{n}-1}$ in any cardinal-preserving generic extension of $\mathbf{L}$.*

Proof. The set $W = \{\langle \nu, f\rangle : \nu \in \mathcal{I} \wedge f \in \mathbb{U}(\nu)\}$ is $\Delta^{\mathrm{HC}}_{\mathfrak{n}-1}$ in $\mathbf{L}$, by Lemma 16, and hence so is $W' = \{\langle \nu, f\rangle : \nu \in \mathcal{I} \wedge f \in \mathbf{Fun} \smallsetminus \mathbb{U}(\nu)\}$. Let $\varphi(\nu, f)$ and $\varphi'(\nu, f)$ be $\Sigma_{\mathfrak{n}-1}$ formulas that define resp. W, W' in HC, in $\mathbf{L}$. Then, in any generic extension of $\mathbf{L}$, $\Gamma_\nu(S)$ is equivalent to $\nu \in \mathcal{I} \wedge S \subseteq \mathbf{Seq} \wedge \forall f \in \mathbf{Fun} \cap \mathbf{L}\, \Psi(\nu, f)$, where $\Psi(\nu, f)$ is the $\Pi_{\mathfrak{n}-1}$ formula

$$((\mathbf{L} \models \varphi(\nu, f)) \implies \mathtt{max}\,(S/f) < \omega) \wedge ((\mathbf{L} \models \varphi'(\nu, f)) \implies \mathtt{max}\,(S/f) = \omega). \qquad \square$$

Theorem 10. *Suppose that, in $\mathbf{L}$, $K \subseteq \mathbb{P}$ is a regular subforcing. Let $G \subseteq \mathbb{P}$ be $\mathbb{P}$-generic over $\mathbf{L}$. Then:*

(i) *if $\nu \in |G \cap K|$, then $S_G(\nu) \in \mathbf{L}[G \cap K]$ and $\Gamma_\nu(S_G(\nu))$ holds, but*
(ii) *if $\nu \notin |G \cap K|$, then $S_G(\nu) \notin \mathbf{L}[G \cap K]$, and there is no sets $S \subseteq \mathbf{Seq}$ in $\mathbf{L}[G \cap K]$ satisfying $\Gamma_\nu(S)$.*

Proof. (i) This is a corollary of Lemma 9(vi).

(ii) Suppose towards the contrary that some $S \in \mathbf{L}[G \cap K]$ satisfies $\Gamma_\nu(S)$. Note that $S \in \mathbf{L}[G \restriction_{\neq \nu}]$ by Lemma 9(iv). Now we can forget about the given set K. It follows from Theorem 5(iii) (with $K = \mathbb{P} \restriction_{\neq \nu}$), that there is a name $\tau \in \mathbf{SN}_{\mathbf{Seq}}(\mathbb{P} \restriction_{\neq \nu})$ such that $S = \tau[G \restriction_{\neq \nu}]$. There is an ordinal $\xi < \omega_1$ satisfying $\tau \in \mathbb{M}_\xi$ and $\tau \in \mathbf{SN}_{\mathbf{Seq}}(\mathbb{P}_\xi \restriction_{\neq \nu})$. Then $S = \tau[G^\xi \restriction_{\neq \nu}]$, where $G^\xi = G \cap \mathbb{P}_\xi$ is $\mathbb{P}_\xi$-generic over $\mathbb{M}_\xi$ by Corollary 6, and hence S belongs to $\mathbb{M}_\xi[G^\xi \restriction_{\neq \nu}]$.

Note that $\mathbb{U}(\nu)$ is uncountable by Corollary 6(ii), and hence $F = \mathbb{U}(\nu) \smallsetminus \mathbb{U}_\xi(\nu)$ is uncountable. Let $f \in F$. Then f is Cohen generic over the model $\mathbb{M}_\xi$ by Corollary 6. On the other hand $G^\xi \restriction_{\neq \nu}$ is $\mathbb{P}_\xi \restriction_{\neq \nu}$-generic over $\mathbb{M}_\xi[f]$ by Theorem 8(iii). Therefore f is Cohen generic over $\mathbb{M}_\xi[G^\xi \restriction_{\neq \nu}]$ as well.

Recall that $S \in \mathbb{M}_\xi[G^\xi \restriction_{\neq \nu}]$ and $\Gamma_\nu(S)$ holds, hence $\mathtt{max}\,(S/f) < \omega$. As f is Cohen generic over $\mathbb{M}_\xi[G^\xi \restriction_{\neq \nu}]$, it follows that there is a string $s \in \mathbf{Seq}$, $s \subset f$, such that S contains no strings extending s. Take any $\mu \in \mathcal{I}$, $j \neq \nu$. By Corollary 6(ii), there exists a function $g \in \mathbb{U}(\mu) \smallsetminus \mathbb{U}_\xi(\mu)$, $g \notin \mathbb{U}(\nu)$, satisfying $s \subset g$. Then, $\mathtt{max}\,(S/g) = \omega$ by $\Gamma_\nu(S)$. However, this is absurd by the choice of s. $\square$

Corollary 7. *Suppose that, in $\mathbf{L}$, $K \subseteq \mathbb{P}$ is a regular forcing. Let $G \subseteq \mathbb{P}$ be a set $\mathbb{P}$-generic over $\mathbf{L}$. Then*

(i) *$|G \cap K|$ is equal to the set $\{\nu \in \mathcal{I} : \mathbf{L}[G \cap K] \models \exists S\, \Gamma_\nu(S)\}$;*
(ii) *it is true in $\mathbf{L}[G \cap K]$ that the set $\{\langle \nu, S\rangle : \Gamma_\nu(S)\}$ is $\Pi^{\mathrm{HC}}_{\mathfrak{n}-1}$;*
(iii) *therefore $|G \cap K|$ is $\Sigma^{\mathrm{HC}}_{\mathfrak{n}}$ in $\mathbf{L}[G \cap K]$.*

Proof. Claim (i) follows from the theorem, because by the regularity we have $G \cap K \in \mathbf{L}[G \restriction_{\neq \nu}]$ for all $\nu \notin |G \cap K|$. Claim (ii) immediately follows from Lemma 17. To prove (iii) note that, by (i) and (ii), it holds in $\mathbf{L}[G \cap K]$ that the set $|G \cap K|$ is defined by a $\Sigma^{\mathrm{HC}}_{\mathfrak{n}}$ formula $\exists S\, \Gamma_\nu(S)$ in HC. $\square$

5. Forcing Approximations

Here we define and study here an important forcing-like relation **forc**. It will give us control over various phenomena of analytic definability in the generic extensions considered.

We continue to assume $\mathbf{V} = \mathbf{L}$ in the ground universe by Assumption 1.

5.1. Models and Absolute Sets

To consider transitive models of weaker theories, we let $\mathbf{ZFC}^-$ be **ZFC** minus the Power Set axiom, with the schema of Collection instead of replacement, and **AC** in the form of well-orderability of every set. See [24] on $\mathbf{ZFC}^-$ in detail.

Let $\mathbf{ZFC}_1^-$ be $\mathbf{ZFC}^-$ plus the axioms $\mathbf{V} = \mathbf{L}$ and "every set is at most countable".

Let $W \subseteq \mathrm{HC}$. By definition, a set $X \subseteq \mathrm{HC}$ is $\Delta_1^{\mathrm{HC}}(W)$ iff there exist a Σ_1 formula $\sigma(x)$ and a Π_1 formula $\pi(x)$, with sets in W as parameters, such that

$$X = \{x \in \mathrm{HC} : \sigma^{\mathrm{HC}}(x)\} = \{x \in \mathrm{HC} : \pi^{\mathrm{HC}}(x)\}, \tag{4}$$

in particular, we have $\sigma^{\mathrm{HC}}(x) \iff \pi^{\mathrm{HC}}(x)$ for all x. However, generally speaking, this does not imply that $X \cap M \in \Delta_1^M(W)$, where $M \in \mathrm{HC}$ is a countable transitive model (CTM). The goal of the next two definitions is to distinguish and formalize this kind of absoluteness.

Definition 17. *If for a given $\Delta_1^{\mathrm{HC}}(W)$ set X, there exists such a pair of formulas, containing only parameters in W and satisfying (4) and $\forall x \in M\,(\sigma^M(x) \iff \pi^M(x))$ for all countable transitive models $M \models \mathbf{ZFC}^-$ containing all parameters that occur in σ and/or in π, then we say that X is absolute $\Delta_1^{\mathrm{HC}}(W)$. In this case, if M is as indicated then the set $X \cap M$ is $\Delta_1^M(W)$ via the same pair of formulas. In particular, any $\Delta_0^{\mathrm{HC}}(W)$ set is absolute $\Delta_1^{\mathrm{HC}}(W)$ by obvious reasons.*

Definition 18. *In continuation of the last definition, a function $f : D \to \mathrm{HC}$, defined on a set $D \subseteq \mathrm{HC}$, is absolute $\Delta_1^{\mathrm{HC}}(W)$ function, if f is absolute $\Delta_1^{\mathrm{HC}}(W)$ as a set of pairs in the sense of Definition 17, and in addition, if $M \models \mathbf{ZFC}^-$ is a CTM and $x \in D \cap M$ then $f(x) \in M$.*

5.2. Formulas

Here we introduce a language that will help us study analytic definability in $\mathbf{P}[U]$-generic extensions, for different systems U, and their submodels.

Definition 19. *Let $\mathcal{L}$ be the 2nd order Peano language, with variables of type 1 over ω^ω. If $K \subseteq \mathbf{P}^*$ then an $\mathcal{L}(K)$ formula is any formula of $\mathcal{L}$, with some free variables of types 0, 1 replaced by resp. numbers in ω and names in $\mathbf{SN}_\omega^\omega(K)$, and some type 1 quantifiers are allowed to have bounding indices B (i.e., $\exists^B$, $\forall^B$) such that $B \subseteq \mathcal{I}$ is finite or countable.*

Typically K will be a regular forcing in Definition 19, in the sense of Definition 8, or a regular subforcing of the form $K[U]$, U being a system.

If φ is a $\mathcal{L}(\mathbf{P}^*)$ formula then let

$$\begin{aligned}
\mathrm{NAM}\,\varphi &= \text{the set of all names } \tau \text{ that occur in } \varphi;\\
|\varphi| &= \textstyle\bigcup_{\tau \in \mathrm{NAM}\,\varphi} |\tau| \qquad \text{(at most countable);}\\
\mathrm{IND}\,\varphi &= \text{the set of all quantifier indices } B \text{ which occur in } \varphi;\\
\|\varphi\| &= |\varphi| \cup \left(\textstyle\bigcup \mathrm{IND}\,\varphi\right) \qquad \text{(at most countable).}
\end{aligned}$$

Note that $|\varphi| \subseteq \|\varphi\| \subseteq \mathcal{I}$ provided φ is an $\mathcal{L}(\mathbf{P}^*)$ formula.

If a set $G \subseteq \mathbf{P}^*$ is minimally φ-generic (i.e., minimally τ-generic w.r.t. every name $\tau \in \mathrm{NAM}\,\varphi$, in the sense of Section 3.5), then let the valuation $\varphi[G]$ be the result of substitution of $\tau[G]$ for any name $\tau \in \mathrm{NAM}\,\varphi$, and changing each quantifier $\exists^B x$, $\forall^B x$ to $\exists\,(\forall)\, x \in \omega^\omega \cap \mathbf{L}[G \restriction B]$ respectively, while index-free type 1 quantifiers are relativized to ω^ω; $\varphi[G]$ is a formula of $\mathcal{L}$ with real parameters, and with some quantifiers of type 1 explicitly relativized to certain submodels of $\mathbf{L}[G]$.

An arithmetic formula in $\mathcal{L}(K)$ is a formula with no quantifiers of type 1 (names in $\mathbf{SN}^{\omega}_{\omega}(K)$ as in Definition 19 are allowed). If $n < \omega$ then let a $\mathcal{L}\Sigma^1_n(K)$, resp., $\mathcal{L}\Pi^1_n(K)$ formula be a formula of the form

$$\exists^\circ x_1\, \forall^\circ x_2 \ldots \forall^\circ (\exists^\circ)\, x_{n-1}\, \exists\, (\forall)\, x_n\, \psi\,, \quad \forall^\circ x_1\, \exists^\circ x_2 \ldots \exists^\circ (\forall^\circ)\, x_{n-1}\, \forall\, (\exists)\, x_n\, \psi$$

respectively, where ψ is an arithmetic formula in $\mathcal{L}(K)$, all variables x_i are of type 1 (over ω^ω), the sign $^\circ$ means that this quantifier can have a bounding index as in Definition 19, and it is required that the rightmost (closest to the kernel ψ) quantifier doesn't have a bounding index.

If in addition $M \models \mathbf{ZFC}^-$ is a transitive model and $U \in M$ a system then define

$\mathcal{L}\Sigma^1_n(K[U], M) =$ all $\mathcal{L}\Sigma^1_n(K)$ formulas φ such that $\text{NAM}\,\varphi \subseteq \mathbf{SN}^{\omega}_{\omega}(K[U]) \cap M$ and all indices $B \in \text{IND}\,\varphi$ belong to M and satisfy $B \subseteq |U|$.

Define $\mathcal{L}\Pi^1_n(K[U], M)$ similarly. All formulas in $\mathcal{L}\Sigma^1_n(K[U], M) \cup \mathcal{L}\Pi^1_n(K[U], M)$ are by definition (finite) strings in M.

5.3. Forcing Approximation

The next definition invents a convenient forcing-type relation **forc** for pairs $\langle M, U\rangle$ in **sJS** and formulas φ in $\mathcal{L}(K[U])$, associated with the truth in $K[U]$-generic extensions of $\mathbf{L}$, where $K \subseteq \mathbf{P}^*$ is a regular forcing. Recall that $K[U] = K \cap \mathbf{P}[U]$ whenever $K \subseteq \mathbf{P}^*$ is a regular forcing and U is a system.

Definition 20 (in $\mathbf{L}$). *We introduce a relation $p\ {}^K\mathbf{forc}^M_U\ \varphi$. First of all,*

(F1) *Writing $p\ {}^K\mathbf{forc}^M_U\ \varphi$, it is assumed that:*

- (a) $\langle M, U\rangle \in \mathbf{sJS}$,
- (b) *$K \subseteq \mathbf{P}^*$ is a regular forcing and an absolute $\Delta^{\text{HC}}_1(M)$ set,*
- (c) *p belongs to $K[U]$ (a regular subforcing of $\mathbf{P}[U]$ by Lemma 7),*
- (d) *φ is a closed formula in $\mathcal{L}\Pi^1_k(K[U], M) \cup \mathcal{L}\Sigma^1_{k+1}(K[U], M)$ for some $k \geq 1$, and each name $\tau \in \text{NAM}\,\varphi$ is $K[U]$-full below p.*

Under these assumptions, the sets U, $K[U]$, p, $\text{NAM}\,\varphi$, $\text{IND}\,\varphi$ belong to M. The property of $K[U]$-fullness in (F1)d is equivalent to just $\mathbf{P}[U]$-fullness, by Corollary 1, since $K[U]$ is a regular subforcing of $\mathbf{P}[U]$ by Lemma 7.

The definition of **forc** *goes on by induction on the complexity of formulas.*

(F2) *If $\langle M, U\rangle \in \mathbf{sJS}$, $p \in K[U]$, and φ is a closed formula in $\mathcal{L}\Pi^1_1(K[U], M)$ (then by definition it has no quantifier indices), then: $p\ {}^K\mathbf{forc}^M_U\ \varphi$ iff (F1) holds and p $K[U]$-forces φ over M in the usual sense. Note that the forcing notion $K[U]$ belongs to M in this case by (F1)b.*

(F3) *If $\varphi(x) \in \mathcal{L}\Pi^1_k(K[U], M)$, $k \geq 1$, then:*

- (a) *$p\ {}^K\mathbf{forc}^M_U\ \exists^B x\, \varphi(x)$ iff there is a name $\tau \in M \cap \mathbf{SN}^{\omega}_{\omega}(K[U] \restriction B)$, $K[U]$-full below p (by (F1)d) and such that $p\ {}^K\mathbf{forc}^M_U\ \varphi(\tau)$.*
- (b) *$p\ {}^K\mathbf{forc}^M_U\ \exists\, x\, \varphi(x)$ iff there is a name $\tau \in M \cap \mathbf{SN}^{\omega}_{\omega}(K[U])$, $K[U]$-full below p (by (F1)d) and such that $p\ {}^K\mathbf{forc}^M_U\ \varphi(\tau)$.*

(F4) *If $k \geq 2$, φ is a closed $\mathcal{L}\Pi^1_k(K[U], M)$ formula, $p \in K[U]$, and (F1) holds, then $p\ {}^K\mathbf{forc}^M_U\ \varphi$ iff we have $\neg\, q\ {}^K\mathbf{forc}^{M'}_{U'}\ \varphi^\neg$ for every pair $\langle M', U'\rangle \in \mathbf{sJS}$ extending $\langle M, U\rangle$, and every condition $q \in K[U']$, $q \leq p$, where $\varphi^\neg$ is the result of canonical conversion of $\neg\,\varphi$ to $\mathcal{L}\Sigma^1_k(K[U], M)$.*

Lemma 18 (in $\mathbf{L}$). *Let K, $\langle M, U\rangle$, p, φ satisfy (F1) of Definition 20. Then:*

(i) *if $p\ {}^K\mathbf{forc}^M_U\ \varphi$, $\langle M, U\rangle \preccurlyeq \langle M', U'\rangle \in \mathbf{sJS}$ and $q \in K[U']$, $q \leq p$, then $q\ {}^K\mathbf{forc}^{M'}_{U'}\ \varphi$;*
(ii) *if $k \geq 2$, φ is $\mathcal{L}\Pi^1_k(K[U], M)$, and $p\ {}^K\mathbf{forc}^M_U\ \varphi$, then $p\ {}^K\mathbf{forc}^M_U\ \varphi^\neg$ fails.*

Thus by the first claim of the lemma **forc** is monotone w. r. t. both the extension of pairs in **sJS** and the strengthening of forcing conditions.

Proof. (i) Let $\varphi = \varphi(\tau_1, \dots, \tau_m)$ be a closed formula in $\mathcal{L}\Pi^1_1(K[U], M)$, where all names $\tau_j \in \mathbf{SN}^\omega_\omega(K[U]) \cap M$ are $K[U]$-full below the condition $p \in K[U]$ considered. Then all names τ_j remain $K[U']$-full below p, and below q as well since $q \le p$, by Corollary 5. Consider a set $G' \subseteq K[U']$, $K[U']$-generic over M' and containing q. We have to prove that $\varphi[G']$ is true in $M'[G']$. Note that the set $G = G' \cap K[U]$ is $K[U]$-generic over M by Corollary 4, and we have $p \in G$. Moreover the valuation $\varphi[G']$ coincides with $\varphi[G]$ since all names in φ belong to $\mathbf{SN}^\omega_\omega(K[U])$. $\varphi[G]$ is true in $M[G]$ as $p \ {}^K\mathbf{forc}^M_U \ \varphi$. It remains to apply Mostowski's absoluteness between the models $M[G] \subseteq M'[G']$.

The inductive steps related to (F3), (F4) of Definition 20 are easy.

Claim (ii) immediately follows from (F4) of Definition 20. □

The next theorem classifies the complexity of **forc** in terms of projective hierarchy. Recall that all formulas in $\bigcup_n \left(\mathcal{L}\Sigma^1_n(K, M) \cup \mathcal{L}\Pi^1_n(K, M)\right)$ are by definition (finite) strings in M. This allows us to consider and analyze sets

$$\begin{aligned}\mathbf{Forc}^K_w(\Sigma^1_n) \quad = \quad \{ \quad & \langle M, U, p, \varphi\rangle : \langle M, U\rangle \in \mathbf{sJS} \wedge w \in M \wedge p \in K[U] \wedge \\ & \varphi \text{ is a closed } \mathcal{L}\Sigma^1_n(K[U], M) \text{ formula} \wedge p \ {}^K\mathbf{forc}^M_U \ \varphi \ \};\end{aligned}$$

and similarly defined $\mathbf{Forc}^K_w(\Pi^1_n)$, where it is assumed that $w \in \omega^\omega$ and $K \subseteq \mathbf{P}^*$ is a regular forcing and an absolute $\Delta^{\mathrm{HC}}_1(w)$ set.

Theorem 11 (in **L**). *Let $w \in \omega^\omega$ and $K \subseteq \mathbf{P}^*$ be a regular forcing and an absolute $\Delta^{\mathrm{HC}}_1(w)$ set. Then:*

(i) *$\mathbf{Forc}^K_w(\Pi^1_1)$ and $\mathbf{Forc}^K_w(\Sigma^1_2)$ are $\Delta^{\mathrm{HC}}_1(w)$;*
(ii) *if $k \ge 2$ then $\mathbf{Forc}^K_w(\Pi^1_k)$ and $\mathbf{Forc}^K_w(\Sigma^1_{k+1})$ are $\Pi^{\mathrm{HC}}_{k-1}(w)$.*

Proof (sketch). Suppose that φ is $\mathcal{L}\Pi^1_1$. Under the assumptions of the theorem, items (F1)a, (F1)c, (F1)d of Definition 20(F1) are $\Delta^{\mathrm{HC}}_1(w)$ relations, (F1)b is automatic, while (F2) is reducible to a forcing relation over M that we can relativize to M. The inductive step goes on straightforwardly using (F3), (F4) of Definition 20. Note that the quantifier over names in (F3) is a bounded quantifier (bounded by M), hence it does not add any extra complexity. □

5.4. Advanced Properties of Forcing Approximations

The following lemma works whenever the domain $K \subseteq \mathbf{P}^*$ (a regular forcing) of conditions p related to the definition of $p \ {}^K\mathbf{forc}^M_U \ \varphi$ is bounded by a set $c \subseteq \mathcal{I}$. (Compare with Theorem 7.)

Lemma 19 (restriction lemma, in **L**). *Suppose that $K, \langle M, U\rangle, p, \varphi$ satisfy (F1) of Definition 20, a set $c \subseteq \mathcal{I}$ is absolute $\Delta^{\mathrm{HC}}_1(M)$, $K \subseteq \mathbf{P}^* \restriction c$, and $p \ {}^K\mathbf{forc}^M_U \ \varphi$. Then $p \ {}^K\mathbf{forc}^M_{U\restriction c} \ \varphi$.*

Note that $|U| \subseteq c$ is not assumed in the lemma. On the other hand, we have $|p| \subseteq c$ by Definition 20(F1)c, because $p \in K[U]$ and $K \subseteq \mathbf{P}^* \restriction c$, and $|\varphi| \subseteq c$ holds because φ is an $\mathcal{L}(K[U])$ formula. In addition, $U \restriction c \in M$ by the choice of c.

Proof. The direction $\Longleftarrow$ immediately follows from Lemma 18(i) since we have $\langle M, U\restriction c\rangle \preccurlyeq \langle M, U\rangle$ by Remark 1 in Section 4.1. Prove the opposite implication by induction.

Case of $\mathcal{L}\Pi^1_1$ formulas: $K[U] = K[U \restriction c]$ under the assumptions of the lemma.

Step $\mathcal{L}\Pi^1_n \to \mathcal{L}\Sigma^1_{n+1}$. Let $\psi(x)$ be a $\mathcal{L}\Pi^1_n(K[U], M)$ formula, and φ be $\exists^B x\, \psi(x)$, $B \subseteq \mathcal{I}$, $B \in M$. If $p \ {}^K\mathbf{forc}^M_U \ \varphi$ then there is a name $\tau \in M \cap \mathbf{SN}^\omega_\omega(K[U] \restriction B)$ such that $p \ {}^K\mathbf{forc}^M_U \ \psi(\tau)$. We conclude that $p \ {}^K\mathbf{forc}^M_{U\restriction c} \ \psi(\tau)$ by the inductive hypothesis. However we have $\mathbf{SN}^\omega_\omega(K[U] \restriction B) = \mathbf{SN}^\omega_\omega(K[U \restriction c] \restriction B)$ since $|K| \subseteq c$. Thus $p \ {}^K\mathbf{forc}^M_{U\restriction c} \ \varphi$. The case φ being $\exists\, x\, \psi(x)$ is similar.

Step $\mathcal{L}\Sigma^1_n \to \mathcal{L}\Pi^1_n$, $n \geq 2$. Let φ be a $\mathcal{L}\Pi^1_n(K[U], M)$ formula. Suppose towards the contrary that $p \ {}^K\mathbf{forc}^M_U \ \varphi$ holds, but $p \ {}^K\mathbf{forc}^M_{U\restriction c} \ \varphi$ fails, so that there exist a pair $\langle M', V \rangle \in \mathbf{sJS}$ and a condition $q \in K[V]$, such that $\langle M, U\restriction c \rangle \preccurlyeq \langle M', V \rangle$, $q \leq p$, and $q \ {}^K\mathbf{forc}^{M'}_V \ \varphi^\neg$. Then $q \ {}^K\mathbf{forc}^{M'}_{V\restriction c} \ \varphi^\neg$ by the inductive hypothesis. Note that $|q| \subseteq c$ by the choice of K, but not necessarily $|V| \subseteq c$.

Define a system $W \in M'$ such that $|W| = (|V| \cap c) \cup (|U| \smallsetminus c)$, $W\restriction(|V| \cap c) = V\restriction(|V| \cap c)$, and $W\restriction(|U| \smallsetminus c) = U\restriction(|U| \smallsetminus c)$. Then $\langle M', V\restriction c \rangle \preccurlyeq \langle M', W \rangle$, therefore still $q \ {}^K\mathbf{forc}^{M'}_W \ \varphi^\neg$ by Lemma 18(i).

Now we claim that $\langle M, U \rangle \preccurlyeq \langle M', W \rangle$. Indeed, suppose that $\nu \in |U|$. If $\nu \notin c$ then $W(\nu) = U(\nu)$. If $\nu \in c$ then $U(\nu) \subseteq V(\nu) = W(\nu)$ by construction. It follows that $|U| \subseteq |W|$, $U \preccurlyeq W$, and $\Delta(U, W) \subseteq \Delta(U\restriction c, V)$—which implies $U \preccurlyeq_M W$, since $\langle M, U\restriction c \rangle \preccurlyeq \langle M', V \rangle$. Thus $\langle M, U \rangle \preccurlyeq \langle M', W \rangle$.

We have $q \leq p$ as well. This contradicts the assumption $p \ {}^K\mathbf{forc}^M_U \ \varphi$ by Lemma 18(ii). □

Lemma 20 (in **L**). *Let K, $\langle M, U \rangle$, p, φ, k satisfy (F1) of Definition 20, $\text{NAM}\,\varphi = \{\tau_1, \ldots, \tau_m\}$, $\mu_1, \ldots, \mu_m$ be another list of names in $\mathbf{SN}^\omega_\omega(K[U]) \cap M$, $K[U]$-full below p and such that τ_ℓ and μ_ℓ are equivalent below p for each $\ell = 1, \ldots, m$. Then $p \ {}^K\mathbf{forc}^M_U \ \varphi(\tau_1, \ldots, \tau_m)$ iff $p \ {}^K\mathbf{forc}^M_U \ \varphi(\mu_1, \ldots, \mu_m)$.*

Proof. It suffices to consider the case of Π^1_1 formulas; the induction steps $\mathcal{L}\Pi^1_k \to \mathcal{L}\Sigma^1_{k+1}$ and $\mathcal{L}\Sigma^1_k \to \mathcal{L}\Pi^1_k$ are rather easy.

Suppose that φ is $\mathcal{L}\Pi^1_1$ and $p \ {}^K\mathbf{forc}^M_U \ \varphi(\tau_1, \ldots, \tau_m)$. Suppose that $G \subseteq K[U]$ is a set $K[U]$-generic over M, and $p \in G$. We claim that $\tau_\ell[G] = \mu_\ell[G]$ for all ℓ; this obviously implies the result required. Suppose that this is not the case. Then, by definition, there exist numbers m and $j \neq k$ and conditions $q \in G \cap (\tau''\langle m, j \rangle)$ and $r \in G \cap (\mu''\langle m, k \rangle)$. Then p, q, r must be compatible (as elements of the same generic set), which is a contradiction. □

Lemma 21 (in **L**). *Suppose that K, $\langle M, U \rangle$, p, φ, k satisfy (F1) of Definition 20, φ is $\mathcal{L}\Pi^1_k(K[U], M)$, $P = \{q \in K[U] : q \leq p\}$, a set $A \in M$, $A \subseteq P$ is a maximal antichain in P, and $q \ {}^K\mathbf{forc}^M_U \ \varphi$ for all $q \in A$. Then $p \ {}^K\mathbf{forc}^M_U \ \varphi$.*

Proof. If φ is a $\mathcal{L}\Pi^1_1$ formula then the result follows from (F2) of Definition 20 and known properties of the ordinary forcing over M. Now let φ be Π^1_k, $k \geq 2$. Suppose towards the contrary that $p \ {}^K\mathbf{forc}^M_U \ \varphi$ fails. Then there exist: a pair $\langle M', U' \rangle \in \mathbf{sJS}$ extending $\langle M, U \rangle$, and a condition $r \in K[U']$, $r \leq p$, such that $r \ {}^K\mathbf{forc}^{M'}_{U'} \ \varphi^\neg$. Note that A remains a maximal antichain in the set $Q = \{q \in \mathbf{P}[U] : q \leq p\}$ (bigger than P above), by Lemma 8. Therefore, A is still a maximal antichain in $Q' = \{q \in \mathbf{P}[U'] : q \leq p\}$, by Theorem 8(i), hence a maximal antichain in $P' = \{q \in K[U'] : q \leq p\}$. It follows that r is compatible in $K[U']$ with at least one condition $q \in A$. However, $r \ {}^K\mathbf{forc}^{M'}_{U'} \ \varphi^\neg$ while $q \ {}^K\mathbf{forc}^M_U \ \varphi$, easily leading to a contradiction with Lemma 19. □

5.5. Transformations and Invariance

Here we show that, under certain assumptions, the transformations of the first two groups defined in Section 3.7 preserve forcing approximations **forc**. This is not an absolutely elementary thing: there is no way to reasonably apply transformations to transitive models M involved in the definition of **forc**. What we can do is to require that the transformations involved belong to the models involved. This leads to certain complications of different sort.

Family 1: permutations. First of all we have to extend the definition of the action of π in Section 3.7 to include formulas. Suppose that $c, c' \subseteq \mathcal{I}$. Define the action of any $\pi \in \text{BIJ}^c_{c'}$ onto formulas φ of $\mathcal{L}(\mathbf{P}^*)$ such that $\|\varphi\| \subseteq c$:

- to get $\pi\varphi$ substitute $\pi \cdot \tau$ for any $\tau \in \text{NAM}\,\varphi$ and $\pi \cdot B$ for any $B \in \text{IND}\,\varphi$.

Lemma 22. *Suppose that $\langle M, U \rangle$, K, p, φ satisfy (F1) of Definition 20, sets $c, c' \subseteq \mathcal{I}$ have equal cardinality and are absolute $\Delta^{\text{HC}}_1(M)$, $\pi \in \text{BIJ}^c_{c'}$ is an absolute $\Delta^{\text{HC}}_1(M)$ function, and $\|\varphi\| \subseteq c$, $|U| \subseteq c$, $K \subseteq \mathbf{P}^*\restriction c$.*

Then, $p\ {}^{K}\mathbf{forc}_{U}^{M}\ \varphi$ *iff* $(\pi \cdot p)\ {}^{\pi \cdot K}\mathbf{forc}_{\pi \cdot U}^{M}\ \pi\varphi$.

Proof. Under the assumptions of the lemma, in particular, the requirement of c, c', π being absolute $\Delta_1^{\mathrm{HC}}(M)$, π acts as an isomorphism on all relevant domains and preserves all relevant relations between the objects involved. Thus $\langle M, \pi \cdot U \rangle, \pi \cdot K, \pi \cdot p, \pi\varphi$ still satisfy Definition 20(F1), and $\|\pi\varphi\| \subseteq c'$, $|\pi \cdot U| \subseteq c'$, $\pi \cdot K \subseteq \mathbf{P}^* \restriction c'$. (For instance, to show that $\pi \cdot U$ still belongs to M, note that the set $|U| \subseteq c$ belongs to M, thus $\pi \restriction |U| \in M$, too, since π is an absolute $\Delta_1^{\mathrm{HC}}(M)$ function.) This allows to prove the lemma by induction on the complexity of φ.

Suppose that φ is a closed formula in $\mathcal{L}\Pi_1^1(K[U], M)$. Then $\pi\varphi$ is a closed formula in $\mathcal{L}\Pi_1^1((\pi \cdot K)[\pi \cdot U], M)$. Then easily $P' = (\pi \cdot K)[\pi \cdot U] = \pi \cdot (K[U]) \subseteq \mathbf{P}^*$ is a set in M order isomorphic to $P = K[U]$ itself by means of the map $p \longmapsto \pi \cdot p$. Moreover a set $G \subseteq P$ is P-generic over M iff $\pi \cdot G$ is, accordingly, P'-generic over M and the valuated formulas $\varphi[G]$ and $(\pi\varphi)[\pi \cdot G]$ coincide. Now the result for Π_1^1 formulas follows from (F2) of Definition 20.

Step $\Pi_n^1 \to \Sigma_{n+1}^1$, $n \geq 1$. Let $\psi(x)$ be a $\mathcal{L}\Pi_n^1(K[U], M)$ formula, and φ be $\exists x\, \psi(x)$. Assume $p\ {}^{K}\mathbf{forc}_{U}^{M}\ \varphi$. By definition there is a name $\tau \in \mathbf{SN}_{\omega}^{\omega}(K[U]) \cap M$ such that $p\ {}^{K}\mathbf{forc}_{U}^{M}\ \psi(\tau)$. Then, by the inductive hypothesis, $\pi \cdot p\ {}^{\pi \cdot K}\mathbf{forc}_{\pi \cdot U}^{M}\ (\pi\psi)(\pi \cdot \tau)$, and hence by definition $\pi \cdot p\ {}^{\pi \cdot K}\mathbf{forc}_{\pi \cdot U}^{M}\ \pi\varphi$.

The case of φ being $\exists^B x\, \psi(x)$ is similar.

Step $\Sigma_n^1 \to \Pi_n^1$, $n \geq 2$. This is somewhat less trivial. Assume that φ is a closed $\mathcal{L}\Pi_n^1(K[U], M)$ formula; all names in φ belong to $\mathbf{SN}_{\omega}^{\omega}(K[U]) \cap M$ and are $K[U]$-full below a given $p \in K[U]$. Then, by rather obvious reasons, $\pi\varphi$ is a closed $\mathcal{L}\Pi_n^1((\pi \cdot K)[\pi \cdot U], M)$ formula, whose all names belong to $\mathbf{SN}_{\omega}^{\omega}((\pi \cdot K)[\pi \cdot U]) \cap M$ and are $(\pi \cdot K)[\pi \cdot U]$-full below $\pi \cdot p$. Suppose that $p\ {}^{K}\mathbf{forc}_{U}^{M}\ \varphi$ **fails.** By definition there exist a pair $\langle M_1, U_1 \rangle \in \mathbf{sJS}$ with $\langle M, U \rangle \preccurlyeq \langle M_1, U_1 \rangle$, and a condition $q \in K[U_1]$, $q \leq p$, such that $q\ {}^{K}\mathbf{forc}_{U_1}^{M_1}\ \varphi^{\neg}$. We can also assume by Lemma 19, that $|U_1| \subseteq c$. Then $(\pi \cdot q)\ {}^{\pi \cdot K}\mathbf{forc}_{\pi \cdot U_1}^{M_1}\ \pi\varphi^{\neg}$ by the inductive hypothesis. Yet the pair $\langle M_1, \pi \cdot U_1 \rangle$ belongs to $\mathbf{sJS}$ and extends $\langle M, \pi \cdot U \rangle$. (As π is absolute $\Delta_1^{\mathrm{HC}}(M)$ and $U \in M$, the restriction $\pi \restriction |U|$ belongs to M.) In addition, $\pi \cdot q \in (\pi \cdot K)[\pi \cdot U_1]$, and $\pi \cdot q \leq \pi \cdot p$. Therefore the statement $(\pi \cdot p)\ {}^{K}\mathbf{forc}_{\pi \cdot U}^{M}\ \pi\varphi$ fails, as required. □

Family 2: Lipschitz transformations. We extend the action of $\alpha \in \mathbf{Lip}^{\mathcal{I}}$ to formulas of $\mathcal{L}(\mathbf{P}^*)$:

- to get $\pi\varphi$ substitute $\pi \cdot \tau$ for any $\tau \in \text{NAM}\, \varphi$ but do not change the quantifier indices B.

Note that the action of any $\alpha \in \mathbf{Lip}^{\mathcal{I}} \cap M$ on systems, conditions, names, and formulas, as defined there, is absolute $\Delta_1^{\mathrm{HC}}(M)$. This allows to prove the next invariance lemma similarly to Lemma 22, which we leave for the reader.

Lemma 23. *Suppose that* $\langle M, U \rangle, K, p, \varphi$ *satisfy* (F1) *of Definition* 20, *and* $\alpha \in \mathbf{Lip}^{\mathcal{I}} \cap M$. *Then* $p\ {}^{K}\mathbf{forc}_{U}^{M}\ \varphi$ *iff* $(\alpha \cdot p)\ {}^{\alpha \cdot K}\mathbf{forc}_{\alpha \cdot U}^{M}\ \alpha\varphi$.

6. Elementary Equivalence Theorem

This section presents further properties of $\mathbb{P}$-generic extensions of $\mathbf{L}$ and their subextensions, including Theorem 13 and its corollaties on the elementary equivalence of different subextensions.

Assumption 2. *We continue to assume* $\mathbf{V} = \mathbf{L}$ *in the ground universe. Below in this section, a number* $\mathtt{n} \geq 2$ *is fixed, and pairs* $\langle \mathbb{M}_\xi, \mathbb{U}_\xi \rangle$, *the system* $\mathbb{U} = \bigvee_{\xi < \omega_1} \mathbb{U}_\xi$, *the forcing notions* $\mathbb{P}_\xi = \mathbf{P}[\mathbb{U}_\xi]$ *and* $\mathbb{P} = \mathbf{P}[\mathbb{U}] = \bigcup_{\xi < \omega_1} \mathbb{P}_\xi$ *are as in Definition* 16 *for this* $\mathtt{n}$.

6.1. Further Properties of Forcing Approximations

Coming back to the complete sequence of pairs $\langle \mathbb{M}_\xi, \mathbb{U}_\xi \rangle$ introduced by Definition 16, we consider the auxiliary forcing relation **forc** with respect to those pairs. We begin with the following definition.

Definition 21 (in $\mathbf{L}$). *Let $K \subseteq \mathbf{P}^*$ be a regular forcing. Recall that*

$$K[\mathbb{U}] = K \cap \mathbb{P} \quad \text{and} \quad K[\mathbb{U}_\xi] = K \cap \mathbf{P}[\mathbb{U}_\xi] = K \cap \mathbb{P}_\xi$$

for any $\xi < \omega_1$. Let $p \,{}^K\mathbf{forc}_\xi\, \varphi$ mean $p \,{}^K\mathbf{forc}_{\mathbb{U}_\xi}^{\mathbb{M}_\xi}\, \varphi$—then by definition K has to be an absolute $\Delta_1^{\mathrm{HC}}(\mathbb{M}_\xi)$ set, by the way. We let $p \,{}^K\mathbf{forc}_\infty\, \varphi$ mean: $\exists \xi < \omega_1 \, (p \,{}^K\mathbf{forc}_\xi\, \varphi)$.

Thus, if $p \,{}^K\mathbf{forc}_\xi\, \varphi$ then definitely K is an absolute $\Delta_1^{\mathrm{HC}}(\mathbb{M}_\xi)$ set, $p \in K[\mathbb{U}_\xi]$, φ is a formula with names in $\mathbb{M}_\xi \cap \mathbf{SN}_\omega^\omega(K[\mathbb{U}_\xi])$ as parameters, all names $\tau \in \mathrm{NAM}\,\varphi$ are $K[\mathbb{U}_\xi]$-full below p, all indices $B \in \mathrm{IND}\,\varphi$ belong to $\mathbb{M}_\xi$. The following is an easy consequence of Lemma 18.

Lemma 24 (in $\mathbf{L}$). *Let $K \subseteq \mathbf{P}^*$ be a regular forcing. Assume that φ is a closed formula in $\mathcal{L}\Pi_k^1(K[\mathbb{U}]) \cup \mathcal{L}\Sigma_{k+1}^1(K[\mathbb{U}])$, $1 \le k$, $p \in K[\mathbb{U}]$. Then:*

(i) *if $p \,{}^K\mathbf{forc}_\xi\, \varphi$ and $\xi \leqslant \zeta < \omega_1$, $q \in K[\mathbb{U}_\zeta]$, $q \leqslant p$, then $q \,{}^K\mathbf{forc}_\zeta\, \varphi$;*
(ii) *$p \,{}^K\mathbf{forc}_\infty\, \varphi$ and $p \,{}^K\mathbf{forc}_\infty\, \varphi^\neg$ contradict to each other;*
(iii) *if φ is a $\mathcal{L}\Pi_k^1(K[\mathbb{U}])$ formula, $A \subseteq Q = \{q \in K[\mathbb{U}] : q \le p\}$ is a maximal antichain in Q, and $q \,{}^K\mathbf{forc}_\infty\, \varphi$ for all $q \in A$, then $p \,{}^K\mathbf{forc}_\infty\, \varphi$.*

Proof. (iii) As A is a countable set, there exists an ordinal $\xi < \omega_1$ such that $q \,{}^K\mathbf{forc}_\xi\, \varphi$ for all $q \in A$. Apply Lemma 21. □

Lemma 25 (in $\mathbf{L}$). *Assume that $K \subseteq \mathbf{P}^*$ is a regular forcing, φ is a closed formula in $\mathcal{L}\Pi_k^1(K[\mathbb{U}]) \cup \mathcal{L}\Sigma_k^1(K[\mathbb{U}])$, $2 \le k < \mathfrak{n}$, $p \in K[\mathbb{U}]$, all names in φ are $K[\mathbb{U}]$-full below p, and finally $w \in \omega^\omega$ and K is absolute $\Delta_1^{\mathrm{HC}}(w)$. Then:*

(i) *there is $q \in K[\mathbb{U}]$, $q \leqslant p$, such that $q \,{}^K\mathbf{forc}_\infty\, \varphi$ or $q \,{}^K\mathbf{forc}_\infty\, \varphi^\neg$;*
(ii) *if φ is $\mathcal{L}\Pi_k^1(K[\mathbb{U}])$, $2 \le k < \mathfrak{n}$, then $p \,{}^K\mathbf{forc}_\infty\, \varphi$ iff there is no condition $q \in K[\mathbb{U}]$, $q \le p$, such that $q \,{}^K\mathbf{forc}_\infty\, \varphi^\neg$.*

Proof. (i) As any name is a countable object, there is an ordinal $\eta < \omega_1$ such that $p \in K[\mathbb{U}_\eta]$, $w \in \mathbb{M}_\eta$, and all names in φ belong to $\mathbb{M}_\eta \cap \mathbf{SN}_\omega^\omega(K[\mathbb{U}_\eta])$; then all names in φ are $K[\mathbb{U}_\eta]$-full below p, of course. As $k < \mathfrak{n}$, the set D of all pairs $\langle M, U\rangle \in \mathbf{sJS}$ that extend $\langle \mathbb{M}_\eta, \mathbb{U}_\eta\rangle$ and there is a condition $q \in K[U]$, $q \leqslant p$, satisfying $q \,{}^K\mathbf{forc}_U^M\, \varphi^\neg$, belongs to $\mathbf{\Sigma}_{n-2}^{\mathrm{HC}}$ by Theorem 11. Therefore, by the n-completeness of the sequence $\{\langle \mathbb{M}_\xi, \mathbb{U}_\xi\rangle\}_{\xi<\omega_1}$, there is an ordinal ζ, $\eta \leqslant \zeta < \omega_1$, such that $\langle \mathbb{M}_\zeta, \mathbb{U}_\zeta\rangle \in D^{\mathtt{solv}}$. (By the way, this is the only use of the $\mathfrak{n}$-completeness!)

We have two cases.

Case 1: $\langle \mathbb{M}_\zeta, \mathbb{U}_\zeta\rangle \in D$. Then there is a condition $q \in K[\mathbb{U}_\zeta]$, $q \le p$, satisfying $q \,{}^K\mathbf{forc}_\zeta\, \varphi^\neg$. However, obviously $q \in K[\mathbb{U}]$.

Case 2: there is no pair $\langle M, U\rangle \in D$ extending $\langle \mathbb{M}_\zeta, \mathbb{U}_\zeta\rangle$. Prove $p \,{}^K\mathbf{forc}_\zeta\, \varphi$. Suppose otherwise. Then by the choice of η and (F4) of Definition 20 there exist a pair $\langle M, U\rangle \in \mathbf{sJS}$ extending $\langle \mathbb{M}_\zeta, \mathbb{U}_\zeta\rangle$, and a condition $q \in K[U]$, $q \le p$, such that $q \,{}^K\mathbf{forc}_U^M\, \varphi^\neg$. Then $\langle M, U\rangle \in D$, a contradiction.

(ii) Suppose that there is no condition $q \in K[\mathbb{U}]$, $q \le p$, with $q \,{}^K\mathbf{forc}_\infty\, \varphi^\neg$. Then by (i) the set $Q = \{q \in K[\mathbb{U}] : q \le p \wedge q \,{}^K\mathbf{forc}_\infty\, \varphi\}$ is dense in $K[\mathbb{U}]$ below p. Let $A \subseteq Q$ be a maximal antichain. It remains to apply Lemma 24(iii). □

6.2. Relations to the Truth in Generic Extensions

According to the next theorem, the truth in the generic extensions considered is connected in the usual way with the relation $\mathtt{forc}_\infty$ up to the $\mathfrak{n}$-th level of analytic hierarchy. Recall that $\mathbf{V} = \mathbf{L}$ is assumed in the ground universe.

Theorem 12. *Assume that, in* $\mathbf{L}$, $K \subset \mathbf{P}^*$ *is a regular forcing,* φ *is a closed formula in* $\mathcal{L}\Pi^1_k(K[\mathbb{U}]) \cup \mathcal{L}\Sigma^1_{k+1}(K[\mathbb{U}])$, $1 \le k \le \mathbb{n}$, *all names in* NAM φ *are* $K[\mathbb{U}]$*-full,* $w \in \omega^\omega$, *and* K *is an absolute* $\Delta^{\mathrm{HC}}_1(w)$ *set. Let* $G \subseteq \mathbb{P}$ *be a* $\mathbb{P}$*-generic set over* $\mathbf{L}$. *Then:*

(i) *if* $p \in G$ *and* $p \,{}^K\mathbf{forc}_\infty\, \varphi$, *then* $\varphi[G]$ *is true in* $\mathbf{L}[G \cap K[\mathbb{U}]]$;
(ii) *conversely, if* $\varphi[G]$ *is true in* $\mathbf{L}[G \cap K[\mathbb{U}]]$ *and strictly* $k < \mathbb{n}$ *holds, then* $\exists\, p \in G \cap K \,(p \,{}^K\mathbf{forc}_\infty\, \varphi)$.

The formulas $\varphi[G]$, $\varphi[G \cap K]$ coincide under the assumptions of the theorem.

Proof. (ii) We argue by induction on the complexity of φ.

The case of $\mathcal{L}\Pi^1_1$ **formulas.** Let φ be a closed formula in $\mathcal{L}\Pi^1_1(K[\mathbb{U}])$. As names in the formulas considered are countable objects, there is an ordinal $\xi < \omega_1$ such that $w \in \mathbb{M}_\xi$ and φ is a $\mathcal{L}\Pi^1_1(K[\mathbb{U}_\xi], \mathbb{M}_\xi)$ formula. As $G \subseteq \mathbb{P}$ is $\mathbb{P}$-generic over $\mathbf{L}$, the smaller set $G_\xi = G \cap K[\mathbb{U}_\xi]$ is $K[\mathbb{U}_\xi]$-generic over $\mathbb{M}_\xi$ by Corollary 4, and the formulas $\varphi[G]$, $\varphi[G_\xi]$ coincide by the choice of ξ. Therefore if $\varphi[G]$ holds in $\mathbf{L}[G \cap K[\mathbb{U}]]$ then $\varphi[G_\xi]$ holds in $\mathbb{M}_\xi[G_\xi]$, by Shoenfield's absoluteness theorem, and hence there is a condition $p \in G_\xi$ which $K[\mathbb{U}_\xi]$-forces φ over $\mathbb{M}_\xi$, that is, $p \,{}^K\mathbf{forc}_\xi\, \varphi$ by (F2) of Definition 20, and finally $p \,{}^K\mathbf{forc}_\infty\, \varphi$, as required. If conversely, $p \in G \cap K[\mathbb{U}]$, ζ, $\xi \le \zeta < \omega_1$, and $p \,{}^K\mathbf{forc}_\zeta\, \varphi$, then by definition p $K[\mathbb{U}_\zeta]$-forces φ over $\mathbb{M}_\xi$. It follows that $\varphi[G_\xi]$ holds in $\mathbb{M}_\xi[G_\xi]$, and hence $\varphi[G]$ holds in $\mathbf{L}[G \cap K[\mathbb{U}]]$ as well by the Shoenfield absoluteness.

Step $\mathcal{L}\Pi^1_k \to \mathcal{L}\Sigma^1_{k+1}$, $k < \mathbb{n}$. Let $\varphi(x)$ be a $\mathcal{L}\Pi^1_k(K[\mathbb{U}])$ formula; let us prove the result for $\exists\, x\, \varphi(x)$. If $p \in G$ and $p \,{}^K\mathbf{forc}_\xi\, \exists\, x\, \varphi(x)$ then by definition there is a name $\tau \in \mathbb{M}_\xi \cap \mathbf{SN}^\omega_\omega(K[\mathbb{U}_\xi])$, $K[\mathbb{U}_\xi]$-full below p, and such that $p \,{}^K\mathbf{forc}_\xi\, \varphi(\tau)$. By Lemma 10, there is a $K[\mathbb{U}_\xi]$-full name $\tau' \in \mathbb{M}_\xi \cap \mathbf{SN}^\omega_\omega(K[\mathbb{U}_\xi])$, equivalent to τ below p. Then $p \,{}^K\mathbf{forc}_\xi\, \varphi(\tau')$ by Lemma 20. Note that τ' is $\mathbb{P}_\xi$-full by Corollary 1, hence $\mathbb{P}$-full by Corollary 6(iv), and $K[\mathbb{U}]$-full, too. It follows that $\varphi(\tau')[G]$ holds in $\mathbf{L}[G \cap K[\mathbb{U}]]$ by the inductive hypothesis, thus $(\exists\, x\, \varphi(x))[G]$ holds in $\mathbf{L}[G \cap K[\mathbb{U}]]$ because $\tau'[G] = \tau[G] \in \mathbf{L}[G \cap K[\mathbb{U}]]$ by the choice of τ.

If conversely $(\exists\, x\, \varphi(x))[G]$ is true in $\mathbf{L}[G \cap K[\mathbb{U}]]$ then by definition there is an element $x \in \mathbf{L}[G \cap K] = \mathbf{L}[G \cap K[\mathbb{U}]]$ such that $\varphi[G](x)$ is true in $\mathbf{L}[G \cap K[\mathbb{U}]]$. By Theorem 5(ii), there is a $K[\mathbb{U}]$-full name $\tau \in \mathbf{SN}^\omega_\omega(K[\mathbb{U}])$ such that $x = \tau[G]$. Thus $\varphi(\tau)[G]$ is true in $\mathbf{L}[G \cap K[\mathbb{U}]]$. Note that τ is $\mathbb{P}$-full as well, by Corollary 1, and hence $K[\mathbb{U}]$-full, too. By the inductive hypothesis, there is a condition $p \in G$ such that $p \,{}^K\mathbf{forc}_\infty\, \varphi(\tau)$. It follows that $p \,{}^K\mathbf{forc}_\infty\, \exists\, x\, \varphi(x)$.

Step $\mathcal{L}\Sigma^1_k \to \mathcal{L}\Pi^1_k$, $2 \le k < \mathbb{n}$. Prove the theorem for a $\mathcal{L}\Pi^1_k(K[\mathbb{U}])$ formula φ, assuming that the result holds for $\varphi^\neg$. If $\varphi[G]$ is false in $\mathbf{L}[G]$ then $\varphi^\neg[G]$ is true. Thus by the inductive hypothesis, there is a condition $p \in G$ such that $p \,{}^K\mathbf{forc}_\infty\, \varphi^\neg$. Then $q \,{}^K\mathbf{forc}_\infty\, \varphi$ for any $q \in G$ is impossible by Lemma 24(ii). Conversely, suppose that $p \,{}^K\mathbf{forc}_\infty\, \varphi$ fails for all $p \in G \cap K$. Then by Lemma 25(i) there is $q \in G \cap K[\mathbb{U}]$ such that $q \,{}^K\mathbf{forc}_\infty\, \varphi^\neg$. It follows that $\varphi^\neg[G]$ is true in $\mathbf{L}[G \cap K[\mathbb{U}]]$ by the inductive hypothesis, therefore $\varphi[G]$ is false.

(i) Let φ be a $\mathcal{L}\Pi^1_{\mathbb{n}}(K[\mathbb{U}])$ formula, $p \in G \cap K[\mathbb{U}]$, $p \,{}^K\mathbf{forc}_\infty\, \varphi$. By Lemma 24(ii), there is no $q \in G \cap K[\mathbb{U}]$ such that $q \,{}^K\mathbf{forc}_\infty\, \varphi^\neg$. However, $\varphi^\neg$ is $\mathcal{L}\Sigma^1_{\mathbb{n}}(K[\mathbb{U}])$, thus $\neg\, \varphi[G]$ in $\mathbf{L}[G \cap K]$ holds by (ii).

Finally prove (i) for a formula $\varphi := \exists\, x\, \psi(x)$, ψ being $\mathcal{L}\Pi^1_{\mathbb{n}}(K[\mathbb{U}])$. Suppose that $p \in G \cap K[\mathbb{U}]$ and $p \,{}^K\mathbf{forc}_\infty\, \varphi$. Then there is a name $\tau \in \mathbf{SN}^\omega_\omega(K[\mathbb{U}])$, $K[\mathbb{U}]$-full below p and such that $p \,{}^K\mathbf{forc}_\infty\, \psi(\tau)$. We can w.l.o.g. assume that τ is totally $K[\mathbb{U}]$-full, by Lemmas 10 and 20. We have to prove that the formula $\psi(\tau)[G]$, that is, $\psi[G](\tau[G])$, holds in $\mathbf{L}[G \cap K]$—then $\varphi[G]$ holds in $\mathbf{L}[G \cap K]$ as well. Suppose to the contrary that $\psi(\tau)[G]$ fails in $\mathbf{L}[G \cap K]$. However, $\psi(\tau)^\neg$ is a Σ^1_n formula. Therefore, by the first claim of the lemma, there is a condition $q \in G \cap K$ such that $q \,{}^K\mathbf{forc}_\infty\, \psi(\tau)^\neg$. However, $p \,{}^K\mathbf{forc}_\infty\, \psi(\tau)$ and p, q are compatible (as they belong to the same generic set). This contradicts Lemma 24(ii). □

6.3. Consequences for the Ordinary Forcing Relation

For any forcing $P \in \mathbf{L}$, we let $\Vdash_P$ be the ordinary P-forcing relation over $\mathbf{L}$ as the ground universe. In particular $\Vdash_{\mathbb{P}}$ is the $\mathbb{P}$-forcing relation over $\mathbf{L}$.

Corollary 8 (in $\mathbf{L}$). *Under the assumptions of Theorem 12, let $p \in K[\mathbb{U}]$. Then:*

(i) *if φ is $\mathcal{L}\Pi^1_k(K[\mathbb{U}])$ or $\mathcal{L}\Sigma^1_{k+1}(K[\mathbb{U}])$ and $p\ {}^K\mathbf{forc}_\infty\ \varphi$, then $p \Vdash_{K[\mathbb{U}]} \varphi$;*
(ii) *if φ is $\mathcal{L}\Pi^1_k(K[\mathbb{U}])$, then $p \Vdash_{K[\mathbb{U}]} \varphi$ iff $\neg\exists q \in K[\mathbb{U}]\ (q \le p \wedge q\ {}^K\mathbf{forc}_\infty\ \varphi^\neg)$;*
(iii) *if $k < \mathfrak{n}$ strictly, φ belongs to $\mathcal{L}\Pi^1_k(K[\mathbb{U}])$ or $\mathcal{L}\Sigma^1_{k+1}(K[\mathbb{U}])$, and $p \Vdash_{K[\mathbb{U}]} \varphi$, then*
$$\exists q \in K[\mathbb{U}]\ (q \le p \wedge q\ {}^K\mathbf{forc}_\infty\ \varphi);$$
(iv) *if $k < \mathfrak{n}$ strictly, φ is $\mathcal{L}\Pi^1_k(K[\mathbb{U}])$, and $p \Vdash_{K[\mathbb{U}]} \varphi$ then $p\ {}^K\mathbf{forc}_\infty\ \varphi$.*

Proof. (i) follows from Theorem 12(i).

(iii) Let $G \subseteq \mathbb{P}$ be $\mathbb{P}$-generic over $\mathbf{L}$, and $p \in G$. If $p \Vdash_{K[\mathbb{U}]} \varphi$ then $\varphi[G]$ is true in $\mathbf{L}[G \cap K[\mathbb{U}]]$, and hence there is $r \in G \cap K$ such that $r\ {}^K\mathbf{forc}_\infty\ \varphi$, by Theorem 12. However, then p, r are compatible (as members of G), hence $q = p \wedge r$ still is a condition, and $q \in K[\mathbb{U}]$.

(iv) If $p\ {}^K\mathbf{forc}_\infty\ \varphi$ fails, then by Lemma 25(ii) there is a condition $q \in K[\mathbb{U}]$, $q \le p$, such that $q\ {}^K\mathbf{forc}_\infty\ \varphi^\neg$. Then $q \Vdash_{K[\mathbb{U}]} \varphi^\neg$ by (i), thus $p \Vdash_{K[\mathbb{U}]} \varphi$ fails.

(ii) Suppose that $q \in K[\mathbb{U}]$, $q \le p$, $q\ {}^K\mathbf{forc}_\infty\ \varphi^\neg$. Then $q \Vdash_{K[\mathbb{U}]} \neg\varphi$ by (i), and hence $p \Vdash_{K[\mathbb{U}]} \varphi$ fails. Now suppose that $p \Vdash_{K[\mathbb{U}]} \varphi$ fails. Then there is a condition $r \in K[\mathbb{U}]$, $r \le p$, $r \Vdash_{K[\mathbb{U}]} \varphi^\neg$. However, then, by (iii), there is a condition $q \in K[\mathbb{U}]$, $q \le r$, $q\ {}^K\mathbf{forc}_\infty\ \varphi^\neg$, as required. □

The next corollary evaluates the complexity of the ordinary forcing relations $\Vdash_{K[\mathbb{U}]}$. The result is related to formulas in classes $\mathcal{L}\Pi^1_{\mathfrak{n}}$ and higher.

Corollary 9 (in $\mathbf{L}$). *Let $\varphi(x_1, \dots, x_m)$ be an $\mathcal{L}(\varnothing)$ formula (that is, no names), and $K \subseteq \mathbf{P}^*$ be a regular forcing. Suppose that $w \in \omega^\omega$, and K is an absolute $\Delta^{\mathrm{HC}}_1(w)$ set. Then:*

(i) *if φ belongs to $\mathcal{L}\Pi^1_k$, $k \ge \mathfrak{n}$, then the following set is $\Pi^{\mathrm{HC}}_{k-1}(w)$:*

$$\begin{aligned}\mathbf{FORC}_K(\varphi) = \{ & \langle p, \tau_1, \dots, \tau_m\rangle : p \in K[\mathbb{U}] \wedge \\ & \tau_1, \dots, \tau_m \in \mathbf{SN}^\omega_\omega(K[\mathbb{U}]) \text{ are } K[\mathbb{U}]\text{-full names} \wedge \\ & p \Vdash_{K[\mathbb{U}]} \varphi(\tau_1, \dots, \tau_m)\ \};\end{aligned}$$

(ii) *similarly, if φ is $\mathcal{L}\Sigma^1_k$, $k > \mathfrak{n}$, then $\mathbf{FORC}_K(\varphi)$ is $\Sigma^{\mathrm{HC}}_{k-1}(w)$.*

Proof. We argue by induction on $k \ge \mathfrak{n}$. Suppose that φ is $\mathcal{L}\Pi^1_{\mathfrak{n}}$ and $\tau_1, \dots, \tau_m \in \mathbf{SN}^\omega_\omega(K[\mathbb{U}])$ are $K[\mathbb{U}]$-full names. It follows from Corollary 8(ii) that $\langle p, \tau_1, \dots, \tau_m\rangle \in \mathbf{FORC}_K(\varphi)$ iff

$$\neg\exists \xi < \omega_1\, \exists q \in K[\mathbb{U}_\xi]\ (q \le p \wedge q\ {}^K\mathbf{forc}^{\mathbb{M}_\xi}_{\mathbb{U}_\xi}\ \varphi^\neg(\tau_1, \dots, \tau_m)).$$

The formula $q\ {}^K\mathbf{forc}^{\mathbb{M}_\xi}_{\mathbb{U}_\xi}\ \varphi^\neg(\tau_1, \dots, \tau_m)$ can be replaced by

$$\langle \mathbb{M}_\xi, \mathbb{U}_\xi, q, \varphi(\tau_1, \dots, \tau_m)\rangle \in \mathbf{Forc}^K_w(\Sigma^1_{\mathfrak{n}})$$

(see a definition in Theorem 11). However, $\mathbf{Forc}^K_w(\Sigma^1_{\mathfrak{n}})$ is $\Delta^{\mathrm{HC}}_{\mathfrak{n}-1}(w)$ by Theorem 11 (even $\Pi^{\mathrm{HC}}_{\mathfrak{n}-2}(w)$ provided $\mathfrak{n} \ge 3$). On the other hand, the maps $\xi \longmapsto \mathbb{M}_\xi$ and $\xi \longmapsto \mathbb{U}_\xi$ are $\Delta^{\mathrm{HC}}_{\mathfrak{n}-1}$ by construction (Definition 16). As K is $\Delta^{\mathrm{HC}}_1(w)$, it easily follows that $\xi \longmapsto K[\mathbb{U}_\xi]$ is $\Delta^{\mathrm{HC}}_{\mathfrak{n}-1}(w)$. We conclude that $\mathbf{FORC}_K(\varphi)$ is $\Pi^{\mathrm{HC}}_{\mathfrak{n}-1}(w)$.

Step $\mathcal{L}\Pi^1_k \to \mathcal{L}\Sigma^1_{k+1}$. Suppose that $\varphi(\vec{\tau})$ is a $\mathcal{L}\Sigma^1_{k+1}$ formula of the form $\exists y\, \psi(y, \vec{\tau})$, where accordingly ψ is $\mathcal{L}\Pi^1_k$. Let us show that simply

$$\langle p, \vec{\tau}\rangle \in \mathbf{FORC}_K(\varphi) \iff \exists \sigma \in \mathbf{SN}^\omega_\omega(K[\mathbb{U}])\, (\langle p, \sigma, \vec{\tau}\rangle \in \mathbf{FORC}_K(\psi)), \tag{5}$$

which obviously suffices to carry out the step $\mathcal{L}\Pi^1_k \to \mathcal{L}\Sigma^1_{k+1}$.

If σ is a name as in the right-hand side then obviously any p forces $\sigma[G] \in \mathbf{L}[G \cap K[\mathbb{U}]]$, and on the other hand by definition $p \Vdash_{K[\mathbb{U}]} \psi(\sigma, \vec{\tau})$. Thus $p \Vdash_{K[\mathbb{U}]} \varphi(\vec{\tau})$, hence, $\langle p, \vec{\tau}\rangle \in \mathbf{FORC}_K(\varphi)$, as required. Now suppose that $p \Vdash_{K[\mathbb{U}]} \varphi(\vec{\tau})$. This means, by definition, that $p \Vdash_{K[\mathbb{U}]} \exists y\, \psi(y, \vec{\tau})$. By Theorem 5(iv), there is a $K[\mathbb{U}]$-full name $\sigma \in \mathbf{SN}^{\omega}_{\omega}(K[\mathbb{U}])$ such that $p \Vdash_{K[\mathbb{U}]} \psi(\sigma, \vec{\tau})$, thus $\langle p, \sigma, \vec{\tau}\rangle \in \mathbf{FORC}_K(\psi)$.

Step $\mathcal{L}\Sigma^1_k \to \mathcal{L}\Pi^1_k$, $k > \mathfrak{n}$. Suppose that φ is a $\mathcal{L}\Pi^1_k$ formula; accordingly, $\varphi^{\neg}$ is $\mathcal{L}\Sigma^1_k$. It is clear that, under the assumptions that $p \in K[\mathbb{U}]$ and $\tau_1, \dots, \tau_m \in \mathbf{SN}^{\omega}_{\omega}(K[\mathbb{U}])$ are $K[\mathbb{U}]$-full names, the following holds:

$$\langle p, \vec{\tau}\rangle \in \mathbf{FORC}_K(\varphi) \iff \neg\, \exists q \in K[\mathbb{U}]\ (q \le p \wedge \langle p, \vec{\tau}\rangle \in \mathbf{FORC}_K(\varphi^{\neg})), \tag{6}$$

which is sufficient to accomplish the step $\mathcal{L}\Sigma^1_k \to \mathcal{L}\Pi^1_k$. □

6.4. Elementary Equivalence Theorem

According to Theorem 10, sets S satisfying $\mathbb{F}_i(S)$ are different for different indices $i \in \mathcal{I}$, and the difference can be determined, in the extensions of the form $\mathbf{L}[G \restriction z]$, at the level $\Pi^{\mathrm{HC}}_{\mathfrak{n}-1}$ by Corollary 7, that is, $\Pi^1_{\mathfrak{n}}$ (see Remark 2 in Section 4.3). On the other hand, the extensions considered remain rather amorphous w. r. t. lower levels of definability, as witnessed by the following key theorem.

Theorem 13. *Suppose that, in* $\mathbf{L}$: $d \subseteq \mathcal{I}$, $w \in \omega^\omega$, *sets* $b, c \subseteq d^{\complement} = \mathcal{I} \smallsetminus d$ *have equal cardinality,* $d^{\complement}$ *is uncountable,* $K \subseteq \mathbf{P}^* \restriction d$ *is a regular forcing,* $\Psi(y)$ *is a* $\Pi^1_{\mathfrak{n}-1}$ *formula with parameters in* $\omega^\omega \cap \mathbf{L}[G \cap K]$, *and* K, b, c, d *are absolute* $\Delta^{\mathrm{HC}}_1(w)$ *sets. Let* $G \subseteq \mathbb{P}$ *be* $\mathbb{P}$*-generic over* $\mathbf{L}$.

Then, if there is a real $y \in \omega^\omega \cap \mathbf{L}[G \cap K, G \restriction b]$ *such that* $\Psi(y)$ *holds in* $\mathbf{L}[G \cap K, G \restriction d^{\complement}]$, *then there exists* $y' \in \mathbf{L}[G \cap K, G \restriction c]$ *such that* $\Psi(y')$ *holds in* $\mathbf{L}[G \cap K, G \restriction d^{\complement}]$.

Recall that $\Delta^{\mathrm{HC}}_1(w)$ means that w is admitted as the only parameter. The assumption that $d^{\complement}$ is uncountable, can be avoided at the cost of extra complications, but the case of $d^{\complement}$ countable is not considered below. The proof makes use of the transformations introduced in Section 3.7.

Proof. As all cardinals are preserved in $\mathbf{L}[G]$, we w. l. o. g. assume that b, c are countably infinite (or finite of equal cardinality) in $\mathbf{L}$. Suppose towards the contrary that

(A) there is a real $y \in \mathbf{L}[G \cap K, G \restriction b]$ such that $\Psi(y)$ holds in $\mathbf{L}[G \cap K, G \restriction d^{\complement}]$, but

(B) there is no $y' \in \mathbf{L}[G \cap K, G \restriction c]$ satisfying $\Psi(y')$ in $\mathbf{L}[G \cap K, G \restriction d^{\complement}]$.

By Theorem 5(ii), for every real parameter z in Ψ there is a $K[\mathbb{U}]$-full name $\tau_z \in \mathbf{SN}^{\omega}_{\omega}(K[\mathbb{U}])$ such that $z = \tau_z[G]$. Replace each parameter z in $\Psi(x)$ by such a name τ_z in $\mathbf{L}$, and let $\psi(x)$ be the $\mathcal{L}\Pi^1_{\mathfrak{n}-1}(K[\mathbb{U}])$ formula obtained. Then $|\psi| \subseteq d$. Further, the set

$$K' = \{p \in \mathbf{P}^* \restriction (d \cup b) : p \restriction d \in K\} = K \times (\mathbf{P}^* \restriction b) \subseteq \mathbf{P}^* \restriction (d \cup b)$$

is a regular forcing, and $\mathbf{L}[G \cap K, G \restriction b] = \mathbf{L}[G \cap K']$. Choose y by (A). Once again, Theorem 5(ii), yields a $K'[\mathbb{U}]$-full name $\tau_y \in \mathbf{SN}^{\omega}_{\omega}(K'[\mathbb{U}])$ such that $y = \tau_y[G]$. The name τ_y is small, hence the set $|\tau_y| \subseteq d \cup b$ is countable (in $\mathbf{L}$). We let $d_0 = |\tau_y| \cap d$; the set $B = d_0 \cup b$ is still countable and $|\tau_y| \subseteq B$. Thus the formula $\exists^B y\, \psi(y)[G]$ is true in $\mathbf{L}[G \cap K, G \restriction d^{\complement}]$.

Now let $Q = \{p \in \mathbf{P}^* : p \restriction d \in K\} = K \times (\mathbf{P}^* \restriction d^{\complement})$. Thus Q is a regular forcing, and $\mathbf{L}[G \cap K, G \restriction d^{\complement}] = \mathbf{L}[G \cap Q] = \mathbf{L}[G \cap Q[\mathbb{U}]]$. Therefore $\exists^B y\, \psi(y)[G]$ is true in $\mathbf{L}[G \cap Q[\mathbb{U}]]$ by the above. It follows by Theorem 12(ii) that there is a condition $p \in G \cap Q$ such that $p \ {}^{Q}\mathbf{forc}_\infty\ \exists^B y\, \psi(y)$, and, by (B), we can also assume that p $Q[\mathbb{U}]$-forces $\neg\, \exists^C y\, \psi(y)$ over $\mathbf{L}$ where $C = d_0 \cup c$. Further, in $\mathbf{L}$, there exists an ordinal $\xi < \omega_1$ such that

$$p \ {}^{Q}\mathbf{forc}^M_U\ \exists^B y\, \psi(y), \tag{7}$$

where $M = \mathbb{M}_\xi$ and $U = \mathbb{U}_\xi$, and in addition the countable sets d_0, b, c belong to M, $w \in M$, $p \in Q[U]$, $d_0 \cup b \cup c \subseteq A = |U|$, and all names in ψ belong to $M \cap \mathbf{SN}^\omega_\omega(K[U])$, so that $\psi(x)$ is a $\mathcal{L}\Pi^1_{\mathfrak{n}-1}(K[U], M)$ formula.

Now we can assume that both sets $|U| \smallsetminus (d \cup b)$ and $|U| \smallsetminus (d \cup c)$ are infinite. (Otherwise take a suitably bigger ξ.) Then there is a bijection $f \in M$, $f : |U| \xrightarrow{\text{onto}} |U|$, such that $f \upharpoonright d$ is the identity and $f[b] = c$. Define a bijection $\pi \in \mathrm{BIJ}^{\mathcal{I}}_{\mathcal{I}}$ such that $\pi \upharpoonright |U|$ coincides with f and $\pi \upharpoonright (\mathcal{I} \smallsetminus |U|)$ is the identity. Let $q = \pi \cdot p$ and $V = \pi \cdot U$. Acting by π on (7), we obtain, by Lemma 22,

$$q \ {}^{Q}\mathbf{forc}^M_V \ \exists^{\mathrm{C}} y\, \psi(y), \tag{8}$$

Comments: 1) $\pi \cdot Q = Q$ since $\pi \upharpoonright d$ is the identity by construction and $K \subseteq \mathbf{P}^* \upharpoonright d$; 2) $\pi \cdot B = \pi[B] = f[B] = C$ by construction; 3) $\pi \cdot \psi(x)$ is $\psi(x)$ because $|\psi| \subseteq d$ and $\pi \upharpoonright d$ is the identity.

Note that $V \in M$ is a system with $|V| = \pi \cdot |U| = |U|$, and $p \in U$, $q \in V$, $U \upharpoonright d = V \upharpoonright d$ and $q \upharpoonright d = p \upharpoonright d$ by the choice of π and f. In addition, U, V are countable systems in $M \models \mathbf{ZFC}^-_1$. Corollary 2 yields a transformation $\boldsymbol{\alpha} \in \mathbf{Lip}^{\mathcal{I}}$ in M such that $|\boldsymbol{\alpha}| = |U| = |V|$, $\boldsymbol{\alpha} \cdot V = U$, conditions $q' = \boldsymbol{\alpha} \cdot q \in Q[U]$ and p are compatible, and $\boldsymbol{\alpha} \upharpoonright d$ is the identity (as $U \upharpoonright d = V \upharpoonright d$ and $p \upharpoonright d = q \upharpoonright d$). However, then $\boldsymbol{\alpha} \cdot Q = Q$, and $\boldsymbol{\alpha}(\exists^{\mathrm{C}} x\, \psi(x))$ coincides with $\exists^{\mathrm{C}} x\, \psi(x)$ since $|\psi| \subseteq d$. Therefore $q' \ {}^{Q}\mathbf{forc}^M_U \ \exists^{\mathrm{C}} y\, \psi(y)$ by (8) and Lemma 23. This implies $q' \ {}^{Q}\mathbf{forc}_\infty \ \exists^{\mathrm{C}} y\, \psi(y)$. We conclude that q' $Q[\mathbb{U}]$-forces $\exists^{\mathrm{C}} y\, \psi(y)$ over $\mathbf{L}$, by Corollary 8(i). However, q' is compatible with p and p forces the negation of this sentence. The contradiction completes the proof. □

Corollary 10. *Under the assumptions of Theorem 13, if c is uncountable in $\mathbf{L}$, then $\mathbf{L}[G \cap K, G \upharpoonright c]$ is an elementary submodel of $\mathbf{L}[G \cap K, G \upharpoonright d^{\mathrm{C}}]$ w.r.t. all $\Sigma^1_{\mathfrak{n}}$ formulas with parameters in $\omega^\omega \cap \mathbf{L}[G \cap K, G \upharpoonright c]$.*

Proof. Prove by induction that if $k \leq \mathfrak{n}$ then $\mathbf{L}[G \cap K, G \upharpoonright c]$ is an elementary submodel of $\mathbf{L}[G \cap K, G \upharpoonright d^{\mathrm{C}}]$ w.r.t. all Σ^1_k formulas with parameters in $\mathbf{L}[G \cap K, G \upharpoonright c]$. If $k = 2$ then the result holds by the Shoenfield absoluteness theorem. It remains to carry out the step $k \to k+1$ ($k < \mathfrak{n}$). Let $\varphi(x)$ be a Π^1_k formula with parameters in $\mathbf{L}[G \cap K, G \upharpoonright c]$; we have to prove the result for the Σ^1_k formula $\exists\, x\, \varphi(x)$, assuming $k < \mathfrak{n}$. First of all, as the cardinals are preserved, there is a set $\delta \in \mathbf{L}$, $\delta \subseteq d^{\mathrm{C}}$, countable in $\mathbf{L}$ and such that all parameters of φ belong to $\mathbf{L}[G \cap K, G \upharpoonright \delta]$. Let $d' = d \cup \delta$ and $K' = \{p \in \mathbf{P}^* \upharpoonright d' : p \upharpoonright d \in K\}$; we can identify K' with $K \times (\mathbf{P}^* \upharpoonright \delta)$, of course. Then, in $\mathbf{L}$, K' is a regular forcing, $K' \subseteq \mathbf{P}^* \upharpoonright d'$, and all parameters of φ belong to $\mathbf{L}[G \cap K']$.

Now suppose that $\exists\, x\, \varphi(x)$ holds in $\mathbf{L}[G \cap K, G \upharpoonright d^{\mathrm{C}}]$, the bigger of the two models of the lemma. Let this be witnessed by a real $x_0 \in \mathbf{L}[G \cap K, G \upharpoonright d^{\mathrm{C}}] = \mathbf{L}[G \cap K', G \upharpoonright (d')^{\mathrm{C}}]$, where $(d')^{\mathrm{C}} = \mathcal{I} \smallsetminus d' = d^{\mathrm{C}} \smallsetminus \delta$, so that $\varphi(x_0)$ holds in the model $\mathbf{L}[G \cap K, G \upharpoonright d^{\mathrm{C}}] = \mathbf{L}[G \cap K', G \upharpoonright (d')^{\mathrm{C}}]$. As the cardinals are preserved, there is a set $b' \in \mathbf{L}$, $b' \subseteq (d')^{\mathrm{C}}$, countably infinite in $\mathbf{L}$ and such that x_0 belongs to $\mathbf{L}[G \cap K', G \upharpoonright b']$. Since c is uncountable, there exists a set $c' \in \mathbf{L}$, $c' \subseteq (d')^{\mathrm{C}} \cap c$, countably infinite in $\mathbf{L}$. By the choice of δ, there is a real $w' \in \omega^\omega \cap \mathbf{L}$ such that the sets K', d', c', b' are absolute $\Delta^{\mathrm{HC}}_1(w')$ in $\mathbf{L}$. By Theorem 13, there is a real $y_0 \in \mathbf{L}[G \cap K', G \upharpoonright c']$ such that $\varphi(y_0)$ holds in $\mathbf{L}[G \cap K', G \upharpoonright (d')^{\mathrm{C}}] = \mathbf{L}[G \cap K, G \upharpoonright d^{\mathrm{C}}]$, and then in $\mathbf{L}[G \cap K, G \upharpoonright c]$ by the inductive assumption. □

Note that if say c is uncountable but b countable, and d is countable, then Theorem 13 fails by means of the formula "there is a real x such that all reals belong to $\mathbf{L}[x, G \cap K]$", and $G \cap K$ is equiconstructible with a real in this case.

Question 1. It would be very interesting to figure out whether Theorem 13 and Corollary 10 hold also for sets b, c not necessarily constructible.

The following corollary presents a partial positive result.

A set $z \subseteq \mathcal{I} = \omega_1^{\mathbf{L}}$ is bounded iff there is $\alpha < \omega_1^{\mathbf{L}}$ such that $z \subseteq \alpha$.

Corollary 11. *Suppose that $G \subseteq \mathbb{P}$ is $\mathbb{P}$-generic over $\mathbf{L}$, and $z \subseteq \mathcal{I}$ is a set unbounded in $\mathcal{I}$, locally constructible in the sense that $z \cap \alpha \in \mathbf{L}$ for all $\alpha \in \mathcal{I}$, and all $\mathbf{L}$-cardinals are preserved in $\mathbf{L}[G \upharpoonright z]$. Then $\mathbf{L}[G \upharpoonright z]$ is elementarily equivalent to $\mathbf{L}[G]$ w.r.t. all $\Sigma^1_{\mathfrak{n}}$ formulas with parameters in $\mathbf{L}[G \upharpoonright z]$.*

Remark: under the assumptions of the corollary, it is not necessary that $\mathbf{L}[G\restriction z]\subseteq\mathbf{L}[G]$, since the set z is not assumed to belong to $\mathbf{L}[G]$, but we necessarily have $\mathbf{L}[G\restriction z]\cap\omega^\omega\subseteq\mathbf{L}[G]\cap\omega^\omega$ by rather obvious reasons.

Proof. Prove by induction that for any $k\le\mathfrak{n}$, $\mathbf{L}[G\restriction z]$ is elementarily equivalent to $\mathbf{L}[G]$ w.r.t. all Σ^1_k formulas with parameters in $\mathbf{L}[G\restriction z]$. For $k=2$ use Shoenfield's absoluteness. To carry out the step $k\to k+1$ ($k<\mathfrak{n}$), let $\varphi:=\exists\, y\,\psi(y)$ be a Σ^1_{k+1} formula with parameters in $\mathbf{L}[G\restriction z]$. Then, by the choice of z, 1) there is a set $d\in\mathbf{L}$, $d\subseteq z$, countable in $\mathbf{L}$ and such that all parameters in φ belong to $\mathbf{L}[G\restriction d]$, and 2) there is a set $e\in\mathbf{L}$, $e\subseteq z\smallsetminus d$, countably infinite in $\mathbf{L}$.

Now suppose that $\exists\, y\,\psi(y)$ is true in $\mathbf{L}[G]$. This is witnessed by a real $y'\in\mathbf{L}[G\restriction(d\cup e')]$ for a set $e'\in\mathbf{L}$, $e'\subseteq\mathcal{I}\smallsetminus d$, countably infinite in $\mathbf{L}$. Then, by Theorem 13 with $K=\mathbf{P}^*\restriction d$, there is a real $y\in\mathbf{L}[G\restriction(d\cup e)]$, hence, $y\in\mathbf{L}[G\restriction z]$, such that $\psi(y)$ is true in $\mathbf{L}[G]$. However, then $\psi(y)$ is true in $\mathbf{L}[G\restriction z]$ by the inductive hypothesis. Hence φ is true in $\mathbf{L}[G\restriction z]$ as well, as required. □

7. Application 1: Nonconstructible Δ^1_n Reals

In this section, we proveTheorems 1 and 2(i). Theorem 1 provides change of definability of reals situated in the ground set universe $\mathbf{L}$, in generic extensions of $\mathbf{L}$. Thus, any real $a\notin\Sigma^1_{\mathfrak{n}}\cup\Pi^1_{\mathfrak{n}}$ in $\mathbf{L}$ can be placed exactly at $\Delta^1_{\mathfrak{n}+1}$ in an appropriate (almost disjoint) extension of $\mathbf{L}$. Theorem 2 contains several results for nonconstructible reals. The proofs of these results will make use of various results in Sections 5 and 6, in particular, a result (Theorem 11) related to definability of relevant forcing relations.

Assumption 3. *We continue to assume* $\mathbf{V}=\mathbf{L}$ *in the ground universe. We fix an integer* $\mathfrak{n}\ge 2$, *for which Theorems 1 and 2 will be proved, and make use of a system* $\mathbb{U}$ *and the forcing notion* $\mathbb{P}=\mathbf{P}[\mathbb{U}]$ *as in Definition 16; both* $\mathbb{U}$ *and* $\mathbb{P}$ *belong to* $\mathbf{L}$.

7.1. Changing Definability of an Old Real

Proof (Theorem 1). Fix a set $b\subseteq\omega$, $b\notin\Sigma^1_n\cup\Pi^1_n$, in $\mathbf{L}$, and define

$$c=\{2k:k\in b\}\cup\{2k+1:k\notin b\}\ \text{ and }\ K=\mathbf{P}^*\restriction c=\{p\in\mathbf{P}^*:|p|\subseteq c\}.$$

Thus $c\subseteq\omega\subseteq\mathcal{I}=\omega_1$, $c\in\mathbf{L}$, $K\subseteq\mathbf{P}^*$ is a regular forcing. Let $G\subseteq\mathbb{P}$ be a $\mathbb{P}$-generic set over $\mathbf{L}$. Then the set $G\cap K=G\restriction c$ is $K[\mathbb{U}]$-generic over $\mathbf{L}$ by Lemma 9(ii), where $K[\mathbb{U}]=K\cap\mathbf{P}[\mathbb{U}]$, as usual.

Define $S(\nu)=S_G(\nu)\subseteq\mathbf{Seq}$ and $a_\nu=a_G(\nu)=\{k\ge1:s_k\in S_G(\nu)\}$ for every ν, as in Definition 9. We assert that the submodel $\mathbf{L}[G\restriction c]=\mathbf{L}[G\cap K]=\mathbf{L}[\{a_m\}_{m\in c}]$ of the whole generic extension $\mathbf{L}[G]$ witnesses Theorem 1. This amounts to the two following claims:

Claim 3. *It is true in* $\mathbf{L}[G\restriction c]$ *that* c *is* $\Sigma^1_{\mathfrak{n}+1}$, *therefore* b *is* $\Delta^1_{\mathfrak{n}+1}$.

Proof. By definition we have $c=|K|=|K\cap G|$. Therefore c is $\Sigma^{\mathrm{HC}}_{\mathfrak{n}}$ in $\mathbf{L}[G\restriction c]$ by Corollary 7(iii), hence $\Sigma^1_{\mathfrak{n}+1}$ (see Remark 2 in Section 4.3), and $b=\{k:2k\in c\}=\{k:2k+1\notin c\}\in\Delta^1_{\mathfrak{n}+1}$, as required. In more detail,

$$\begin{aligned} c &= \{m:S_G(m)\in\mathbf{L}[G\restriction c]\} &&= \{m:\mathbf{L}[G\restriction c]\models\exists\, S\,\Gamma_m(S)\},\ \text{hence}\\ a &= \{k:S_G(2k)\in\mathbf{L}[G\restriction c]\} &&= \{k:\mathbf{L}[G\restriction c]\models\exists\, S\,\Gamma_{2k}(S)\}\\ &= \{k:S_G(2k+1)\notin\mathbf{L}[G\restriction c]\} &&= \{k:\mathbf{L}[G\restriction c]\models\neg\exists\, S\,\Gamma_{2k+1}(S)\}\end{aligned}$$

by Theorem 10, and it remains to apply Lemma 17. □

Claim 4. *In* $\mathbf{L}[G\restriction c]$: *if* $x\subseteq\omega$ *is* $\Sigma^1_{\mathfrak{n}}$ *then* $x\in\mathbf{L}$ *and* x *is* $\Sigma^1_{\mathfrak{n}}$ *in* $\mathbf{L}$.

Proof (Claim 4). Let $x=\{m:\varphi(m)\}$ in $\mathbf{L}[G\restriction c]$, where $\varphi(m)$ is a $\Sigma^1_{\mathfrak{n}}$ formula. Define $c'=\omega$, $K'=\mathbf{P}^*\restriction\omega$, and $K'[U]=K'\cap\mathbf{P}[U]$, as usual. Prove that

$$m\in x\iff\exists\,\langle M,U\rangle\in\mathbf{sJS}\,\exists\, p\in K'[U]\,(p\ {}^{K'}\!\mathbf{forc}^M_U\ \varphi(m)).\tag{9}$$

The right-hand side of (9) is relativized to $\mathbf{L}$ and is Σ^1_n in $\mathbf{L}$ by Theorem 11. Thus (9) implies Claim 4.

To verify $\Longrightarrow$ in (9), suppose that $m \in x$, that is, $\varphi(m)$ holds in $\mathbf{L}[G \restriction c] = \mathbf{L}[G \cap K]$. Then by Theorem 12(ii) there is a condition $p \in G \cap K$ such that $p\ {}^K\mathbf{forc}_\infty\ \varphi(m)$, that is, $p\ {}^K\mathbf{forc}^M_U\ \varphi(m)$, where $M = \mathbb{M}_\xi$, $U = \mathbb{U}_\xi$ for some $\xi < \omega_1$. As $\mathbb{M}_\xi = M \models \mathbf{ZFC}^-_1$, M contains c, c', and the increasing bijection $\pi \in \mathrm{BIJ}^c_{c'}$. It follows that $q\ {}^{K'}\mathbf{forc}^M_{U'}\ \varphi(m)$, by Lemma 22, where $U' = \pi \cdot U$ and $q = \pi \cdot p$, as obviously $\pi \cdot K = K'$. This implies the right-hand side of (9).

To verify $\Longleftarrow$, let $\langle M', U' \rangle \in \mathbf{sJS}$, $p' \in K'[U']$, and $p'\ {}^{K'}\mathbf{forc}^{M'}_{U'}\ \varphi(m)$. Suppose towards the contrary that $\varphi(m)$ fails in $\mathbf{L}[G \cap K]$, so that there is a condition $q \in G \cap K$ such that $q \Vdash_{K[\mathbb{U}]} \neg\, \varphi(m)$. Then $q \in K[\mathbb{U}]$ (since $G \subseteq \mathbb{P}$), and hence there is an ordinal $\xi < \omega_1$ such that $q \in K[\mathbb{U}_\xi]$, $\omega \cup |U'| \subseteq |\mathbb{U}_\xi|$ and $M' \subseteq \mathbb{M}_\xi$. Then still $p'\ {}^{K'}\mathbf{forc}^{\mathbb{M}_\xi}_{U'}\ \varphi(m)$ by Lemma 18, and Lemma 22 implies $p\ {}^K\mathbf{forc}^{\mathbb{M}_\xi}_U\ \varphi(m)$, where $p = \pi^{-1} \cdot p'$ and $U = \pi^{-1} \cdot U'$. (By obvious reasons, $K = \pi^{-1} \cdot K'$.) Note that $|U| \subseteq |\mathbb{U}_\xi|$ by the choice of ξ. Therefore, we can define a system $V \in \mathbb{M}_\xi$ such that $V \restriction |U| = U$ and $V(\nu) = \mathbb{U}_\xi(\nu)$ for all $\nu \notin |U|$. Then obviously $\langle \mathbb{M}_\xi, U \rangle \preccurlyeq \langle \mathbb{M}_\xi, V \rangle$, therefore $p\ {}^K\mathbf{forc}^{\mathbb{M}_\xi}_V\ \varphi(m)$.

Now, V and $\mathbb{U}_\xi$ are countable systems in $\mathbb{M}_\xi$ with $|V| = |\mathbb{U}_\xi|$ and $p \in K[V]$ but $q \in K[\mathbb{U}_\xi]$. Corollary 2 yields a transformation $\boldsymbol{\alpha} \in \mathbf{Lip}^{\mathcal{I}}$ in M such that $|\boldsymbol{\alpha}| \subseteq c$, $\boldsymbol{\alpha} \cdot V = \mathbb{U}_\xi$, and conditions $r = \boldsymbol{\alpha} \cdot p \in K[\mathbb{U}_\xi]$ and q are compatible. Then $r\ {}^K\mathbf{forc}^{\mathbb{M}_\xi}_{\mathbb{U}_\xi}\ \varphi(m)$ by Lemma 23. (Comment: $\boldsymbol{\alpha}\varphi$ is φ, and $\boldsymbol{\alpha} \cdot K = K$ because regular forcings of the form $K = \mathbf{P}^* \restriction c$ are invariant w.r.t. the transformations in $\mathbf{Lip}^{\mathcal{I}}$.) Thus $r\ {}^K\mathbf{forc}_\infty\ \varphi(m)$, and hence $r \Vdash_{K[\mathbb{U}]} \varphi(m)$ by Corollary 8(i). However, r is compatible with q, and q forces the opposite, a contradiction. This ends the proof of (9). (Claim 4) □

(Theorem 1) □

7.2. Nonconstructible Δ^1_{n+1} Real, Part 1

Here we begin the proof of Theorem 2(i). Suppose that a set $G \subseteq \mathbb{P}$ is $\mathbb{P}$-generic over $\mathbf{L}$. Define $S(\nu) = S_G(\nu) \subseteq \mathbf{Seq}$ and $a_\nu = a_G(\nu) = \{k \geq 1 : \mathbb{s}_k \in S_G(\nu)\}$ for every ν as in Definition 9. Emulating the construction in Section 7.1, put

$$z = z_G = \{0\} \cup \{2k+2 : k \in a_0\} \cup \{2k+1 : k \notin a_0\}. \tag{10}$$

The sets $S_G(\nu)$ and a_ν do not belong to $\mathbf{L}$, accordingly, $z = z_G \in \mathbf{L}[a_0] \smallsetminus \mathbf{L}$—unlike c in Section 7.1. Nevertheless, we are going to prove that the extension $\mathbf{L}[G \restriction z] = \mathbf{L}[\{a_m\}_{m \in z}]$ witnesses Theorem 2(i) with $a = a_0$.

Note that the setup here is not exactly the same as in the proof of Theorem 1 in Section 7.1 since the set z does not belong to $\mathbf{L}$, the ground universe. Therefore we cannot treat $\mathbf{P}^* \restriction z$ as a forcing in $\mathbf{L}$. Instead of $\mathbf{P}^* \restriction z$, we make use of the set K of all conditions $p \in \mathbf{P}^* \restriction \omega$ such that for any $k \geq 1$:

(A) if $2k \in |p|$ then $\mathbb{s}_k \in S_p(0)$;
(B) if $2k - 1 \in |p|$ then $\mathbb{s}_k \in F^\vee_p(0) \smallsetminus S_p(0)$—and hence $2k \notin |p|$ by (A).

as well as the related set $K[\mathbb{U}] = K \cap \mathbb{P} = K \cap \mathbf{P}[\mathbb{U}]$.

Lemma 26. *K is a regular forcing in $\mathbf{L}$. If $G \subseteq \mathbb{P}$ is $\mathbb{P}$-generic over $\mathbf{L}$ then $G \cap K = G \cap K[\mathbb{U}]$ is a set $K[\mathbb{U}]$-generic over $\mathbf{L}$ and $\mathbf{L}[G \cap K] = \mathbf{L}[G \restriction z_G]$.*

Proof. The nontrivial item of the regularity property here is (4) of Definition 8. If $p \in \mathbf{P}^*$ then define $p^* \in \mathbf{P}^*$ to be equal to p everywhere except for $S_{p^*}(0) = S_p(0) \cup S$, where S consists of all strings $s = \mathbb{s}_k$ such that 1) $2k \in |p|$ or $2k-1 \in |p|$, and 2) $s \notin F^\vee_p(0)$ (to make sure that $p^* \leq p$). Now we let d contain 0, all numbers $2k \in |p^*|$ such that $\mathbb{s}_k \in S_{p^*}(0)$, and all numbers $2k-1 \in |p^*|$ such that $\mathbb{s}_k \in F^\vee_p(0) \smallsetminus S_{p^*}(0)$. (Compare to Example 2 in Section 3.2!)

The rest of the lemma follows from Lemma 9. □

Thus extensions of the form $\mathbf{L}[G \restriction z_G]$ considered here are exactly $K[\mathbb{U}]$-generic extensions of $\mathbf{L}$. To check that those extensions satisfy Theorem 2(i), we prove the following Claims 5 and 6. The first of them is entirely similar to Claim 3, so the proof is omitted (left to the reader).

Claim 5. *It is true in $\mathbf{L}[G \restriction z]$ that z is $\Sigma^1_{\mathfrak{n}+1}$, therefore a_0 is $\Delta^1_{\mathfrak{n}+1}$.*

Claim 6. *In $\mathbf{L}[G \restriction z]$, if $x \subseteq \omega$ is $\Sigma^1_{\mathfrak{n}}$, then $x \in \mathbf{L}$ and x is $\Sigma^1_{\mathfrak{n}}$ in $\mathbf{L}$.*

The proof of this claim involves the following lemma.

Lemma 27 (proved below in Section 7.3). *Suppose that $\langle M, U \rangle \in \mathbf{sJS}$, $p \in K[U]$, $q \in K[\mathbb{U}]$. Let Φ be any closed parameter-free $\Sigma^1_{\mathfrak{n}}$ formula. Then it is impossible that simultaneously $q \Vdash_{K[\mathbb{U}]} \neg\Phi$ and $p \ {}^K\mathbf{forc}^M_U\ \Phi$.*

Proof (Claim 6 from the lemma). Assume that $x = \{m : \varphi(m)\}$ in $\mathbf{L}[G \restriction c] = \mathbf{L}[G \cap K]$, where $\varphi(m)$ is a $\Sigma^1_{\mathfrak{n}}$ formula. We claim that then

$$m \in x \iff \exists \langle M, U \rangle \in \mathbf{sJS}\, \exists p \in K[U]\ (p\ {}^K\mathbf{forc}^M_U\ \varphi(m)). \tag{11}$$

This proves Claim 6, of course, by Theorem 11. Now let us check (11) itself; this will be similar to the proof of (9) in Section 7.1.

Assume that $m \in x$, that is, $\varphi(m)$ holds in $\mathbf{L}[G \cap K]$. By Theorem 12(ii) there is a condition $p \in G \cap K$ such that $p\ {}^K\mathbf{forc}_\infty\ \varphi(m)$, that is, $p\ {}^K\mathbf{forc}^M_U\ \varphi(m)$, where $M = \mathbb{M}_\xi$, $U = \mathbb{U}_\xi$, $\xi < \omega_1$. However, this implies the right-hand side of (9).

Now assume that $\langle M, U \rangle \in \mathbf{sJS}$, $p \in K[U]$, and $p\ {}^K\mathbf{forc}^M_U\ \varphi(m)$. Suppose towards the contrary that $\varphi(m)$ is false in $\mathbf{L}[G \cap K]$, so that there is a condition $q \in G \cap K$ such that $q \Vdash_{K[\mathbb{U}]} \neg\varphi(m)$. However, this contradicts Lemma 27. (Claim 6 and Theorem 2(i) modulo Lemma 27) □

7.3. Nonconstructible $\Delta^1_{\mathfrak{n}+1}$ Real, Part 2

We continue the proof of Theorem 2(i).

The proof of Lemma 27 that follows below makes use of transformations in BIJ^ω_ω (bijections of ω) and those in the set $\mathbf{Lip}^\omega = \{\boldsymbol{\alpha} \in \mathbf{Lip}^{\mathcal{I}} : |\boldsymbol{\alpha}| \subseteq \omega\}$, essentially the ω-product of $\mathbf{Lip}$. Yet this will be somewhat more complicated than the proof of Theorem 1 above, because in this case K is not preserved under the action of arbitrary transformations in BIJ^ω_ω and $\mathbf{Lip}^\omega$. This is why we have to consider certain combinations of those transformations.

Namely consider superpositions of the form $\sigma = \pi \circ \boldsymbol{\alpha}$, where $\pi \in \text{BIJ}^\omega_\omega$ and $\boldsymbol{\alpha} \in \mathbf{Lip}^\omega$. (Such σ acts so that $\sigma \cdot x = \pi \cdot (\boldsymbol{\alpha} \cdot x)$ for any applicable object x.)

Remark 4. The set Σ of all σ of this form is a group under the superposition, because the transformations of the two families considered commute so that $\boldsymbol{\alpha} \circ \pi = \pi \circ \boldsymbol{\alpha}'$, where $\boldsymbol{\alpha}' = \pi \cdot \boldsymbol{\alpha}$, that is, $\alpha'_k = \alpha_{\pi(k)}$ for all k.

Definition 22. *A transformation $\sigma = \pi \circ \boldsymbol{\alpha} \in \Sigma$ is K-preserving, if $p \in K \iff \sigma \cdot p \in K$ for all $p \in \mathbf{P}^* \restriction \omega$.*

Not all $\pi \in \text{BIJ}^\omega_\omega$ are K-preserving, and neither is any $\boldsymbol{\alpha} \in \mathbf{Lip}^\omega$ with $\alpha_0 \neq$ the identity. Yet there are plenty of K-preserving transformations in Σ.

Lemma 28. *Let U, V be countable systems with $|U| = |V| = \omega$, and $p \in K[U]$, $q \in K[V]$. There is a K-preserving transformation $\sigma = \pi \circ \boldsymbol{\alpha} \in \Sigma$ such that $\sigma \cdot U = V$, and the conditions $\sigma \cdot p$ and q are compatible.*

Proof. First of all, Lemma 5 yields a transformation $\alpha_0 \in \mathbf{Lip}$ such that $\alpha_0 \cdot U(0) = V(0)$ and the conditions $\alpha_0 \cdot p(0)$ and $q(0)$ (in P^*) are compatible. Define $\boldsymbol{\alpha} = \{\alpha_i\}_{i \in \omega} \in \mathbf{Lip}^\omega$ so that α_0 has just been defined, and $\alpha_k =$ the identity for all $k > 0$. Note that α_0 is a $\subseteq$-preserving bijection of the set $\mathbf{Seq}$ of all non-empty strings of integers. Let $f : \omega \xrightarrow{\text{onto}} \omega$ be the associated permutation of integers, so that $f(k) = n$ iff $\alpha_0(s_k) = s_n$ (and $f(0) = 0$). Define $\pi \in \mathrm{BIJ}^{\omega}_{\omega}$ so that $\pi(0) = 0$ and then $\pi(2k+2) = 2f(k) + 2$ and $\pi(2k+1) = 2f(k) + 1$. It is quite obvious that $\rho = \pi \circ \alpha$ is K-preserving. Let $U' = \rho \cdot U$ and $p' = \rho \cdot p$. Thus U' is a countable system with $|U'| = \omega$, $p' \in K[U']$, and in addition $U'(0) = V(0)$ and the conditions $p'(0) = \alpha_0 \cdot p(0)$ and $q(0)$ are compatible.

It follows from Lemma 5 that there is a transformation $\gamma = \{\gamma_\nu\}_{\nu<\omega} \in \mathbf{Lip}^{\omega}$ such that γ_0 is the identity (and hence γ is K-preserving) and for any $k \geq 1$ we have $\gamma_k \cdot U'(k) = V(k)$ and $\gamma \cdot p'(k)$ is compatible with $q(k)$. We conclude that the transformation $\sigma = \gamma \circ \rho = \gamma \circ \pi \circ \alpha$ is K-preserving, $V = \gamma \cdot U' = \sigma \cdot U$, and the condition $\gamma \cdot p' = (\gamma \circ \pi \circ \alpha) \cdot p$ is compatible with q. Then, we have $\sigma \in \Sigma$ by Remark 4 in Section 7.3. □

Proof (Lemma 27). Suppose towards the contrary that both $q \Vdash_{K[\mathbb{U}]} \neg\,\Phi$ and $p \;{}^{K}\mathbf{forc}^{M}_{U}\; \Phi$. By the way we can w.l.o.g. assume that $|U| \subseteq \omega$, by Lemma 19, and moreover, that $|U| = \omega$ exactly. (Otherwise extend U by $U(\nu) = Q$ for all $\nu \in \omega \smallsetminus |U|$, where Q = all eventually-0 functions $f \in \mathbf{Fun}$.)

There is an ordinal $\xi < \omega_1$ such that $q \in K[\mathbb{U}_\xi]$, $\omega \subseteq |\mathbb{U}_\xi|$, and $M \subseteq \mathbb{M}_\xi$. Let $V = \mathbb{U}_\xi \restriction \omega$. Note that $|q| \subseteq \omega$ since $K \subseteq \mathbf{P}^* \restriction \omega$. Thus $q \in K[V]$. Apply Lemma 28 in $\mathbb{M}_\xi$. It gives a K-preserving transformation $\sigma = \alpha \circ \pi \in \mathbb{M}_\xi$ such that $\sigma \cdot U = V$ and the conditions $r = \sigma \cdot p$ and q (both in $K[V]$) are compatible. On the other hand, we have $r \;{}^{K}\mathbf{forc}^{\mathbb{M}_\xi}_{V}\; \Phi$ by Lemmas 22 and 23, and hence $r \;{}^{K}\mathbf{forc}^{\mathbb{M}_\xi}_{\mathbb{U}_\xi}\; \Phi$ by Lemma 18, that is, $r \;{}^{K}\mathbf{forc}_{\infty}\; \Phi$. Thus $r \Vdash_{K[\mathbb{U}]} \Phi$ by Corollary 8(i). However, r is compatible with q, and q forces the opposite, a contradiction. (Lemma 27) (Claim 6) (Theorem 2(i)) □

8. Application 2: Nonconstructible Self-Definable Δ^1_n Reals

Note that the set a as in Theorem 2(i) is definable in the generic extension of $\mathbf{L}$, considered in Section 7.2, by means of other reals in that extension, including those which do not necessarily belong to $\mathbf{L}[a]$. Claim (ii) of Theorem 2 achieves the same effect with the advantage that a is definable inside $\mathbf{L}[a]$.

The key idea (originally from [9] Section 4) can be explained as follows. Recall that a set of the form $a_0 = a_G(0)$ was made definable in a generic extension of the form $\mathbf{L}[G \restriction z_G]$ by means of the presence/absense of other sets of the form $S_G(\nu)$, $\nu < \omega$, in $\mathbf{L}[G \restriction z]$, see Sections 7.2 and 7.3. Our plan will now be to make each of the according sets $a_G(\nu) \in \mathbf{L}[G \restriction z]$ (note that $a_G(\nu) \subseteq \omega \smallsetminus \{0\}$, see Definition 9), as well as the whole sequence of them, $\Delta^1_{\mathtt{n}+1}$-definable in $\mathbf{L}[G \restriction z]$. In order to do this, we need to develop a suitable coding construction.

Assumption 4. *We continue to assume* $\mathbf{V} = \mathbf{L}$ *in the ground universe. We fix an integer* $\mathtt{n} \geq 2$, *for which Theorem 1(ii) will be proved, and make use of a system* $\mathbb{U}$ *and the forcing notion* $\mathbb{P} = \mathbf{P}[\mathbb{U}]$ *as in Definition 16; both* $\mathbb{U}$ *and* $\mathbb{P}$ *belong to* $\mathbf{L}$.

8.1. Nonconstructible Self-Definable $\Delta^1_{\mathtt{n}+1}$ Reals: The Model

Here we begin **the proof of Theorem 2(ii).** Recall that $\omega^{\omega} = \{s_k : k < \omega\}$ is a fixed recursive enumeration of strings of natural numbers, such that $s_0 = \Lambda$, the empty string, and $s_k \subseteq s_{k'} \Longrightarrow k \leq k'$. Let $\ell^k_i = \mathtt{num}\,(s_k {}^\frown i)$, thus $s_{\ell^k_i} = s_k {}^\frown i$. Then we have:

- Each set $L(k) = \{\ell^k_i : i < \omega\} \subseteq \omega$ is countably infinite, $k < \mathtt{min}_i\, \ell^k_i$, $k \neq k' \Longrightarrow L(k) \cap L(k') = \varnothing$ and $i \neq j \Longrightarrow \ell^k_i \neq \ell^k_j$, and finally each $m \geq 1$ is equal to ℓ^k_i for exactly one pair of indices of $i, k < \omega$.

Define a partial order $\ll$ on ω so that $i \ll k$ iff $s_i \subset s_k$. Obviously $k \ll \ell^k_i$ for all $i, k \in \omega$, and 0 is the $\ll$-least element.

For any sequence $\vec{a} = \{a_k\}_{k<\omega}$ of sets $a_k \subseteq \omega$, we define a set $\zeta_{\vec{a}} \subseteq \omega$ so that:

1) $0 \in \zeta_{\vec{a}}$;

2) if $k \in \zeta_{\vec{a}}$ then, for every i: if $i \in a_k$ then $\ell^k_{2i} \in \zeta_{\vec{a}}$ and $\ell^k_{2i+1} \notin \zeta_{\vec{a}}$, but if $i \notin a_k$ then $\ell^k_{2i} \notin \zeta_{\vec{a}}$ and $\ell^k_{2i+1} \in \zeta_{\vec{a}}$;
3) if $k \notin \zeta_{\vec{a}}$ then $\ell^k_i \notin \zeta_{\vec{a}}$ for all i.

The next theorem obviously implies Theorem 2(ii).

Theorem 14. *Let $G \subseteq \mathbb{P}$ be $\mathbb{P}$-generic over $\mathbf{L}$. Define $\vec{a}[G] = \{a_G(i)\}_{i<\omega}$ and $\zeta = \zeta_{\vec{a}[G]} \subsetneq \omega$. Then $\mathbf{L}[\zeta] = \mathbf{L}[G \restriction \zeta]$, and it holds in $\mathbf{L}[\zeta]$ that:*

(i) *ζ is $\Delta^1_{\mathfrak{n}+1}$;*
(ii) *if $x \subseteq \omega$ is $\Sigma^1_{\mathfrak{n}}$, then $x \in \mathbf{L}$ and x is $\Sigma^1_{\mathfrak{n}}$ in $\mathbf{L}$.*

Proof (will continue towards the end of Section 7). Our arguments will be a more elaborate version of arguments in Sections 7.2, 7.3. We'll make use of the set K of all conditions $p \in \mathbf{P}^* \restriction \omega$ such that for all i and k:

(A) if $\ell^k_{2i} \in |p|$ then $s_i \in S_p(k)$;
(B) if $\ell^k_{2i+1} \in |p|$ then $s_i \in F^{\vee}_p(k) \smallsetminus S_p(k)$—and hence $\ell^k_{2i} \notin |p|$ by (A).

(compare to (A), (B) in Section 7.2), and the related set $K[\mathbb{U}] = K \cap \mathbb{P}$.

Lemma 29. *K is a regular forcing in $\mathbf{L}$. If $G \subseteq \mathbb{P}$ is a set $\mathbb{P}$-generic over $\mathbf{L}$ then $G \cap K = G \cap K[\mathbb{U}]$ is $K[\mathbb{U}]$-generic over $\mathbf{L}$, $|G \cap K| = \zeta_{\vec{a}[G]}$, and accordingly $\mathbf{L}[G \cap K] = \mathbf{L}[G \restriction \zeta_{\vec{a}[G]}] = \mathbf{L}[\zeta_{\vec{a}[G]}]$.*

Proof. As above, the nontrivial item of the regularity property is (4) of Definition 8. Suppose that $p \in \mathbf{P}^*$. Then $|p| \subseteq \omega$ is finite. Let δ be the least $\lll$-initial segment of ω satisfying $|p| \subseteq \delta$; δ is finite as well. Define $p^* \in \mathbf{P}^*$ so that $|p^*| = \delta$ and $F_{p^*}(k) = F_p(k)$ for all k, but the sets $S_{p^*}(k)$ may be strictly bigger than the corresponding sets $S_p(k)$. The definition of $S_{p^*}(k)$ goes on by $\lll$-inverse induction on $k \in \delta$. If $k \in \delta$ is $\lll$-maximal in δ then obviously $k \in |p|$, and we put $S_{p^*}(k) = S_p(k)$. Assume that $k \in \delta$ is not $\lll$-maximal in δ, and the value of $p^*(\ell^k_m) = \langle S_{p^*}(\ell^k_m); F_p(\ell^k_m)\rangle$ is defined for all m such that $\ell^k_m \in \delta$. Put $S_{p^*}(k) = S_p(k) \cup S$, where S consists of all strings $s = s_i$ such that

(a) $\ell^k_{2i+1} \in |p^*| = \gamma$ or $\ell^k_{2i} \in |p^*|$, and
(b) $s \notin F^{\vee}_p(k)$ (to make sure that $p^* \leq p$).

By definition, $|p^*| = \delta$, and if $i, k \in \omega$ and at least one of the numbers ℓ^k_{2i+1}, ℓ^k_{2i} belongs to δ, then the string s_i belongs to $F^{\vee}_{p^*}(k) \cup S_{p^*}(k)$.

Now we define a set $d \subseteq \delta$ so that the decision whether a number $k \in \delta$ belongs to d is made by direct $\lll$-induction. We put $0 \in d$. Suppose that some $k \in \delta$ already belongs to d. We define:

(1) $\ell^k_{2i} \in d$, if $\ell^k_{2i+1} \in \delta$ and $s_i \in S_{p^*}(k)$;
(2) $\ell^k_{2i+1} \in d$, if $\ell^k_{2i} \in \delta$ and $s_i \in F^{\vee}_{p^*}(k) \smallsetminus S_{p^*}(k)$.

A simple verification that p^* and d satisfy Definition 8(4) is left to the reader.

Further, the set $G \cap K = G \cap K[\mathbb{U}]$ is $K[\mathbb{U}]$-generic by Lemma 9(ii).

By definition if $k \in \zeta_{\vec{a}[G]}$ then $a_G(k) = \{i : \ell^k_{2i} \in \zeta_{\vec{a}[G]}\} = \{i : \ell^k_{2i+1} \notin \zeta_{\vec{a}[G]}\} \in \mathbf{L}[\zeta_{\vec{a}[G]}]$, therefore $G \restriction \zeta_{\vec{a}[G]} \in \mathbf{L}[\zeta_{\vec{a}[G]}]$ and $\mathbf{L}[G \restriction \zeta_{\vec{a}[G]}] = \mathbf{L}[\zeta_{\vec{a}[G]}]$.

Now to prove $\mathbf{L}[G \cap K[\mathbb{U}]] = \mathbf{L}[G \restriction \zeta_{\vec{a}[G]}]$ it remains to show that $|G \cap K| = \zeta_{\vec{a}[G]}$—then use Lemma 9(iii). Note that both $|p|$ for any $p \in K$ and $\zeta_{\vec{a}[G]}$ are $\lll$-initial segments. Thus it suffices to check that if $k \in |G \cap K| \cap \zeta_{\vec{a}[G]}$ then

$$\ell^k_{2i+1} \in |G \cap K| \iff \ell^k_{2i+1} \in \zeta_{\vec{a}[G]} \quad \text{and} \quad \ell^k_{2i} \in |G \cap K| \iff \ell^k_{2i} \in \zeta_{\vec{a}[G]}.$$

Prove, e.g., the first equivalence. Suppose that $\ell^k_{2i+1} \in |G \cap K|$. Then $\ell^k_{2i+1} \in |p|$ for some $p \in K$ in G, and we have $s_i \in F^{\vee}_p(k) \smallsetminus S_p(k)$ by (B), so that $s_i \notin S_G(k)$ and accordingly $i \notin a_G(k)$, thus by definition $\boldsymbol{\ell}^k_{2i+1} \in \zeta_{\vec{a}[G]}$. Suppose conversely that $\boldsymbol{\ell}^k_{2i+1} \in \zeta_{\vec{a}[G]}$. Then by definition $i \notin a_G(k)$, hence $\mathbb{s}_i \notin G_G(k)$. This must be forced by some $p \in K \cap G$, and, as $k \in |G \cap K|$, we can assume that $k \in |p|$. However, in this case forcing $\mathbb{s}_i \notin G_G(k)$ means by necessity that just $\mathbb{s}_i \in F^{\vee}_p(k) \smallsetminus S_p(k)$, so there exists a stronger condition $p' \in K \cap G$ with $\boldsymbol{\ell}^k_{2i+1} \in |p'|$. We conclude that $\boldsymbol{\ell}^k_{2i+1} \in |G \cap K|$. (Lemma) □

It follows that $\zeta_{\vec{a}[G]}$ is $\Sigma^1_{\mathtt{n}+1}$ in $\mathbf{L}[G \restriction \zeta]$ by Corollary 7. On the other hand, by definition, if $k \in \zeta_{\vec{a}[G]}$, then, for any k, we have $\boldsymbol{\ell}^k_{2i} \in \zeta_{\vec{a}[G]}$ iff $\boldsymbol{\ell}^k_{2i+1} \notin \zeta_{\vec{a}[G]}$. This easily leads to a Π^1_{n+1} definition of $\zeta_{\vec{a}[G]}$. Thus $\zeta_{\vec{a}[G]}$ is $\Delta^1_{\mathtt{n}+1}$ in $\mathbf{L}[G \restriction \zeta]$, and hence we have claim (i) of Theorem 14. The proof of claim (ii) follows in the next two subsections.

Remark 5. A slightly more elaborate argument, like in the end of Section 4 in [9], shows that even more $\{\zeta_{\vec{a}[G]}\}$ is a $\Pi^1_{\mathtt{n}}$ singleton in $\mathbf{L}[\zeta_{\vec{a}[G]}]$ since $\zeta_{\vec{a}[G]}$ is equalto the only set $\zeta \subseteq \omega$ in $\mathbf{L}[\zeta_{\vec{a}[G]}]$ satisfyings the following requirements:

(a) $0 \in \zeta$, and if $k \notin \zeta$ then $\boldsymbol{\ell}^k_{2i} \notin \zeta$ and $\boldsymbol{\ell}^k_{2i} \notin \zeta$ for all i;
(b) if $k \in \zeta$ then we have $\boldsymbol{\ell}^k_{2i} \in \zeta$ iff $\boldsymbol{\ell}^k_{2i+1} \notin \zeta$ for every i, and
(c) if $k \in \zeta$ then the set $S_{\zeta k} = \{\mathbb{s}_i : \boldsymbol{\ell}^k_{2i} \in \zeta\}$ satisfies $\Gamma_k(S_{\zeta k})$.

The conjunction of them amounts to a $\Pi^1_{\mathtt{n}}$ definition of $\{\zeta\}$ in $\mathbf{L}[\zeta]$.

8.2. Key Lemma

As in Section 7.2, Claim (ii) of Theorem 14 is a consequence of the following lemma (the key lemma from the title), the proof of which will end the proof of theorems 14 and 2(ii).

Lemma 30 (in $\mathbf{L}$). *Suppose that $\langle M, U \rangle \in \mathbf{sJS}$, $p \in K[U]$, $q \in K[\mathbb{U}]$. Let Φ be any closed parameter-free $\Sigma^1_{\mathtt{n}}$ formula. Then it is impossible that simultaneously $q \Vdash_{K[\mathbb{U}]} \neg\,\Phi$ and $p \;{}^K\mathbf{forc}^M_U\; \Phi$.*

Following Definition 22, a transformation $\sigma \in \Sigma$ (see Remark 4 in Section 7.3 on Σ) is called K-preserving, if $p \in K \iff \sigma \cdot p \in K$ for all $p \in \mathbf{P}^* \restriction \omega$. Clearly the regular forcing K here is different (and way more complex in some aspects) than K in Section 7.3. The following lemma is analogous to Lemma 28.

Lemma 31 (in $\mathbf{L}$). *Suppose that U, V are countable systems with $|U| = |V| = \omega$, and $p \in K[U], q \in K[V]$. Then there is a K-preserving transformation $\sigma \in \Sigma$ such that $\sigma \cdot U = V$, and the conditions $\sigma \cdot p$ and q are compatible.*

Proof. The proof resembles the proof of Lemma 28, but is somewhat more complicated. Essentially, we'll have a ramified ω-long iteration in which the construction employed in Lemma 28 will be just one step. We define $\ll$-cones $C_k = \{i \in \omega : k \ll i\}$ and $C'_k = C_k \cup \{k\}$ for any $k \in \omega$.

Claim 7. *If $\boldsymbol{\alpha} = \{\alpha_k\}_{k<\omega} \in \mathbf{Lip}^\omega$, $k_0 \in \omega$, and α_k is the identity for each $k \neq k_0$ then there is a bijection $\pi = \pi[\alpha_{k_0}] \in \mathrm{BIJ}^\omega_\omega$, recursive in $\boldsymbol{\alpha}$, $\ll$-preserving, and such that $\pi(k) = k$ for all $k \notin C_{k_0}$ and $\pi \circ \boldsymbol{\alpha}$ is K-preserving.*

Proof. Note that α_{k_0} is a $\subseteq$-preserving bijection of the set $\mathbf{Seq}$ of all finite non-empty strings of integers. Let $f = f_{\alpha_{k_0}} : \omega \overset{\text{onto}}{\longrightarrow} \omega$ be the associated permutation of integers, so that $f(i) = j$ iff $\alpha_0(\mathbb{s}_i) = \mathbb{s}_j$. Let the transformation $\pi = \pi[\alpha_{k_0}]$ be the identity outside of the strict $\ll$-cone C_{k_0}; in particular, $\pi(k_0) = k_0$. Beyond this, put $\pi(\boldsymbol{\ell}^{k_0}_{2i}) = \boldsymbol{\ell}^{k_0}_{2f(i)}$ and $\pi(\boldsymbol{\ell}^{k_0}_{2i+1}) = \boldsymbol{\ell}^{k_0}_{2f(i)+1}$ for all i. Now, if $k \in C_{k_0}$ and $\pi(k) = k'$ is defined then put $\pi(\boldsymbol{\ell}^k_{2m}) = \boldsymbol{\ell}^{k'}_{2m}$ for all m. (Claim) □

8.3. Matching Permutation

Now, in continuation of the proof of Lemma 31, given any $\boldsymbol{\alpha} \in \mathbf{Lip}^{\omega}$ we outline a construction of a permutation $\Pi \in \text{BIJ}_{\omega}^{\omega}$ such that the superposition $\boldsymbol{\alpha} \circ \Pi$ is K-preserving. Suppose that $\boldsymbol{\alpha} = \{\alpha_k\}_{k<\omega} \in \mathbf{Lip}^{\omega}$. We define

(I) a sequence of numbers k_m, $m < \omega$, such that $k_0 = 0$ and, for any m, k_{m+1} is the least (in the usual order of ω) $\ll$-minimal element of $\omega \smallsetminus d_m$, where $d_m = \{k_i : i \le m\}$,—then $\bigcup_m d_m = \omega$ and each d_m is a $\ll$-initial segment of ω;
(II) for every m, a transformation $\boldsymbol{\alpha}^m = \{\alpha_k^m\}_{k<\omega} \in \mathbf{Lip}^{\omega}$, such that α_k^m is the identity for all $k \ne k_m$ but $\alpha_{k_m}^m = \alpha_{k_m}$, and a matching permutation $\pi^m = \pi[\alpha_{k_m}^m] \in \text{BIJ}_{\omega}^{\omega}$ by Claim 7 — thus π^m is the identity outside of the cone C_{k_m};
(III) a K-preserving superposition $\boldsymbol{\rho}_m = \pi^m \circ \boldsymbol{\alpha}^m$, equal to the identity outside of the extended $\ll$-cone $C'_{k_m} = C_{k_m} \cup \{k_m\}$, in the sense that if U is a system with $|U| = \omega$, or a condition $p \in \mathbf{P}^*$ satisfies $|p| \subseteq \omega$, then $(\boldsymbol{\rho}_m \cdot U)(k) = U(k)$ and $(\boldsymbol{\rho}_m \cdot p)(k) = p(k)$ for all $k \in \omega \smallsetminus C'_{k_m}$.

The whole sequence of transformations is thereby specified by the choice of the components $\alpha_{k_m}^m \in \mathbf{Lip}$, $m \in \omega$; we address this issue below. Now put

$$T_m = \boldsymbol{\rho}_m \circ \cdots \circ \boldsymbol{\rho}_2 \circ \boldsymbol{\rho}_1 \circ \boldsymbol{\rho}_0 \in \Sigma, \quad \Pi_m = \pi_m \circ \cdots \circ \pi_2 \circ \pi_1 \circ \pi_0 \in \text{BIJ}_{\omega}^{\omega}. \tag{12}$$

Claim 8. (i) *the sets $D_m = (\Pi_m)^{-1}(d_m)$ satisfy $\bigcup_m D_m = \omega$;*
(ii) *If $m \le i$ and $k \in D_m$ then $\Pi_i(k) = \Pi_m(k)$;*
(iii) *there is a single permutation $\Pi \in \text{BIJ}_{\omega}^{\omega}$ such that $\Pi(k) = \Pi_m(k) = \Pi_i(k)$ whenever $i \ge m$ and $k \in D_m$.*

Proof. (i) Suppose that $k < \omega$ belongs to some D_m. Prove that any number $j = \boldsymbol{\ell}_{2i}^k$ or $j = \boldsymbol{\ell}_{2i+1}^k$, $i < \omega$, also belongs to some $D_{m'}$. By definition $k' = \Pi_m(k) \in d_m$. The number $j' = \Pi_m(j)$ either belongs to d_m, QED, or is $\ll$-minimal in $\omega \smallsetminus d_m$. In the latter case, we have $\neg\, k_{m'} \ll j'$ for all $m' > m$, and hence $\Pi_{m'}(j)$ is equal to j' for every $m' > m$. Take $m' > m$ big enough for $j' \in d_{m'}$; then $j \in D_{m'}$.

To prove (ii) apply assumption (II) above. Finally (iii) easily follows from items (i), (ii). □

The transformation Π as in item (iii) of the claim can be understood as the infinite superposition $\cdots \circ \pi_m \circ \cdots \circ \pi_2 \circ \pi_1 \circ \pi_0$.

Claim 9. *Suppose that $m \le i$, U is a system, $|U| = \omega$, and $p \in \mathbf{P}^*$, $|p| \subseteq \omega$. Then $(T_i \cdot U)(k_m) = ((\boldsymbol{\alpha} \circ \Pi) \cdot U)(k_m)$ and $(T_i \cdot p)(k_m) = ((\boldsymbol{\alpha} \circ \Pi) \cdot p)(k_m)$.*

Proof. By Claim 8(ii), there is an index $j \in D_m$ such that $k_m = \Pi(j) = \Pi_i(j)$ for all $i \ge m$. Thus $(T_i \cdot U)(k_m)$ is equal to $\alpha_{k_m}^m \cdot U(j) = \alpha_{k_m}^m \cdot ((T_i \cdot U)(k_m))$.

The argument for p is similar. (Claim) □

It follows that the superposition $\boldsymbol{\alpha} \circ \Pi \in \Sigma$ is K-preserving. Indeed, since sets $|p|$ are finite, if $p \in K$ then there is m such that $|p| \subseteq d_m \cap D_m$. However, then $(\boldsymbol{\alpha} \circ \Pi) \cdot p = T_i \cdot p$ by Claim 9, and on the other hand T_i is K-preserving as a finite superposition of K-preserving transformations $\boldsymbol{\rho}^m$.

8.4. Final Argument

Now let U, V, p, q be as in Lemma 31. To accomplish the proof of Lemma 31, we note that the construction of $\boldsymbol{\alpha}^m, \pi^m, \boldsymbol{\rho}_m$ depends on α_{k_m} rather than on $\boldsymbol{\alpha} = \{\alpha_k\}_{k<\omega} \in \mathbf{Lip}^{\omega}$ as a whole. This enables us to carry out the following definition of $\alpha_{k_m} \in \mathbf{Lip}$ $(m \in \omega)$ by induction on m.

Definition 23. *Choose, using Lemma 5, a transformation $\alpha_{k_0} \in \mathbf{Lip}$ such that $\alpha_{k_0} \cdot U(k_0) = V(k_0)$ and the conditions $\alpha_{k_0} \cdot p(k_0)$ and $q(k_0)$ (in P^*) are compatible.*

Now suppose that transformations $\alpha_{k_0}, \ldots, \alpha_{k_m} \in \mathbf{Lip}$ have been defined, and define $\alpha_{k_{m+1}} \in \mathbf{Lip}$. Note that k_{m+1} is a $\ll$-minimal element in $\omega \smallsetminus d_m$, where $d_m = \{k_0, \ldots, k_m\}$, as above. First of all if $\mu \le m$ then define:

- $\boldsymbol{\alpha}^\mu = \{\alpha^\mu_k\}_{k<\omega} \in \mathbf{Lip}^\omega$ *so that* $\alpha^\mu_{k_\mu} = \alpha_{k_\mu}$*, but* α^μ_k *is the identity, whenever* $k \neq k_\mu$*;*
- $\pi^\mu = \pi[\alpha^\mu_{k_\mu}] \in \mathrm{BIJ}^\omega_\omega$ *as in assumption (II) of Section 8.3 — thus* π^μ *is the identity outside of* C_{k_μ}*;*
- *a K-preserving superposition* $\boldsymbol{\rho}_\mu = \pi^\mu \circ \boldsymbol{\alpha}^\mu$*, equal to the identity outside of the extended cone* C'_{k_μ}*, as in assumption (III) of Section 8.3.*

Define Π_m *and* T_m *by (12) above. Put* $U^m = T_m \cdot U$ *and* $p^m = T_m \cdot p$*. By Lemma 5, there is a transformation* $\alpha_{k_{m+1}} \in \mathbf{Lip}$ *such that* $\alpha_{k_{m+1}} \cdot U^m(k_{m+1}) = V(k_{m+1})$ *and the conditions* $\alpha_{k_{m+1}} \cdot p^m(k_{m+1})$ *and* $q(k_{m+1})$ *are compatible.*

After we have defined $\alpha_{k_m} \in \mathbf{Lip}$ by induction on m, let's take the transformation $\boldsymbol{\alpha} = \{\alpha_k\}_{k<\omega} \in \mathbf{Lip}^\omega$ as the input of the construction in Section 8.3. The latter gives us a permutation $\Pi \in \mathrm{BIJ}^\omega_\omega$ of Claim 8, such that the superposition $\sigma = \boldsymbol{\alpha} \circ \Pi \in \Sigma$ is K-preserving. It remains to check that 1) $\sigma \cdot U = V$ and that 2) $\sigma \cdot p$ and q are compatible conditions.

To prove 1), consider any $k = k_{m+1} \in \omega$. (The argument will also work for the case $m = -1$, that is, $k = 0$.) By definition, we have

$$V(k_{m+1}) = \alpha_{k_{m+1}} \cdot U^m(k_{m+1}) = (\boldsymbol{\alpha}^{m+1} \cdot U^m)(k_{m+1}),$$

and hence, as obviously $\pi^{m+1}(k_{m+1}) = k_{m+1}$,

$$V(k_{m+1}) = ((\pi^{m+1} \circ \boldsymbol{\alpha}^{m+1} \circ T_m) \cdot U))(k_{m+1}) = (T_{m+1} \cdot U))(k_{m+1}),$$

therefore $V(k_{m+1}) = ((\boldsymbol{\alpha} \circ \Pi) \cdot U)(k_{m+1}) = (\sigma \cdot U)(k_{m+1})$ by Claim 9, as required. (Lemma 31) □

Proof (Lemma 30)**.** Similar to the proof of Lemma 27, but using Lemma 31 just proved. □

(Theorem 14) (Theorem 2(ii)) □

9. Application 3: Nonconstructible Σ^1_n Reals

Here we prove Theorem 3.

Assumption 5. *We continue to assume* $\mathbf{V} = \mathbf{L}$ *in the ground universe. We fix an integer* $\mathtt{n} \geq 2$*, for which Theorem 3 will be proved, and make use of a system* $\mathbb{U}$ *and the forcing notion* $\mathbb{P} = \mathbf{P}[\mathbb{U}]$ *as in Definition 16; both* $\mathbb{U}$ *and* $\mathbb{P}$ *belong to* $\mathbf{L}$.

9.1. Nonconstructible $\Sigma^1_{\mathtt{n}+1}$ Reals: The Model

The most obvious idea as of getting an extension required is to slightly modify the proof of Theorem 2(ii) in the following direction. Suppose that $G \subseteq \mathbb{P}$ be $\mathbb{P}$-generic over $\mathbf{L}$, and let $S_G(\nu)$ and $a_\nu = a_G(\nu) = \{k \geq 1 : \mathtt{s}_k \in S(i)\}$ be defined as in Definition 9. We proved (see the proof of Theorem 2(i) above) that if

$$z = \{0\} \cup \{2k+2 : k \in a_0\} \cup \{2k+1 : k \notin a_0\}$$

by (10) of Section 7.2 then the set a_0 is $\Delta^1_{\mathtt{n}+1}$ in $\mathbf{L}[G \restriction z]$, and the part $\{2k+2 : k \in a_0\}$ of z is responsible for a_0 being $\Sigma^1_{\mathtt{n}+1}$ in $\mathbf{L}[G \restriction z]$ (by means of the equality $a_0 = \{k : \exists S \Vdash_{2k+2}(S)\}$) while the part $\{2k+1 : k \notin a_0\}$ is responsible for a_0 being Π^1_{n+1} in $\mathbf{L}[G \restriction z]$ (by means of the equality $a_0 = \{k : \neg \exists S \Vdash_{2k+1}(S)\}$). As now the second part is not needed, one might hope that if y is defined by

$$y = y_G := \{0\} \cup a_0 = \{0\} \cup a_G(0) \tag{13}$$

then $\mathbf{L}[G \restriction y]$ will be a model for Theorem 3. At least a_0 will be $\Sigma^1_{\mathtt{n}+1}$ in $\mathbf{L}[G \restriction y]$ by exactly the same reasons. However we have not been able to prove the second part of the theorem, i.e., that all reals $\Delta^1_{\mathtt{n}+1}$ in $\mathbf{L}[G \restriction y]$ belong to $\mathbf{L}$. The point of difficulty is the following hypothesis:

Conjecture 1. Under the assumptions above, if $m \notin y = y_G$ then any parameter-free $\Sigma^1_{\mathfrak{n}+1}$ formula true in $\mathbf{L}[G\restriction y]$ is true in $\mathbf{L}[G\restriction y, a_m]$ as well.

We definitely cannot expect the conjecture to be true for formulas with parameters in $\mathbf{L}[G\restriction y]$ (the smaller model) since if $p \in \mathbf{L}[G\restriction y]$, $p \subseteq \omega$ codes the sequence $\{a_i\}_{i\in y}$ then $\mathbf{Fun} \subseteq \mathbf{L}[p]$ is true in $\mathbf{L}[G\restriction y]$ but false in $\mathbf{L}[G\restriction y, a_m]$.

We have a near-counterexample to Conjecture 1: the formula $\exists\, x\, (\mathbb{\Gamma}_0(x) \wedge \mathbf{Fun} \subseteq \mathbf{L}[x])$ of class $\Sigma^1_{\mathfrak{n}+1}$ (assuming $\mathfrak{n} \geq 3$) holds in $\mathbf{L}[a_0]$ and fails in $\mathbf{L}[a_0, a_1]$. The set $y = \{a_0\}$ is definitely not of the form (13), so this is not literally a counterexample, yet it casts doubts on the approach based on (13).

Now we describe the extension involved in the proof of Theorem 3.

The model we define will be a submodel of the whole extension $\mathbf{L}[G]$, where G is $\mathbb{P}$-generic over $\mathbf{L}$, and a set y of (13) is involved in the definition. We let

$$Y = Y_G = y_G \cup (\mathcal{I} \smallsetminus \omega) = \{0\} \cup a_0 \cup (\mathcal{I} \smallsetminus \omega)\,, \tag{14}$$

where $a_0 = a_G(0)$ (then $Y \in \mathbf{L}[a_0] \smallsetminus \mathbf{L}$) and y_G is defined by (13). The goal is to prove that $\mathbf{L}[G\restriction Y]$ witnesses Theorem 3 with $a = a_0$. The task splits in two claims:

Claim 10. *In $\mathbf{L}[G\restriction Y]$, y is $\Sigma^1_{\mathfrak{n}+1}$, therefore a_0 is $\Sigma^1_{\mathfrak{n}+1}$ as well.*

Claim 11. *In $\mathbf{L}[G\restriction Y]$, if $x \subseteq \omega$ is $\Delta^1_{\mathfrak{n}+1}$ then $x \in \mathbf{L}$ and x is $\Delta^1_{\mathfrak{n}+1}$ in $\mathbf{L}$.*

Claim 10 is established just as similar claims above, so we leave it for the reader.

Let us concentrate on Claim 11. We make use of the set K_0 of all conditions $p \in \mathbf{P}^*\restriction\omega$ such that

$$\text{if } k \geq 1 \text{ and } k \in |p|\text{, then } \mathfrak{s}_k \in S_p(0) \;\; (= \text{Example 2 in Section 3.2}); \tag{15}$$

as well as the related sets: $K = K_0 \times (\mathbf{P}^*\restriction(\mathcal{I} \smallsetminus \omega)) = \{p \in \mathbf{P}^* : p\restriction\omega \in K_0\}$, $K_0[\mathbb{U}] = K_0 \cap \mathbb{P}$, and accordingly $K[\mathbb{U}] = K \cap \mathbb{P}$.

Lemma 32. *It is true in $\mathbf{L}$ that: K_0 and K are regular forcings and absolute Δ^{HC}_1 sets, and if $z \subseteq \mathcal{I}$ contains 0 then the restrictions $K\restriction z$, $K_0\restriction z$ are regular forcings, too.*

If $G \subseteq \mathbb{P}$ is a set $\mathbb{P}$-generic over $\mathbf{L}$ then $G \cap K = G \cap K[\mathbb{U}]$ is a set $K[\mathbb{U}]$-generic over $\mathbf{L}$, $G \cap K_0 = G \cap K_0[\mathbb{U}]$ is a set $K_0[\mathbb{U}]$-generic over $\mathbf{L}$, and

$$\mathbf{L}[G \cap K_0] = \mathbf{L}[G\restriction y_G]\,, \quad \mathbf{L}[G \cap K] = \mathbf{L}[G\restriction Y_G] = \mathbf{L}[G\restriction y_G, G\restriction(\mathcal{I} \smallsetminus \omega)]\,.$$

Proof. To check (4) of Definition 8 for K_0 see Example 2 in Section 3.2. To prove, that the set $K_0\restriction z = \{p \in K_0 : |p| \subseteq z\}$ ($z \in \mathbf{L}$, $z \subseteq \omega$) is regular, argue as in Example 2 in Section 3.2. The rest of the lemma is easy: apply Lemma 9. □

9.2. Key Lemma

Here we establish the following key lemma. Recall that sets y_G, Y_G are defined by (13) and (14).

Lemma 33. *Suppose that $G \subseteq \mathbb{P}$ is $\mathbb{P}$-generic over $\mathbf{L}$, and $y_G = \{0\} \cup a_G(0)$, $y \subseteq \omega$, the symmetric difference $\delta = y \,\Delta\, y_G$ is finite, and $0 \notin \delta$. Then the models $\mathbf{L}[G\restriction Y_G] = \mathbf{L}[G\restriction y_G, G\restriction(\mathcal{I} \smallsetminus \omega)]$ and $\mathbf{L}[G\restriction y, G\restriction(\mathcal{I} \smallsetminus \omega)]$ are $K[\mathbb{U}]$-generic extensions of $\mathbf{L}$, elementarily equivalent w.r.t. all $\Sigma^1_{\mathfrak{n}}$ formulas with parameters in the common part $\mathbf{L}[G\restriction(y_G \cap y), G\restriction(\mathcal{I} \smallsetminus \omega)]$ of the two models.*

Proof. That $\mathbf{L}[G\restriction y_G, G\restriction(\mathcal{I} \smallsetminus \omega)] = \mathbf{L}[G \cap K[\mathbb{U}]]$ is a $K[\mathbb{U}]$-generic extension of $\mathbf{L}$ follows from Lemma 32. Consider $\mathbf{L}[G\restriction y, G\restriction(\mathcal{I} \smallsetminus \omega)]$, the other model.

Let $u = y \smallsetminus y_G$ and $v = y_G \smallsetminus y$; thus $\delta = u \cup v$. Then $v \subseteq a_G(0)$ but $u \cap a_G(0) = \varnothing$ by the definition of y_G. In other words, the finite disjoint sets $S^u = \{\mathfrak{s}_k : k \in u\}$ and $S^v = \{\mathfrak{s}_k : k \in v\}$

satisfy $S^v \subseteq S_G(0)$ but $S^u \cap S_G(0) = \varnothing$. It follows that there is a condition $p \in G \cap K[\mathbb{U}]$ such that $|p| = \{0\}$, $S^v \subseteq S_p(0)$, and $S^u \subseteq F_p^{\vee}(0) \smallsetminus S_p(0)$. We can increase $F_p(0)$ if necessary for $S_p(0) \subseteq F_p^{\vee}(0)$ (a technical requirement) to hold.

Now let q be a condition obtained by the following modification of p: still $|q| = \{0\}$ and $F_q(0) = F_p(0)$ (therefore, q belongs to $K[\mathbb{U}]$ together with p), and $S_q(0) = (S_p(0) \cup S^u) \smallsetminus S^v$. It is clear that $S_q(0) \subseteq F_q^{\vee}(0) = F_p^{\vee}(0)$, so p, q satisfy (3) in Section 3.7. Therefore the map (Definition 12)

$$H_q^p : P = \{p' \in \mathbf{P}^* : p' \leq p\} \stackrel{\text{onto}}{\longrightarrow} Q = \{q' \in \mathbf{P}^* : q' \leq q\}$$

is an order isomorphism of P onto Q by Theorem 6, acting so that:

($*$) if $p' \in P$ then $q' = H_q^p(p')$ satisfies $|p'| = |q'|$, $p'(i) = q'(i)$ for all $i \neq 0$, and even $F_{q'}(0) = F_{p'}(0)$, but $S_{q'}(0) = (S_{p'}(0) \cup S^u) \smallsetminus S^v$.

We conclude that H_q^p also is an order isomorphism of $P \cap \mathbb{P}$ onto $Q \cap \mathbb{P}$ by ($*$), and hence the set $H = \{H_q^p(p') : p' \in G\} \subseteq Q$ is $\mathbb{P}$-generic over $\mathbf{L}$. Moreover it follows from ($*$) that $S_H(i) = S_G(i)$ and $a_H(i) = a_G(i)$ for all $i > 0$, but $S_H(0) = (S_G(0) \cup S^u) \smallsetminus S^v$ and $a_H(0) = (a_G(0) \cup u) \smallsetminus v$. Therefore $y_H = (y_G \cup u) \smallsetminus v = y$, thus $\mathbf{L}[G \restriction y, G \restriction (\mathcal{I} \smallsetminus \omega)]$ is a $K[\mathbb{U}]$-generic extension of $\mathbf{L}$.

As for the elementary equivalence claim, note first of all that the common part $\mathbf{L}[G \restriction (y_G \cap y), G \restriction (\mathcal{I} \smallsetminus \omega)]$ of the two models also is a $K[\mathbb{U}]$-generic extension of $\mathbf{L}$ by the above. (Take $y_G \cap y$ as a new y.) Thus in fact it suffices to prove that under the assumptions of the theorem if $j \in \omega \smallsetminus y_G$ then $\mathbf{L}[G \restriction y_G, a_G(j), G \restriction (\mathcal{I} \smallsetminus \omega)]$ is an elementary extension of $\mathbf{L}[G \restriction y_G, G \restriction (\mathcal{I} \smallsetminus \omega)]$ w.r.t. all $\Sigma^1_{\mathfrak{n}}$ formulas.

Let Φ be a closed $\Sigma^1_{\mathfrak{n}}$ formula with parameters in $\mathbf{L}[G \restriction y_G, G \restriction (\mathcal{I} \smallsetminus \omega)]$. It can be deduced, using either Theorem 5(ii) or directly the CCC property of $\mathbb{P}$ (Theorem 4) that there is an ordinal γ, $\omega \leq \gamma < \omega_1$, such that all parameters of Φ belong to $\mathbf{L}[G \restriction y_G, G \restriction h]$, where $h = \gamma \smallsetminus \omega$.

Put $d = \gamma \smallsetminus \{j\}$; the sets $b = \mathcal{I} \smallsetminus \gamma$, $c = b \cup \{j\}$ have cardinality ω_1, and $Y = h \cup b$ while $Y \cup \{j\} = h \cup c$. It follows from Lemma 32 that $K' = K \restriction d$ is a regular forcing, and in fact $G \restriction \gamma \subseteq K'$ since $j \notin y_G$. Moreover, by definition all of K_0, K, K', d, b, c are absolute $\Delta_1^{\mathrm{HC}}(w)$ sets in $\mathbf{L}$ for some $w \in \omega^\omega$. Therefore by Corollary 10 Φ is simultaneously true in $\mathbf{L}[G \cap K', G \restriction b]$ and in $\mathbf{L}[G \cap K', G \restriction c]$. However,

$$\mathbf{L}[G \cap K', G \restriction b] = \mathbf{L}[G \cap K_0, G \restriction (\gamma \smallsetminus \omega), G \restriction (\mathcal{I} \smallsetminus \gamma)] = \mathbf{L}[y_G, G \restriction (\mathcal{I} \smallsetminus \omega)],$$

and similarly $\mathbf{L}[G \cap K', G \restriction c] = \mathbf{L}[y_G, a_G(j), G \restriction (\mathcal{I} \smallsetminus \omega)]$, as required. □

9.3. Second Key Lemma

In continuation of the proof of Claim 11, we establish another key lemma (Lemma 35). Suppose that

(I) $G \subseteq \mathbb{P}$ is $\mathbb{P}$-generic over $\mathbf{L}$, $x \subseteq \omega$, $x \in \mathbf{L}[G \restriction Y_G]$, and $\varphi(m)$, $\psi(m)$ are parameter-free $\Sigma^1_{\mathfrak{n}+1}$ formulas that give a $\Delta^1_{\mathfrak{n}+1}$ definition for $x = \{m \in \omega : \varphi(m)\} = \{m : \neg\, \psi(m)\}$ in $\mathbf{L}[G \restriction Y_G]$.

Thus it is true in $\mathbf{L}[G]$ that "the equivalence $\forall m\, (\varphi(m) \iff \neg\, \psi(m))$ holds in the model $\mathbf{L}[G \restriction Y_G]$". It follows that there is a condition $p_0 \in G$ with

(II) $p_0 \Vdash_{\mathbb{P}}$ "$\mathbf{L}[\underline{G} \restriction Y_{\underline{G}}] \models \forall m\, (\varphi(m) \iff \neg\, \psi(m))$".

Lemma 34. *If $p_0 \in G$ satisfies (II) then so does $p_0 \restriction \{0\}$.*

Proof. We assume w.l.o.g. that $0 \in |p_0|$. Let $u = |p_0| \smallsetminus \{0\}$. In the context of Theorem 7, put $d = \mathcal{I}$, $c = \omega \smallsetminus u$, and $K' = K_0 \restriction c$ (a regular forcing by Lemma 32). Then $Y_G = (\mathcal{I} \smallsetminus \omega) \cup y_G = (\mathcal{I} \smallsetminus \omega) \cup (y_G \cap c) \cup (y_G \cap u)$, hence

$$\mathbf{L}[G \restriction Y_G] = \mathbf{L}[G \restriction (\mathcal{I} \smallsetminus \omega)] \cup \mathbf{L}[G \restriction (y_G \cap c)] \cup \mathbf{L}[G \restriction (y_G \cap u)].$$

Here $\mathcal{I} \smallsetminus \omega \subseteq d \smallsetminus c$ is constructible while $y_G \cap u \subseteq d \smallsetminus c$ is finite and hence constructible as well. We conclude by Theorem 7(i) that $p_0 \upharpoonright c$ $\mathbb{P}$-forces "$\mathbf{L}[\underline{G} \upharpoonright Y_{\underline{G}}] \models \forall m\,(\varphi(m) \iff \neg\,\psi(m))$". However, $c \cap |p_0| = \{0\}$, so we are done. □

Following the lemma, fix a condition $p_0 \in G$ satisfying $|p_0| = \{0\}$ and (II).

Lemma 35. *Assume* (I) *and* (II) *above. Let $m < \omega$. Then the sentence $\varphi(m)$ is $K[\mathbb{U}]$-decided by p_0 : either $p_0 \Vdash_{K[\mathbb{U}]} \varphi(m)$ or $p_0 \Vdash_{K[\mathbb{U}]} \neg\,\varphi(m)$.*

Proof. It will be technically easier to establish the result in the following form equivalent to the original form by Theorem 5(i):

1°: *the sentence "$\mathbf{L}[\underline{G} \upharpoonright Y_{\underline{G}}] \models \varphi(m)$" is $\mathbb{P}$-decided by p_0.*

Assume that this fails; then there exist two conditions $p, q \in \mathbb{P}$ stronger than p_0 and satisfying:

2°: $q \Vdash_{\mathbb{P}}$ "$\mathbf{L}[\underline{G} \upharpoonright Y_{\underline{G}}] \models \varphi(m)$" and $p \Vdash_{\mathbb{P}}$ "$\mathbf{L}[\underline{G} \upharpoonright Y_{\underline{G}}] \models \neg\,\varphi(m)$".

We can assume that $|p| = |q| = \{0\}$; otherwise apply Lemma 34 to formulas φ and $\neg\,\varphi$. Strengthening p, q, if necessary, we can w.l.o.g. assume that

(a) $F_q(0) = F_p(0)$ and $S_p(0) \cup S_q(0) \subseteq F_p^{\vee}(0) = F_q^{\vee}(0)$. (= (3) in Section 3.8.)

Working towards a contradiction, we w.l.o.g. assume that, in addition to (a), the following holds:

(b) the symmetric difference $S_p(0) \,\Delta\, S_q(0)$ contains a single element $s \in \mathbf{Seq}$.

(Any pair of conditions $p, q \leq p_0$ satisfying (a) can be connected by a finite chain of conditions in which any two neighbours satisfy (b) and are $\leq p_0$.)

Thus suppose that $p, q \leq p_0$, $|p| = |q| = \{0\}$, (a), (b), 2° hold; the goal is to infer a contradiction. The associated transformation H_q^p (Definition 12) maps $P = \{p' \in \mathbb{P} : p' \leq p\}$ onto $Q = \{q' \in \mathbb{P} : q' \leq q\}$ order-preservingly by Theorem 6. Let $G \subseteq P$ be a set $\mathbb{P}$-generic over $\mathbf{L}$ and containing p. Then $H = \{H_q^p(p') : p' \in G\} \subseteq Q$ is $\mathbb{P}$-generic as well, $q \in H$, and hence $\mathbf{L}[H \upharpoonright Y_H] \models \varphi(m)$, while $\mathbf{L}[H \upharpoonright Y_G] \models \neg\,\varphi(m)$ by 2°.

Case 1: $S_p(0) = S_q(0) \cup \{s\}$, where $s = \mathbb{s}_\ell \in \mathbf{Seq} \smallsetminus S_q(0)$. Then the map H_q^p acts so that $q' = H_q^p(p')$ is defined by $|p'| = |q'| \supseteq |p| = |q|$, $p'(\nu) = q'(\nu)$ for all $\nu \in \mathcal{I}, \nu \neq 0$, $F_{q'}(0) = F_{p'}(0)$, but $S_{p'}(0) = S_{q'}(0) \cup \{s\}$. It follows that $S_H(\nu) = S_G(\nu)$ for all $\nu \neq 0$ but $S_G(0) = S_H(0) \cup \{s\}$. Thus $a_G(\nu) = a_H(\nu)$ for $\nu \neq 0$ but $a_G(0) = a_H(0) \cup \{\ell\}$ since $s = \mathbb{s}_\ell$. In other words, $a_G(0) = a_H(0) \cup \{\ell\}$, therefore $y_G = y_H \cup \{\ell\}$ and $\mathbf{L}[G \upharpoonright Y_G] = \mathbf{L}[H \upharpoonright Y_H, a_H(\ell)]$. It follows from Lemma 33 that any $\Sigma^1_{\mathfrak{n}+1}$ formula true in $\mathbf{L}[H \upharpoonright Y_H]$ remains true in $\mathbf{L}[G \upharpoonright Y_G]$. In particular, $\mathbf{L}[G \upharpoonright Y_G] \models \varphi(m)$, a contradiction.

Case 2: $S_q(0) = S_p(0) \cup \{s\}$, where $s = \mathbb{s}_\ell \in \mathbf{Seq} \smallsetminus S_p(0)$. Then, similarly to the above, $a_G(\nu) = a_H(\nu)$ for $\nu \neq 0$, but $a_H(0) = a_G(0) \cup \{\ell\}$. Therefore, $y_H = y_G \cup \{\ell\}$ and $\mathbf{L}[H \upharpoonright Y_H] = \mathbf{L}[G \upharpoonright Y_G, a_G(\ell)]$. Thus any $\Pi^1_{\mathfrak{n}+1}$ formula true in $\mathbf{L}[H \upharpoonright Y_H]$ remains true in $\mathbf{L}[G \upharpoonright Y_G]$ by Lemma 33. Apply this to the formula $\neg\,\psi(m)$, equivalent to $\varphi(m)$ in both models by (II) above. (Note that $p, q \leq p_0$, hence $p_0 \in G \cap H$.) We have $\mathbf{L}[G \upharpoonright Y_G] \models \varphi(m)$, a contradiction. (Lemma 35) □

9.4. Final Argument

Here we finish the proof of both Claim 11 in Section 9.1 and Theorem 3. Suppose that $G \subseteq \mathbb{P}$ is a set $\mathbb{P}$-generic over $\mathbf{L}$, $Y = Y_G$, and a set $x \subseteq \omega$ in $\mathbf{L}[G \upharpoonright Y]$, formulas φ, ψ, and a condition p_0 satisfy assumptions (I), (II) above. Then, by Lemma 35,

$$x = \{m < \omega : p_0 \Vdash_{K[\mathbb{U}]} \varphi(m)\} = \{m : A(p_0, m)\}, \tag{16}$$

where, in $\mathbf{L}$, $A \subseteq K[\mathbb{U}] \times \omega$ is a $\Sigma^{\mathrm{HC}}_{\mathfrak{n}}$ set such that $A(p, m) \iff p \Vdash_{K[\mathbb{U}]} \varphi(m)$ for all $p \in K[\mathbb{U}]$ and m (Corollary 9). It follows that, in $\mathbf{L}$, x is $\Sigma^{\mathrm{HC}}_{\mathfrak{n}}(p_0)$, hence $\Sigma^1_{\mathfrak{n}+1}(w)$ (see Remark 2 in Section 4.3), where $w \in \mathbf{L} \cap \omega^\omega$ is a suitable code of p_0.

To eliminate p_0, consider the set Q of all conditions $p \in K[\mathbb{U}]$ such that $|p| = |p_0|$ and $S_p(\nu) = S_{p_0}(\nu)$ for all $\nu \in |p| = |p_0|$. Note that $K[\mathbb{U}] = K \cap \mathbb{P}$ is a set of the same complexity as $\mathbb{P}$, that is, $\Delta^{\text{HC}}_{\mathfrak{n}-1}$, and hence so is Q because $|p_0|$ and all $S_{p_0}(\nu)$, $\nu \in |p_0|$ are finite sets. It follows that Q is $\Delta^{\text{HC}}_{\mathfrak{n}-1}$.

We now claim that, in $\mathbf{L}$, $x = \{m \in \omega : \exists\, p \in Q\, A(p,m)\}$; this obviously yields x being lightface Σ^1_{n+1} in $\mathbf{L}$. Indeed $\subseteq$ follows by taking $p = p_0 \in Q$ and applying (16). Now suppose that $p \in Q$ and $A(p,m)$, that is, $p \Vdash_{K[\mathbb{U}]} \varphi(m)$. Recall that p_0 decides $\varphi(m)$ by Lemma 35. However, $p_0 \Vdash_{K[\mathbb{U}]} \neg\, \varphi(m)$ is impossible since any condition in Q is compatible with p_0. Therefore $p_0 \Vdash_{K[\mathbb{U}]} \varphi(m)$ as required. Thus $x \in \Sigma^1_{\mathfrak{n}+1}$ in $\mathbf{L}$ is established.

That the complementary set $\omega \smallsetminus x$ is $\Sigma^1_{\mathfrak{n}+1}$ as well is verified the same way, using the formula ψ instead of φ. (Theorem 3)

10. Conclusions and Some Further Results

With proofs of the main theorems accomplished, in this final section some further results are briefly discussed, which we plan to achieve and publish elsewhere.

10.1. Separation

This is another application of submodels of the same basic model. Recall that given a class $\mathbb{K}$ of pointsets, the separation principle $\mathbb{K}$-**Sep** claims that any two disjoint $\mathbb{K}$-sets in the same space can be separated by a set in $\mathbb{K} \cap \mathbb{K}^{\complement}$, where $\mathbb{K}^{\complement}$ consists of all complements of $\mathbb{K}$-sets. The separation principle was introduced by N. Luzin. Luzin proved (see [25]) that $\mathbf{\Sigma}^1_1$-**Sep** holds, and then P. Novikov [26,27] demonstrated that $\mathbf{\Pi}^1_1$-**Sep** fails, while at the second projective level, the other way around, $\mathbf{\Pi}^1_2$-**Sep** holds but $\mathbf{\Sigma}^1_2$-**Sep** fails.

As for higher projective levels, the separation problem belongs to a considerable list of problems related to the projective hieharchy in Luzin's book [25], Chapter V. Further development of set theory showed that Luzin's problems are very hard to solve. Some of them are now known to be independent of the Zermelo–Fraenkel set theory **ZFC**, while some others are still open in different aspects, but it is known that adding Gödel's axiom of constructibility $\mathbf{V} = \mathbf{L}$ solves most of them. In particular, $\mathbf{V} = \mathbf{L}$ implies [28,29] that $\mathbf{\Pi}^1_n$-**Sep** holds but $\mathbf{\Sigma}^1_n$-**Sep** fails for all $n \geq 3$—similarly to the classical case $n = 2$. It follows that the statement $\forall\, n \geq 3\, (\mathbf{\Pi}^1_n\text{-}\mathbf{Sep} \wedge \neg\, \mathbf{\Sigma}^1_n\text{-}\mathbf{Sep})$ is consistent with **ZFC**, and the problem is then to find a model in which we have $\mathbf{\Sigma}^1_n$-**Sep** and/or $\neg\, \mathbf{\Pi}^1_n$-**Sep** (opposite to the state of affairs in $\mathbf{L}$) for one or several or all indices $n \geq 3$. This was the content of problems P 3029 and 3030 in the survey [8] of early years of forcing.

This turns out a very difficult question, and still open in its general forms, especially w. r. t. $\mathbf{\Sigma}^1_n$-**Sep**. (Compare to Problem 9 in [30], Section 9.) As for the $\neg\, \mathbf{\Pi}^1_n$-**Sep** side, there are indications in the set-theoretic literature, that generic extensions, where both $\mathbf{\Sigma}^1_n$-**Sep** and $\mathbf{\Pi}^1_n$-**Sep** fail, are constructed by L. Harrington for $n = 3$ (see 5B.3 in [6]) and for arbitrary $n \geq 3$ (see [8] and [31], p. 230). These results were indeed announced in Harrington's handwritten notes (Addendum A in [32]), with brief outline of some key arguments related mainly to case $n = 3$ and based on almost-disjoint forcing. There are no such results in Harrington's published works, assumed methods in their principal part (arbitrary n) are not used even for any other results, and separability theorems in this context are not considered. An article by Harrington, entitled "Consistency and independence results in descriptive set theory", which apparently might have contained these results, was announced in the References list in [31], to appear in *Ann. of Math.*, 1978, but in fact it has never been published.

The following conjecture concludes Addendum A of Harrington's note [32]:

> *In fact (we believe) there is a model of* **ZFC** *in which Separation fails for all of the following at once*: $\mathbf{\Sigma}^1_n$, $\mathbf{\Pi}^1_n$, $3 \leq n < \omega$, $\mathbf{\Sigma}^m_n$, $\mathbf{\Pi}^m_n$, $1 \leq n < \omega$, $2 \leq m < \omega$. ($\mathbf{\Sigma}^m_n$, $\mathbf{\Pi}^m_n$ are classes arising in the type-theoretic hierarchy).

The hypothesis is partially confirmed by the following our theorem (to appear elsewhere).

Theorem 15 (originally Harrington [32]). *If $\mathfrak{n} \geq 2$ then there is a generic extension of* $\mathbf{L}$ *in which* $\mathbf{\Pi}^1_{\mathfrak{n}+1}$-**Sep** *and* $\mathbf{\Sigma}^1_{\mathfrak{n}+1}$-**Sep** *fail, and moreover*

(i) *there exist disjoint $\Pi^1_{\mathfrak{n}+1}$ sets of reals unseparable by disjoint $\mathbf{\Sigma}^1_{\mathfrak{n}+1}$ sets,*
(ii) *there exist disjoint $\Sigma^1_{\mathfrak{n}+1}$ sets of reals unseparable by disjoint $\mathbf{\Pi}^1_{\mathfrak{n}+1}$ sets.*

Moreover there is a generic extension of $\mathbf{L}$ *in which* (i) *and* (i) *simultaneously hold for all $\mathfrak{n} \geq 2$.*

Note that generic models are defined in [33] in which both $\mathbf{\Sigma}^1_3$-**Sep** and $\mathbf{\Pi}^1_3$-**Sep** fail. We used different technique in [33], mostly related to Jensen's minimal Π^1_2 singleton forcing [10] and its iterated forms (see [34–36]) rather than the almost-disjoint forcing as in this paper.

10.2. Projections of Uniform Sets

In his monograph [25] (pp. 276–291) Nikolas Luzin formulated a number of problems about the structure of the projective classes $\mathbf{\Sigma}^1_n$, $\mathbf{\Pi}^1_n$, $\mathbf{\Delta}^1_n$ (or $\mathbf{A}_n$, $\mathbf{CA}_n$, $\mathbf{B}_n$ in the old notational system). Their general meaning was to extend the results obtained by Luzin himself and P. S. Novikov for classes $\mathbf{\Sigma}^1_1$, $\mathbf{\Pi}^1_1$, $\mathbf{\Delta}^1_1$ (level $n = 1$ of the projective hierarchy) to higher levels. Among these problems, the following stands out, along with the separation problem discussed above:

Projection problem: given $n \geq 2$, find out the nature of projections of uniform (planar) $\mathbf{\Pi}^1_n$ sets in comparison with the class $\mathbf{\Sigma}^1_{n+1}$ of arbitrary projections of $\mathbf{\Pi}^1_n$ sets and with the narrower class $\mathbf{\Delta}^1_n$. (A planar set is *uniform*, if it intersects every vertical line at no more than one point.)

Further research has shown the key importance of structural theorems on projective classes for the development of descriptive set theory. For example, separation principles play essential role in research on subsystems of second-order arithmetic, in particular, in the context of reverse mathematics [5].

If $n = 1$ then every $\mathbf{\Sigma}^1_2$ set is equal to the projection of a uniform $\mathbf{\Pi}^1_1$ set by the Novikov–Kondo–Addison uniformization theorem [6, 4E.4]. Under $\mathbf{V} = \mathbf{L}$, the uniformization theorem fails for classes $\mathbf{\Pi}^1_n$, $n \geq 2$, but nevertheless it is known that if $n \geq 2$ then every $\mathbf{\Sigma}^1_{n+1}$ set is equal to the projection of a uniform $\mathbf{\Pi}^1_n$ set [37]. The next theorem (to appear elsewhere) demonstrates that this property is violated in suitable generic models.

Theorem 16. *If $\mathfrak{n} \geq 2$ then there is a generic extension of* $\mathbf{L}$ *in which*:

(i) *there is a $\Sigma^1_{\mathfrak{n}+1}$ set not equal to the projection of a uniform $\mathbf{\Pi}^1_{\mathfrak{n}+1}$ sets,*
(ii) *there is a $\Delta^1_{\mathfrak{n}+1}$ set not equal to the projection of a uniform $\mathbf{\Pi}^1_{\mathfrak{n}}$ set.*

10.3. Harvey Friedman's Δ^1_n Problem

Problem 87 in [38] requires to prove that for each $n > 2$ there is a model of

$$\mathbf{ZFC} + \text{"the constructible reals are precisely the } \Delta^1_n \text{ reals"}. \tag{17}$$

It is noted in the very end of [38] that Harrington had solved this problem affirmatively. Indeed, a sketch is given in the same handwritten notes [32], of a generic extension of $\mathbf{L}$, in which it is true that $\omega^\omega \cap \mathbf{L} = \Delta^1_3$, as well as a few sentences added as how Harrington planned to get a model in which $\omega^\omega \cap \mathbf{L} = \Delta^1_n$ holds for a given (arbitrary) $n \geq 3$, and a model in which $\omega^\omega \cap \mathbf{L} = \Delta^1_\infty$, where $\Delta^1_\infty = \bigcup_n \Delta^1_n$ (all analytically definable reals). This positively solves Problem 87, including the case $n = \infty$. Full proofs have never been published except for an independent proof of the consistency of $\omega^\omega \cap \mathbf{L} = \Delta^1_\infty$ in [39]. Our plan will be to restore Harrington's proof of the next theorem elsewhere.

Theorem 17 (originally Harrington [32]). (i) *If $\mathfrak{n} \geq 2$ then there is a generic extension of* $\mathbf{L}$ *in which it is true that $\omega^\omega \cap \mathbf{L} = \Delta^1_{\mathfrak{n}+1}$.*
(ii) *There is a generic extension of* $\mathbf{L}$ *in which it is true that $\omega^\omega \cap \mathbf{L} = \Delta^1_\infty$.*

Friedman concludes [38] with a modified version of the above problem, given as Problem 87′: find a model of

$$\mathbf{ZFC} + \text{"for any reals } x, y, \text{ we have: if } x \in \mathbf{L}[y] \text{ then } x \text{ is } \Delta^1_3 \text{ in } y\text{"}. \tag{18}$$

This was solved in the positive by David [40], yet so far it is unknown whether this result generalizes to higher classes Δ^1_n, $n \geq 4$, or Δ^1_∞. We also note that problems (17) and (18) were known in the early years of forcing, see, e.g., problems P 3110, 3111, 3112 in [8].

10.4. Axiom Schemata in 2nd Order Arithmetic

Different axiomatic systems in second-order arithmetic $\mathbf{Z}_2$ is widely represented in modern research, in particular, in the context of reverse mathematics and other sections of proof theory. See e.g., Simpson [5] (Part B), and numerous articles, and from older sources—for example, Kreisel [41], where the choice of subsystems is called the central problem. These systems are obtained by joining a particular combination of comprehension schema **CA**, countable choice **AC**, dependent choice **DC**, transfinite induction **TI** and recursion **TR** , etc., to the basic theory, say $\mathbf{ACA}_0$. The schemata can be specifieded by the complexity of the core formula in the Kleene hierarchy, as well as by allowing or prohibiting parameters. (For the importance of parameters, see [41], section III.)

The relationships between the subsystems have been actively studied. In particular, it is known that Σ^1_{n+1}-**CA** is strictly stronger, than Σ^1_n-**CA**, and the same for **AC** and **DC**. Proofs of these results in e.g., [5, Chapter VII] use the fact that the schema at a higher quantifier level allows to get strictly more countable ordinals, than the schema at a lower level, but in essence, it is utilized that the $(n+1)$th level schema proves the consistency of the nth level schema.

A few more complex results are known, where the compared systems are equiconsistent, despite the increase in quantifier complexity in the schemata, so the consistency argument doesn't work. It such a case one has to resort to set theoretic methods. This is the old result of A. Levy [42] that Σ^1_3-**AC** does not follow from **CA**, as well as a recent theorem in [43] saying that Σ^1_3-**DC** does not follow from **AC**; both are obtained using complex generic models of **ZF** without the full axiom of choice. The task of our further research in this direction will be to prove consistency theorems that demonstrate the importance of both the quantifier complexity and the presence of parameters in the $\mathbf{Z}_2$ schemata.

Theorem 18 (to appear elsewhere). *If* $\mathtt{n} \geq 2$, *then the theory* $\mathbf{ACA}_0 + \mathbf{CA}^* + \boldsymbol{\Sigma}^1_{\mathtt{n}}$-**CA** *does not imply* $\Sigma^1_{\mathtt{n}+1}$-**CA** (*unless inconsistent, of course*).

Here **CA*** is the parameter-free part of the comprehension schema **CA**. Thus, both the quantifier complexity and the presence of parameters are essential for the deductive power of the comprehension schema in second-order arithmetic.

Theorem 19 (to appear elsewhere). *If* $\mathtt{n} \geq 2$, *then the theory* $\mathbf{ACA}_0 + \mathbf{CA} + \mathbf{AC} + \boldsymbol{\Sigma}^1_{\mathtt{n}}$-**DC** *does not imply* $\Sigma^1_{\mathtt{n}+1}$-**DC** (*unless inconsistent*).

Remark 6. We are grateful to one of the reviewers for pointing out possible connections of our research with some questions of fuzzy set theory [44,45], yet this issue cannot be considered for a short time.

Supplementary Materials: The following are available online at http://www.mdpi.com/2227-7390/8/6/910/s1.

Author Contributions: Conceptualization, V.K. and V.L.; methodology, V.K. and V.L.; validation, V.K. and V.L.; formal analysis, V.K. and V.L.; investigation, V.K. and V.L.; writing original draft preparation, V.K.; writing review and editing, V.K.; project administration, V.L.; funding acquisition, V.L. All authors have read and agreed to the published version of the manuscript.

Funding: This research was funded by Russian Foundation for Basic Research RFBR grant number 18-29-13037.

Acknowledgments: We thank the anonymous reviewers for their thorough review and highly appreciate the comments and suggestions, which significantly contributed to improving the quality of the publication.

Conflicts of Interest: The authors declare no conflict of interest. The funders had no role in the design of the study; in the collection, analyses, or interpretation of data; in the writing of the manuscript, or in the decision to publish the results.

References

1. Hadamard, J.; Baire, R.; Lebesgue, H.; Borel, E. Cinq lettres sur la théorie des ensembles. *Bull. Soc. Math. Fr.* **1905**, *33*, 261–273. [CrossRef]
2. Antos, C.; Friedman, S.; Honzik, R.; Ternullo, C. (Eds.) *The Hyperuniverse Project and Maximality;* Birkhäuser: Basel, Switzerland, 2018; p. XI, 265.
3. Friedman, S.D.; Schrittesser, D. *Projective Measure without Projective Baire;* Memoirs of the American Mathematical Society; AMS: Providence, RI, USA, 2020.
4. Friedman, S.D.; Hyttinen, T.; Kulikov, V. *Generalized Descriptive Set Theory and Classification Theory;* Memoirs of the American Mathematical Society; AMS: Providence, RI, USA, 2014; Volume 1081, p. 80.
5. Simpson, S.G. *Subsystems of Second Order Arithmetic*, 2nd ed.; Perspectives in Logic; Cambridge University Press: Cambridge, UK; ASL: Urbana, IL, USA, 2009; p. 444.
6. Moschovakis, Y.N. *Descriptive Set Theory;* Studies in Logic and the Foundations of Mathematics; North-Holland: Amsterdam, The Netherlands, 1980; Volume 100, p. 637.
7. Shoenfield, J.R. The problem of predicativity. In *Essays Found. Math., dedicat. to A. A. Fraenkel on his 70th Anniv.;* Bar-Hillel, Y., Ed.; North-Holland: Amsterdam, The Netherlands, 1962; pp. 132–139.
8. Mathias, A.R.D. Surrealist landscape with figures (a survey of recent results in set theory). *Period. Math. Hung.* **1979**, *10*, 109–175. [CrossRef]
9. Jensen, R.B.; Solovay, R.M. Some applications of almost disjoint sets. In *Math. Logic Found. Set Theory, Proc. Int. Colloqu. Jerusalem 1968;* Studies in Logic And the Foundations of Mathematics; Bar-Hillel, Y., Ed.; North-Holland: Amsterdam, The Netherlands, 1970; Volume 59, pp. 84–104.
10. Jensen, R.B. Definable sets of minimal degree. In *Math. Logic Found. Set Theory, Proc. Int. Colloqu. Jerusalem 1968;* Studies in Logic And the Foundations of Mathematics; Bar-Hillel, Y., Ed.; North-Holland: Amsterdam, The Netherlands, 1970; Volume 59, pp. 122–128.
11. Friedman, S.D. Provable Π^1_2-singletons. *Proc. Am. Math. Soc.* **1995**, *123*, 2873–2874.
12. Harrington, L.; Kechris, A.S. Π^1_2 singletons and $0^\#$. *Fundam. Math.* **1977**, *95*, 167–171.
13. Stanley, M.C. A Π^1_2 singleton incompatible with $0^\#$. *Ann. Pure Appl. Logic* **1994**, *66*, 27–88.
14. Kanovei, V.G. On the nonemptiness of classes in axiomatic set theory. *Math. USSR, Izv.* **1978**, *12*, 507–535. [CrossRef]
15. Enayat, A. On the Leibniz – Mycielski axiom in set theory. *Fundam. Math.* **2004**, *181*, 215–231. [CrossRef]
16. Abraham, U. Minimal model of "$\aleph_1^L$ is countable" and definable reals. *Adv. Math.* **1985**, *55*, 75–89. [CrossRef]
17. Kanovei, V.; Lyubetsky, V. Definable E_0 classes at arbitrary projective levels. *Ann. Pure Appl. Logic* **2018**, *169*, 851–871. [CrossRef]
18. Kanovei, V.; Lyubetsky, V. Definable minimal collapse functions at arbitrary projective levels. *J. Symb. Log.* **2019**, *84*, 266–289. [CrossRef]
19. Kanovei, V.; Lyubetsky, V. Non-uniformizable sets with countable cross-sections on a given level of the projective hierarchy. *Fundam. Math.* **2019**, *245*, 175–215. [CrossRef]
20. Barwise, J. (Ed.) Handbook of Mathematical Logic. In *Studies in Logic and the Foundations of Mathematics;* North-Holland: Amsterdam, The Netherlands 1977; Volume 90.
21. Jech, T. *Set Theory;* The Third Millennium Revised and Expanded Edition; Springer-Verlag: Berlin, Germany, 2003; p. 769.
22. Kanovei, V.G. Projective hierarchy of Luzin: Modern state of the theory. In *Handbook of Mathematical Logic. Part 2. Set Theory. (Spravochnaya Kniga Po Matematicheskoj Logike. Chast' 2. Teoriya Mnozhestv);* Transl. from the English; Barwise, J., Ed.; Nauka: Moscow, Russia, 1982; pp. 273-364.
23. Jensen, R.B.; Johnsbraten, H. A new construction of a non-constructible Δ^1_3 subset of ω. *Fundam. Math.* **1974**, *81*, 279–290. [CrossRef]

24. Gitman, V.; Hamkins, J.D.; Johnstone, T.A. What is the theory ZFC without power set? *Math. Log. Q.* **2016**, *62*, 391–406. [CrossRef]
25. Lusin, N. *LeÇons Sur Les Ensembles Analytiques Et Leurs Applications*; Gauthier-Villars, Paris, France, 1930.
26. Novikoff, P. Sur les fonctions implicites mesurables B. *Fundam. Math.* **1931**, *17*, 8–25, . [CrossRef]
27. Novikoff, P. Sur la séparabilité des ensembles projectifs de seconde classe. *Fundam. Math.* **1935**, *25*, 459–466, . [CrossRef]
28. Addison, J.W. Separation principles in the hierarchies of classical and effective descriptive set theory. *Fundam. Math.* **1959**, *46*, 123–135. [CrossRef]
29. Addison, J.W. Some consequences of the axiom of constructibility. *Fundam. Math.* **1959**, *46*, 337–357. [CrossRef]
30. Friedman, S.D. Constructibility and class forcing. In *Handbook of Set Theory*; Springer: Berlin, Germany, 2010; Volume 3, pp. 557–604.
31. Hinman, P.G. *Recursion-Theoretic Hierarchies;* Perspectives in Mathematical Logic; Springer: Berlin, Germany, 1978; p. 480.
32. Harrington, L. The constructible reals can be anything. Preprint dated May 1974 with several addenda dated up to October 1975: (A) Models where Separation principles fail, May 74; (B) Separation without Reduction, April 75; (C) The constructible reals can be (almost) anything, Part II, May 75.
33. Kanovei, V.; Lyubetsky, V. Counterexamples to countable-section Π^1_2 uniformization and Π^1_3 separation. *Ann. Pure Appl. Logic* **2016**, *167*, 262–283. [CrossRef]
34. Kanovei, V. Non-Glimm-Effros equivalence relations at second projective level. *Fund. Math.* **1997**, *154*, 1–35.
35. Kanovei, V. An Ulm-type classification theorem for equivalence relations in Solovay model. *J. Symb. Log.* **1997**, *62*, 1333–1351. [CrossRef]
36. Kanovei, V. On non-wellfounded iterations of the perfect set forcing. *J. Symb. Log.* **1999**, *64*, 551–574. [CrossRef]
37. Kanovei, V.G. N. N. Luzin's problems on imbeddability and decomposability of projective sets. *Math. Notes* **1983**, *32*, 490–499. [CrossRef]
38. Friedman, H. One hundred and two problems in mathematical logic. *J. Symb. Log.* **1975**, *40*, 113–129. [CrossRef]
39. Kanovei, V.G. The set of all analytically definable sets of natural numbers can be defined analytically. *Math. USSR Izvestija* **1980**, *15*, 469–500. [CrossRef]
40. David, R. Δ^1_3 reals. *Ann. Math. Logic* **1982**, *23*, 121–125. [CrossRef]
41. Kreisel, G. A survey of proof theory. *J. Symb. Log.* **1968**, *33*, 321–388, . [CrossRef]
42. Levy, A. Definability in axiomatic set theory II. In *Math. Logic Found. Set Theory, Proc. Int. Colloqu. Jerusalem 1968*; Bar-Hillel, Y., Ed.; North-Holland: Amsterdam, Netherlands, 1970; pp. 129–145.
43. Friedman, S.D.; Gitman, V.; Kanovei, V. A model of second-order arithmetic satisfying AC but not DC. *J. Math. Log.* **2019**, *19*, 39. Id/No 1850013. [CrossRef]
44. Burrascano, P.; Callegari, S.; Montisci, A.; Ricci, M.; Versaci, M. (Eds.) *Ultrasonic Nondestructive Evaluation Systems*; Springer International Publishing: Berlin, Germany, 2015; p. XXXIII, 324.
45. Versaci, M. Fuzzy approach and Eddy currents NDT/NDE devices in industrial applications. *Electron. Lett.* **2016**, *52*, 943–945. [CrossRef]

mathematics

MDPI

Article

On the Δ^1_n Problem of Harvey Friedman

Vladimir Kanovei *,† **and Vassily Lyubetsky** †

Institute for Information Transmission Problems of the Russian Academy of Sciences (Kharkevich Institute), 127051 Moscow, Russia; lyubetsk@iitp.ru

* Correspondence: kanovei@iitp.ru

† These authors contributed equally to this work.

Received: 2 August 2020; Accepted: 23 August 2020; Published: 1 September 2020

Abstract: In this paper, we prove the following. If $n \geq 3$, then there is a generic extension of **L**, the constructible universe, in which it is true that the set $\mathscr{P}(\omega) \cap \mathbf{L}$ of all constructible reals (here—subsets of ω) is equal to the set $\mathscr{P}(\omega) \cap \Delta^1_n$ of all (lightface) Δ^1_n reals. The result was announced long ago by Leo Harrington, but its proof has never been published. Our methods are based on almost-disjoint forcing. To obtain a generic extension as required, we make use of a forcing notion of the form $\mathbb{Q} = \mathbb{C} \times \prod_\nu \mathbb{Q}_\nu$ in **L**, where $\mathbb{C}$ adds a generic collapse surjection b from ω onto $\mathscr{P}(\omega) \cap \mathbf{L}$, whereas each $\mathbb{Q}_\nu$, $\nu < \omega_2^{\mathbf{L}}$, is an almost-disjoint forcing notion in the ω_1-version, that adjoins a subset S_ν of $\omega_1^{\mathbf{L}}$. The forcing notions involved are independent in the sense that no $\mathbb{Q}_\nu$-generic object can be added by the product of $\mathbb{C}$ and all $\mathbb{Q}_\xi$, $\xi \neq \nu$. This allows the definition of each constructible real by a Σ^1_n formula in a suitably constructed subextension of the $\mathbb{Q}$-generic extension. The subextension is generated by the surjection b, sets $S_{\omega \cdot k+j}$ with $j \in b(k)$, and sets S_ξ with $\xi \geq \omega \cdot \omega$. A special character of the construction of forcing notions $\mathbb{Q}_\nu$ is **L**, which depends on a given $n \geq 3$, obscures things with definability in the subextension enough for vice versa any Δ^1_n real to be constructible; here the method of *hidden invariance* is applied. A discussion of possible further applications is added in the conclusive section.

Keywords: Harvey Friedman's problem; definability; nonconstructible reals; projective hierarchy; generic models; almost-disjoint forcing

MSC: 03E15; 03E35

Dedicated to the 70-th Anniversary of A. L. Semenov.

1. Introduction

Problem 87 in Harvey Friedman's treatise *One hundred and two problems in mathematical logic* [1] requires proof that for each n in the domain $2 < n \leq \omega$ there is a model of

$$\mathbf{ZFC} + \text{“the constructible reals are precisely the } \Delta^1_n \text{ reals”}. \tag{1}$$

(For $n \leq 2$ this is definitely impossible e.g., by the Shoenfield's absoluteness theorem.) This problem was generally known in the early years of forcing, see, e.g., problems 3110, 3111, 3112 in an early survey [2] (the original preprint of 1968) by Mathias. At the very end of [1], it is noted that Leo Harrington had solved this problem affirmatively. For a similar remark, see [2] (p. 166), a comment to P 3110. And indeed, Harrington's handwritten notes [3] (pp. 1–4) contain a sketch of a generic extension of **L**, based on the almost-disjoint forcing of Jensen and Solovay [4], in which it is true that $\omega^\omega \cap \mathbf{L} = \Delta^1_3$. Then a few sentences are added on page 5 of [3], which explain, as how Harrington planned to get a model in which $\omega^\omega \cap \mathbf{L} = \Delta^1_n$ holds for a given (arbitrary) natural index $n \geq 3$, and a model in which

$\omega^\omega \cap \mathbf{L} = \Delta^1_\infty$ where $\Delta^1_\infty = \bigcup_n \Delta^1_n$ (all analytically definable reals). This positively solves Problem 87, including the case $n = \infty$. Different cases of higher order definability are observed in [3] as well.

Yet no detailed proofs have ever emerged in Harrington's published works. An article by Harrington, entitled "Consistency and independence results in descriptive set theory", which apparently might have contained these results among others, was announced in the References list in Peter Hinman's book [5] (p. 462) to appear in *Ann. of Math.*, 1978, but in fact, this or similar article has never been published by Harrington.

One may note that finding a model for (1) belongs to the "definability of definable" type of mathematical problems, introduced by Alfred Tarski in [6], where the definability properties of the set D_{1M}, of all sets $x \subseteq \omega$ definable by a parameter-free type-theoretic formula with quantifiers bounded by type M, are discussed for different values of $M < \omega$. In this context, case $n = \infty$ in (1) is equivalent to case $M = 1$ in the Tarski problem, whereas cases $n < \infty$ in (1) can be seen as refinements of case $m = 1$ in the Tarski problem, because classes Δ^1_n are well-defined subclasses of $D_{11} = \bigcup_{n<\omega} \Delta^1_n$.

The goal of this paper is to present a complete proof of the following part of Harrington's statement that solves the mentioned Friedman's problem. No such proof has been known so far in mathematical publications, and this is the **motivation** for our research.

Theorem 1 (Harrington)**.** *If $2 \le \mathfrak{n} < \infty$ then there is a generic extension of $\mathbf{L}$ in which it is true that the constructible reals are precisely the $\Delta^1_{\mathfrak{n}+1}$ reals.*

The Δ^1_∞ case of Harrington's result, as well as different results related to Tarski's problems in [6], will be subject of a forthcoming publication.

This paper is dedicated to the proof of Theorem 1. This will be another application of the technique introduced in our previous paper [7] in this Journal, and in that sense this paper is a continuation and development of the research started in [7]. However, the problem considered here, i.e., getting a model for (1), is different from and irreducible to the problems considered in [7] and related to definability and constructability of individual reals. Subsequently the technique applied in [7] is considerably modified and developed here for the purposes of this new application. In particular, as the models involved here by necessity satisfy $\omega_1^{\mathbf{L}} < \omega_1$ (unlike the models considered in [7], which satisfy the equality $\omega_1^{\mathbf{L}} = \omega_1$), the almost-disjoint forcing is combined with a cardinal collapse forcing in this paper. And hence we will have to substantially deviate from the layout in [7], towards a modification that shifts the whole almost-disjoint machinery from ω to ω_1.

Section 2: we set up the almost-disjoint forcing in the ω_1-version. That is, we consider the sets $\mathbf{SEQ} = (\omega_1)^{<\omega_1}$ and $\mathbf{FUN} = (\omega_1)^{\omega_1}$ in $\mathbf{L}$, the constructible universe, and, given $u \subseteq \mathbf{FUN}$, we define a forcing notion $Q[u]$ which adds a set $G \subseteq \mathbf{SEQ}$ such that if $f \in \mathbf{FUN}$ in $\mathbf{L}$ then G covers f iff $f \notin u$, where covering means that $f \upharpoonright \xi \in G$ for unbounded-many $\xi < \omega_1^{\mathbf{L}}$. We also consider two types of transformations related to forcing notions of the form $Q[u]$.

Section 3. We let $\mathcal{I} = \omega_2^{\mathbf{L}}$ be the index set. Arguing in $\mathbf{L}$, we consider *systems* $U = \{U(\nu)\}_{\nu \in \mathcal{I}}$, where each $U(\nu) \subseteq \mathbf{FUN}$ is dense. Given such U, the *product almost-disjoint forcing* $\mathbf{Q}[U] = \mathbb{C} \times \prod_{\nu \in \mathcal{I}^+} Q[U(\nu)]$ (with finite support) is defined in $\mathbf{L}$, where $\mathbb{C} = (\mathscr{P}(\omega))^{<\omega}$ is a version of Cohen's collapse forcing. Such a forcing notion adjoins a generic map $\boldsymbol{b}_G : \omega \overset{\text{onto}}{\longrightarrow} \mathscr{P}(\omega) \cap \mathbf{L}$ to $\mathbf{L}$, and adds an array of sets $G(\nu) \subseteq \mathbf{SEQ}$ (where $\nu \in \mathcal{I}$) as well, so that each $G(\nu)$ is a $Q[U(\nu)]$-generic set over $\mathbf{L}$. We also investigate the structure of related product-generic extensions and their subextensions, and transformations of forcing notions of the form $\mathbf{Q}[U]$.

Section 4. Given $\mathfrak{n} \ge 2$, we define a system $\mathbb{U} \in \mathbf{L}$ as above, which has some remarkable properties, in particular, (1) being $Q[\mathbb{U}(\nu)]$-generic is essentially a $\Pi^1_{\mathfrak{n}}$ property in all suitable generic extensions, (2) if $\nu \in \mathcal{I}$ and $G \subseteq \mathbf{Q}[\mathbb{U}]$ is generic over $\mathbf{L}$, then the extension $\mathbf{L}[\boldsymbol{b}_G, \{G(\nu')\}_{\nu' \ne \nu}]$ contains no $Q[\mathbb{U}(\nu)]$-generic reals, and (3) all submodels of $\mathbf{L}[G]$ of certain kind are elementarily equivalent w.r.t. $\Sigma^1_{\mathfrak{n}}$ formulas. The latter property is summarized in the key technical instrument,

Theorem 4 (the elementary equivalence theorem), whose proof is placed in a separate Section 6. To prove Theorem 1, we make use of a related generic extension $\mathbf{L}[\boldsymbol{b}_G, \{G(\nu)\}_{\nu \in W[G]}]$, where

$$W[G] = w[G] \cup W = \{\omega \cdot k + 2^j : j \in \boldsymbol{b}_G(k)\} \cup \{\omega \cdot k + 3^j : j, k < \omega\} \cup \{\nu \in \mathcal{I} : \nu \geq \omega^2\}$$

(see Lemma 23), and $\cdot$ is the ordinal multiplication. The first term in $W[G]$ provides a suitable definition of each set $x = \boldsymbol{b}_G(k) \in \mathbf{L}$ in the model $\mathbf{L}[\boldsymbol{b}_G, \{G(\nu)\}_{\nu \in W[G]}]$, namely

$$\boldsymbol{b}_G(k) = \{j : \text{there exists a } Q[U(\nu)]\text{-generic set over } \mathbf{L}\},$$

while the second and third terms in $W[G]$ are added for technical reasons. The proof itself goes on in Section 4.5, modulo Theorem 4.

We introduce *forcing approximations* in Section 5, a forcing-like relation used to prove the elementary equivalence theorem. Its key advantage is the invariance under some transformations, including the permutations of the index set $\mathcal{I}$, see Section 5.4. The actual forcing notion $\mathbb{Q} = \mathbf{Q}[\mathbb{U}]$ is absolutely not invariant under permutations, but the $\mathfrak{n}$-completeness property, maintained through the inductive construction of $\mathbb{U}$ in $\mathbf{L}$, allows us to prove that the auxiliary forcing relation is connected to the truth in $\mathbb{Q}$-generic extensions exactly as the true $\mathbb{Q}$-forcing relation does—up to the level $\Sigma^1_{\mathfrak{n}}$ of the projective hierarchy (Lemma 33). We call this construction *the hidden invariance technique* (see Section 6.1).

Finally, Section 6 presents the proof of the elementary equivalence theorem, with the help of forcing approximations, and hence completes the proof of Theorem 1.

The flowchart can be seen in Figure 1 on page 3. And we added the index and contents as Supplementary Materials for easy reading.

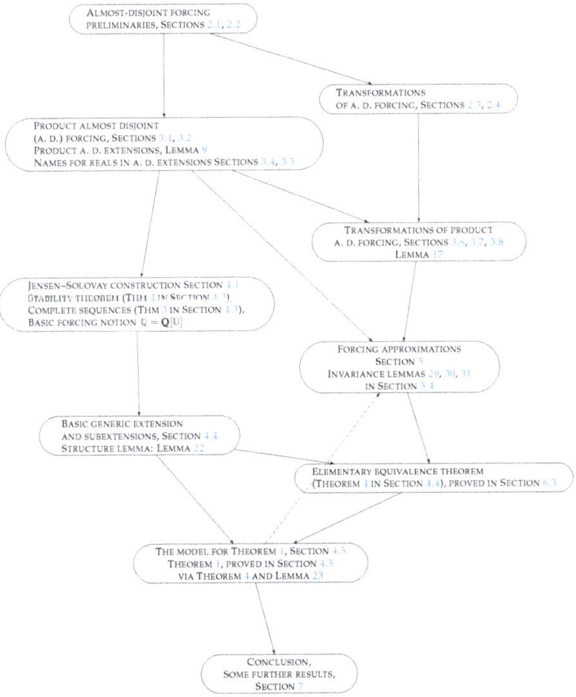

Figure 1. Flowchart.

2. Almost-Disjoint Forcing

Almost-disjoint forcing as a set theoretic tool was invented by Jensen and Solovay [4]. It has been applied in many research directions in modern set theory, in particular, in our paper [7] in this Journal. Here we make use of a considerably different version of the almost-disjoint forcing technique, which, comparably to [7], (1) considers countable cardinality instead of finite cardinalities in some key positions, (2) accordingly considers cardinality ω_1 instead of countable cardinality. In particular, sequences of finite length change to those of length $< \omega_1$. And so on.

Assumption 1. *During arguments in this section, we assume that the ground set universe is* $\mathbf{L}$*, the constructible universe. Recall that in* $\mathbf{L}$, $\mathrm{HC} = \mathbf{L}_{\omega_1}$ *and* $\mathrm{H}\omega_2 = \mathbf{L}_{\omega_2}$. $\square$

For the sake of brevity, we call *ω_1-size sets* those X satisfying $\mathtt{card}\, X \le \omega_1$.

2.1. Almost-Disjoint Forcing: ω_1-Version

This subsection contains a review the basic notation related to almost-disjoint forcing in the ω_1- version. Arguing in $\mathbf{L}$, we put $\mathbf{FUN} = {\omega_1}^{\omega_1}$ = all ω_1 sequences of ordinals $< \omega_1$.

- A set $X \subseteq \mathbf{FUN}$ is *dense* iff for any $s \in \mathbf{SEQ}$ there is $f \in X$ such that $s \subset f$.
- We let $\mathbf{SEQ} = {\omega_1}^{<\omega_1} \smallsetminus \{\Lambda\}$, the set of all non-empty *sequences* s of ordinals $< \omega_1$, of length $\mathtt{lh}\, s = \mathtt{dom}\, s < \omega_1$. We underline that Λ, the empty sequence, does not belong to $\mathbf{SEQ}$.
- If $S \subseteq \mathbf{SEQ}$, $f \in \mathbf{FUN}$ then let $S/f = \mathtt{sup}\{\xi < \omega_1 : f \upharpoonright \xi \in S\}$. If S/f is unbounded in ω_1 then say that S *covers* f, otherwise S *does not cover* f.

The following or very similar version of the almost-disjoint forcing was defined by Jensen and Solovay in [4] ([§ 5]). Its goal can be formulated as follows: given a set $u \subseteq \mathbf{FUN}$ in the ground universe, find a generic set $S \subseteq \mathbf{SEQ}$ such that the equivalence

$$f \in u \iff S \text{ does not cover } f \tag{2}$$

holds for each $f \in \mathbf{FUN}$ in the ground universe.

Definition 1 (in $\mathbf{L}$). *Q^* is the set of all pairs $p = \langle S_p ; F_p \rangle$ of finite sets $F_p \subseteq \mathbf{FUN}$, $S_p \subseteq \mathbf{SEQ}$. Elements of Q^* will be called (forcing) conditions. If $p \in Q^*$ then put*

$$F_p^{\vee} = \{f \upharpoonright \xi : f \in F_p \wedge 1 \le \xi < \omega_1\},$$

a tree in $\mathbf{SEQ}$. If $p, q \in Q^*$ then let $p \wedge q = \langle S_p \cup S_q ; F_p \cup F_q \rangle$; a condition in Q^*.

Let $p, q \in Q^*$. Define $q \leqslant p$ (that is, q is stronger as a forcing condition) iff $S_p \subseteq S_q$, $F_p \subseteq F_q$, and the difference $S_q \smallsetminus S_p$ does not intersect $F_p^{\vee}$, i.e., $S_q \cap F_p^{\vee} = S_p \cap F_p^{\vee}$. Clearly, we have $q \leqslant p$ iff $S_p \subseteq S_q$, $F_p \subseteq F_q$, and $S_q \cap F_p^{\vee} = S_p \cap F_p^{\vee}$. $\square$

Lemma 1 (in $\mathbf{L}$). *Conditions $p, q \in Q^*$ are compatible in Q^* iff (1) $S_q \smallsetminus S_p$ does not intersect $F_p^{\vee}$, and (2) $S_p \smallsetminus S_q$ does not intersect $F_q^{\vee}$. Therefore any $p, q \in P^*$ are compatible in P^* iff $p \wedge q \leqslant p$ and $p \wedge q \leqslant q$.*

Proof. If (1), (2) hold then $p \wedge q \leqslant p$ and $p \wedge q \leqslant q$, thus p, q are compatible. $\square$

If $u \subseteq \mathbf{FUN}$ then put $Q[u] = \{p \in Q^* : F_p \subseteq u\}$.

Any conditions $p, q \in Q[u]$ are compatible in $Q[u]$ iff they are compatible in Q^* iff the condition $p \wedge q = \langle S_p \cup S_q ; F_p \cup F_q \rangle \in Q[u]$ satisfies both $p \wedge q \leqslant p$ and $p \wedge q \leqslant q$. Therefore, we can say that conditions $p, q \in Q^*$ are compatible (or incompatible) without an explicit indication which forcing notion $Q[u]$ containing p, q is considered.

Lemma 2 (in $\mathbf{L}$). *If $u \subseteq \mathbf{FUN}$ and $A \subseteq Q[u]$ is an antichain then $\mathtt{card}\, A \le \omega_1$.*

Proof. Suppose towards the contrary that $\mathtt{card}\, A > \omega_1$. If $p \neq q$ in A are incompatible then obviously $S_p \neq S_q$. Yet $\{S_p : p \in Q^*\}$ = all finite subsets of $\mathbf{SEQ}$, is a set of cardinality ω_1, a contradiction. □

2.2. Almost-Disjoint Generic Extensions

To work with $\mathbf{L}$-sets $\mathbf{FUN}$ and $\mathbf{SEQ}$ in generic extensions of $\mathbf{L}$, possibly in those obtained by means of cardinal collapse, we let

$$\mathbf{FUN}^{\mathbf{L}} = (\omega_1^{\mathbf{L}})^{\omega_1^{\mathbf{L}}} \cap \mathbf{L} \quad \text{and} \quad \mathbf{SEQ}^{\mathbf{L}} = ((\omega_1^{\mathbf{L}})^{<\omega_1^{\mathbf{L}}} \cap \mathbf{L}) \smallsetminus \{\Lambda\} \tag{3}$$

—in other words, $\mathbf{FUN}^{\mathbf{L}}$ and $\mathbf{SEQ}^{\mathbf{L}}$ are just $\mathbf{FUN}$ and $\mathbf{SEQ}$ defined in $\mathbf{L}$.

Lemma 3. *Suppose that in $\mathbf{L}$, $u \subseteq \mathbf{FUN}$ is dense. Let $G \subseteq Q[u]$ be a set $Q[u]$-generic over $\mathbf{L}$. We define $S_G = \bigcup_{p \in G} S_p$; thus $S_G \subseteq \mathbf{SEQ}^{\mathbf{L}}$. Then*

(i) *if $f \in \mathbf{FUN}^{\mathbf{L}}$ then $f \in u$ iff S_G does not cover f ;*

(ii) *if $p \in Q[u]$ then $p \in G$ iff $S_p \subseteq S_G \wedge (S_G \smallsetminus S_p) \cap F_p^{\vee} = \varnothing$.*

(iii) $\mathbf{L}[G] = \mathbf{L}[S_G]$;

(iv) *if $f \in \mathbf{FUN}^{\mathbf{L}} \smallsetminus u$ then $X_f = \{\xi < \omega_1^{\mathbf{L}} : f \restriction \xi \in S_G\}$ is a cofinal subset of $\omega_1^{\mathbf{L}}$ of order type ω ;*

(v) $\omega_1^{\mathbf{L}[G]} = \omega_2^{\mathbf{L}}$.

Proof. (i) Consider any $f \in u$. We claim that $D_f = \{p \in P[u] : f \in F_p\}$ is dense in $P[u]$. (Indeed if $q \in P[u]$ then define $p \in P[u]$ by $S_p = S_q$ and $F_p = F_q \cup \{f\}$; we have $p \in D_f$ and $p \leqslant q$.) It follows that $D_f \cap G \neq \varnothing$. Choose any $p \in D_f \cap G$; we have $f \in F_p$. Each condition $r \in G$ is compatible with p, therefore, by Lemma 1, $S_r / f \subseteq S_p / f$. We conclude that $S_G / f = S_p / f$.

Now assume that $f \notin u$. The set $D_{fl} = \{p \in P[u] : \sup(S_p / f) > l\}$ is dense in $P[u]$ for any $l < \omega$. (Let $q \in P[u]$. Then F_q is finite. There exists $m > l$ with $f \restriction m \notin F_q^{\vee}$, since $f \notin u$. Define a condition p by $F_p = F_q$ and $S_p = S_q \cup \{f \restriction m\}$; we have $p \in D_{fl}$ and $p \leqslant q$.) Pick, by the density, any $p \in D_{fl} \cap G$. Then $\sup(S_G / f) > l$. We conclude that S_G / f is infinite because l is arbitrary.

(ii) Let $p \in G$. Then obviously $s_p \subseteq S_G$. If there exists $s \in (S_G \smallsetminus S_p) \cap F_p^{\vee}$ then $s \in S_q$ for some $q \in G$. Then conditions p, q are incompatible by Lemma 1, which is a contradiction.

Now assume that $p \in P[u] \smallsetminus G$. There is a condition $q \in G$ incompatible with p. We have two cases by Lemma 1. First, there is some $s \in (S_q \smallsetminus S_p) \cap F_p^{\vee}$. Then $s \in S_G \smallsetminus S_p$, so p is not compatible with S_G. Second, there is some $s \in (S_p \smallsetminus S_q) \cap F_q^{\vee}$. In this case, $s \notin S_r$ holds for any condition $r \le q$. It follows that $s \notin S_G$, hence $S_p \not\subseteq S_G$, and p cannot be compatible with S_G.

Further it follows from (ii) that $G = \{p \in P[u] : s_p \subseteq S_G \wedge (S_G \smallsetminus s_p) \cap F_p^{\vee} = \varnothing\}$, hence, we have (iii). Claim (v) is an immediate corollary of (iv) since $\omega_2^{\mathbf{L}}$ remains a cardinal in $\mathbf{L}[G]$ by Lemma 2.

Finally, to prove (iv) let $f \in \mathbf{FUN}^{\mathbf{L}} \smallsetminus u$ and $\lambda < \omega_1^{\mathbf{L}}$. The set $D_{f\lambda}$ of all conditions $p \in Q[u]$, such that $f \restriction \lambda \subset g$ for some $g \in S_p$, is dense in $\mathbf{Q}[u]$. Therefore G contains some $p \in D_{f\lambda}$. Let this be witnessed by some $g \in S_p$. Now, if $\xi < \lambda$ belongs to X_f, so that $s = f \restriction \xi \in S_G$, then s must belong to S_p by (ii), therefore ξ belongs to the finite set $\{\mathtt{lh}\, s : s \in S_p\}$. Thus, $X_f \cap \lambda$ is finite. That $X_f \cap \omega_1^{\mathbf{L}}$ is infinite follows from (i) (recall that $f \notin u$). □

Now we consider two types of transformations related to the forcing notion Q^*.

2.3. Lipschitz Transformations

We argue in **L**. Let $\mathbf{LIP}$ be the group of all $\subseteq$-automorphisms of $\mathbf{SEQ}$, called *Lipschitz transformations*. Any $\lambda \in \mathbf{LIP}$ preserves the *length* $\mathtt{lh}$ of sequences, i.e., $\mathtt{lh}\,s = \mathtt{lh}\,(\lambda \cdot s)$ for all $s \in \mathbf{SEQ}$. Any transformation $\lambda \in \mathbf{LIP}$ acts on:

- sequences $s \in \mathbf{SEQ}$: by $\lambda \cdot s = \lambda(s)$;
- functions $f \in \mathbf{FUN}$: by $\lambda \cdot f \in \mathbf{FUN}$ and $(\lambda \cdot f) \restriction \xi = \lambda \cdot (f \restriction \xi)$ for all $\xi < \omega_1$;
- sets $S \subseteq \mathbf{SEQ}$, $F \subseteq \mathbf{FUN}$: by $\lambda \cdot S = \{\lambda \cdot s : s \in S\}$, $\lambda \cdot F = \{\lambda \cdot f : f \in F\}$;
- conditions $p \in Q^*$: by $\lambda \cdot p = \langle \lambda \cdot S_p ; \lambda \cdot F_p \rangle \in Q^*$.

Lemma 4 (routine). *The action of any $\lambda \in \mathbf{LIP}$ is an order-preserving automorphism of Q^*. If $u \subseteq \mathbf{FUN}$ and $p \in Q[u]$ then $\lambda \cdot p \in Q[\lambda \cdot u]$.* □

We proceed with an important existence lemma. If $f \neq g$ belongs to $\mathbf{FUN}$ then let $\beta(f,g)$ be equal to the least ordinal $\beta < \omega_1$ such that $f(\beta) \neq g(\beta)$ (or, similarly, the largest ordinal β with $f \restriction \beta = g \restriction \beta$). Say that sets $X, Y \subseteq \mathbf{FUN}$ are *intersection-similar*, or *i-similar* for brevity, if there is a bijection $b : X \xrightarrow{\text{onto}} Y$ such that $\beta(f,g) = \beta(b(f), b(g))$ for all $f \neq g$ in X—such a bijection b will be called *an i-similarity bijection*.

Lemma 5. *Suppose that $u, v \subseteq \mathbf{FUN}$ are ω_1-sizesets, dense in $\mathbf{FUN}$. Then u, v are i-similar. Moreover, if $X \subseteq u$, $Y \subseteq v$ are finite and i-similar then*

(i) *there is an i-similarity bijection $b : u \xrightarrow{\text{onto}} v$ such that $b[X] = Y$,*

(ii) *there exists a transformation $\lambda \in \mathbf{LIP}$ such that $\lambda \cdot u = v$ and $\lambda \cdot X = Y$.*

Proof. The key argument is that if $A \subseteq u$, $B \subseteq v$ are at most countable, $b : A \xrightarrow{\text{onto}} B$ is an i-similarity bijection, and $f \in u \smallsetminus A$, then by the density of v there is $g \in v \smallsetminus B$ such that the extended map $b \cup \{\langle f, g \rangle\} : A \cup \{f\} \xrightarrow{\text{onto}} B \cup \{g\}$ is still an i-similarity bijection. This allows proof of (i), iteratively extending an initial i-similarity bijection $b_0 : X \xrightarrow{\text{onto}} Y$ by a ω_1-step back-and-forth argument involving eventually all elements $f \in u$ and $g \in v$, to an i-similarity bijection $u \xrightarrow{\text{onto}} v$ required. See the proof of Lemma 5 in [7] for more detail.

To get (ii) from (i), consider any sequence $s \in \mathbf{SEQ}$. Let $\beta = \mathtt{lh}\,s$. As u is dense, there exist $f, f' \in u$ such that $\beta(f, f') = \beta$ and $s \subset f$, $s \subset f'$. Put $g = b(f)$, $g' = b(f')$. Then still $\beta(g, g') = \beta$, hence $g \restriction \beta = g' \restriction \beta$. Therefore, we can define $\lambda(s) = g \restriction \beta = g' \restriction \beta$. □

2.4. Substitution Transformations

We continue to argue in **L**. Assume that conditions $p, q \in Q^*$ satisfy

$$F_p = F_q \quad \text{and} \quad S_p \cup S_q \subseteq F_p^{\vee} = F_q^{\vee}. \tag{4}$$

We define a transformation h_{pq} acting as follows.

If $p = q$ then define $h_{pq}(r) = r$ for all $r \in Q^*$, the identity.

Suppose that $p \neq q$. Then p, q are incompatible by (4) and Lemma 1. Define $d_{pq} = \{r \in Q^* : r \leqslant p \vee r \leqslant q\}$, the *domain* of h_{pq}. Let $r \in d_{pq}$. We put $h_{pq}(r) = r' := \langle S_{r'}, F_{r'} \rangle$, where $F_{r'} = F_r$ and

$$S_{r'} = \begin{cases} (S_r \smallsetminus S_p) \cup S_q & \text{in case} \quad r \leqslant p, \\ (S_r \smallsetminus S_q) \cup S_p & \text{in case} \quad r \leqslant q. \end{cases} \tag{5}$$

Thus, assuming (4), the difference between S_r and $S_{r'}$ lies entirely within the set $X = F_p^{\vee} = F_q^{\vee}$, so that if $r \leqslant p$ then $S_r \cap X = S_p$ but $S_{r'} \cap X = S_q$, while if $r \leqslant q$ then $S_r \cap X = S_q$ but $S_{r'} \cap X = S_p$.

Lemma 6. (i) *If $u \subseteq$ **FUN** is dense and $p_0, q_0 \in Q[u]$ then there exist conditions $p, q \in Q[u]$ with $p \leqslant p_0$, $q \leqslant q_0$, satisfying* (4).

(ii) *Let $p, q \in Q^*$ satisfy* (4). *If $p = q$ then h_{pq} is the identity transformation. If $p \neq q$ then h_{pq} an order automorphism of $d_{pq} = \{r \in Q^* : r \leqslant p \vee r \leqslant q\}$, satisfying $h_{pq}(p) = q$ and $h_{pq} = (h_{pq})^{-1} = h_{qp}$.*

(iii) *If $u \subseteq$ **FUN** and $p, q \in Q[u]$ satisfy* (4) *then h_{pq} maps the set $Q[u] \cap d_{pq}$ onto itself order-preserving.*

Proof. (i) By the density of u there is a finite set $F \subseteq$ **FUN** satisfying $F_p \cup F_q \subseteq F$ and $S_p \cup S_q \subseteq F^\vee = \{f \restriction \xi : f \in F \wedge 1 \leq \xi < \omega_1\}$. Put $p = \langle S_p, F \rangle$ and $q = \langle S_q, F \rangle$. Claims (ii), (iii) are routine. □

Please note that unlike the Lipschitz transformations above, transformations of the form h_{pq}, called *substitutions* in this paper, act within any given forcing notion of the form $Q[u]$ by claim (iii) of the lemma, and hence the forcing notions of the form $Q[u]$ considered are sufficiently homogeneous.

3. Almost-Disjoint Product Forcing

Here we review the structure and basic properties of product almost-disjoint forcing and corresponding generic extensions in the ω_1-version. There is an important issue here: a forcing $\mathbb{C}$, which collapses ω_1 to ω, enters as a factor in the product forcing notions considered.

3.1. Product Forcing

In **L**, we define $\mathbb{C} = \mathscr{P}(\omega)^{<\omega}$, the set of all finite sequences of subsets of ω, an ordinary forcing to collapse $\mathscr{P}(\omega) \cap \mathbf{L}$ down to ω. We will make use of an ω_2-product of Q^* with $\mathbb{C}$ as an extra factor. (In fact, $\mathbb{C}$ can be eliminated since Q^* collapses $\omega_1^{\mathbf{L}}$ anyway by Lemma 3 (v). Yet the presence of $\mathbb{C}$ somehow facilitates the arguments since $\mathbb{C}$ has a more transparent forcing structure.)

Technically, we put $\mathcal{I} = \omega_2$ (in **L**) and consider the index set $\mathcal{I}^+ = \mathcal{I} \cup \{-1\}$. Let $\mathbf{Q}^*$ be the finite-support product of $\mathbb{C}$ and $\mathcal{I}$ copies of Q^* (Definition 1 in Section 2.1), ordered componentwise. That is, $\mathbf{Q}^*$ consists of all maps p defined on a finite set $\mathtt{dom}\, p = |p|^+ \subseteq \mathcal{I}^+$ so that $p(\nu) \in Q^*$ for all $\nu \in |p| := |p|^+ \smallsetminus \{-1\}$, and if $-1 \in |p|^+$ then $\boldsymbol{b}_p := p(-1) \in \mathbb{C}$. If $p \in \mathbf{Q}^*$ then put $F_p(\nu) = F_{p(\nu)}$ and $S_p(\nu) = S_{p(\nu)}$ for all $\nu \in |p|$, so that $p(\nu) = \langle S_p(\nu); F_p(\nu) \rangle$.

We order $\mathbf{Q}^*$ componentwise: $p \leqslant q$ (p is stronger as a forcing condition) iff $|q|^+ \subseteq |p|^+$, $\boldsymbol{b}_q \subseteq \boldsymbol{b}_p$ in case $-1 \in |q|^+$, and $p(\nu) \leqslant q(\nu)$ in Q^* for all $\nu \in |q|$. Put

$$F_p^\vee(\nu) = F_{p(\nu)}^\vee = \{f \restriction \xi : f \in F_p(\nu) \wedge 1 \leq \xi < \omega_1\}.$$

In particular, $\mathbf{Q}^*$ contains *the empty condition* $\odot \in \mathbf{Q}^*$ satisfying $|\odot|^+ = \varnothing$; obviously $\odot$ is the $\leqslant$-least (and weakest as a forcing condition) element of $\mathbf{Q}^*$.

Because of the factor $\mathbb{C}$, it takes some effort to define $p \wedge q$ for $p, q \in \mathbf{Q}^*$, and only assuming that $\boldsymbol{b}_p, \boldsymbol{b}_q$ are compatible, i.e., $\boldsymbol{b}_p \subseteq \boldsymbol{b}_q$ or $\boldsymbol{b}_q \subseteq \boldsymbol{b}_p$. In such a case define $p \wedge q \in \mathbf{Q}^*$ as follows. First, $|p \wedge q|^+ = |p|^+ \cup |q|^+$. If $\nu \in |p|^+ \smallsetminus |q|^+$ then put $(p \wedge q)(\nu) = p(\nu)$, and similarly if $\nu \in |q|^+ \smallsetminus |p|^+$ then $(p \wedge q)(\nu) = q(\nu)$. Now suppose that $\nu \in |p|^+ \cap |q|^+$.

If $\nu \neq -1$ then $(p \wedge q)(\nu) = p(\nu) \wedge q(\nu)$ in the sense of Definition 1 in Section 2.1.

If $\nu = -1 \in |p|^+ \cap |q|^+$, then, by the compatibility, either $\boldsymbol{b}_p \subseteq \boldsymbol{b}_q$—and then define $\boldsymbol{b}_{p \wedge q} = \boldsymbol{b}_q$, or $\boldsymbol{b}_q \subseteq \boldsymbol{b}_p$—and then accordingly $\boldsymbol{b}_{p \wedge q} = \boldsymbol{b}_p$.

Lemma 7. *Let $p, q \in \mathbf{Q}^*$ be compatible. Then $(p \wedge q) \in \mathbf{Q}^*$, $(p \wedge q) \leqslant p$, $(p \wedge q) \leqslant q$, and if $r \in \mathbf{Q}^*$, $r \leqslant p$, $r \leqslant q$, then $r \leqslant (p \wedge q)$.* □

3.2. Systems

Arguing in **L**, we consider certain subforcings of the total product forcing notion $\mathbf{Q}^*$.

Let a *system* be any map $U : |U| \to \mathscr{P}(\mathbf{FUN})$ such that $|U| \subseteq \mathcal{I}$, each set $U(\nu)$ $(\nu \in |U|)$ is dense in $\mathbf{FUN}$, and the *components* $U(\nu) \subseteq \mathbf{FUN}$ $(\nu \in |U|)$ are pairwise disjoint.

- A system U is *small*, if both $|U|$ and each set $U(\nu)$ $(\nu \in |U|)$ has cardinality $\le \omega_1$.
- If U, V are systems, $|U| \subseteq |V|$, and $U(\nu) \subseteq V(\nu)$ for all $\nu \in |U|$, then say that V *extends* U, in symbol $U \preccurlyeq V$.
- If $\{U_\xi\}_{\xi<\lambda}$ is a $\preccurlyeq$-increasing sequence of systems then define a system $U = \bigvee_{\xi<\lambda} U_\xi$ by $|U| = \bigcup_{\xi<\lambda} |U_\xi|$ and $U(\nu) = \bigcup_{\xi<\lambda, \nu \in |U_\xi|} U_\xi(\nu)$ for all $\nu \in |U|$.
- If U is a system, then $\mathbf{Q}[U]$ is the finite-support product of $\mathbb{C}$ and sets $Q[U(\nu)]$, $\nu \in |U|$, i.e.,

$$\mathbf{Q}[U] = \{p \in \mathbf{Q}^* : |p| \subseteq |U| \wedge \forall \nu \in |p| \, (F_p(\nu) \subseteq U(\nu))\}.$$

Suppose that $c \subseteq \mathcal{I}^+$. If $p \in \mathbf{Q}^*$ then define $p' = p \restriction c \in \mathbf{Q}^*$ so that $|p'|^+ = c \cap |p|^+$ and $p'(\nu) = p(\nu)$ whenever $\nu \in |p'|^+$. A special case: if $\nu \in \mathcal{I}^+$ then let $p \restriction_{\neq\nu} = p \restriction (|p|^+ \smallsetminus \{\nu\})$. Similarly, if U is a system then define a system $U' = U \restriction c$ so that $|U'| = c \cap |U|$ and $U'(\nu) = U(\nu)$ whenever $\nu \in |U'|$. A special case: if $\nu \in \mathcal{I}^+$ then let $U \restriction_{\neq\nu} = U \restriction (|p| \smallsetminus \{\nu\})$. And if $Q \subseteq \mathbf{Q}^*$ then let $Q \restriction c = \{p \in Q : |p|^+ \subseteq c\}$ (will usually coincide with $\{p \restriction c : p \in Q\}$.

Writing $p \restriction c$, $U \restriction c$ etc., it is not assumed that $c \subseteq |p|^+$.

Lemma 8 (in **L**). *If U is a system and $A \subseteq \mathbf{Q}[U]$ is an antichain then* $\mathtt{card}\, A \le \omega_1$.

Proof. Suppose that $\mathtt{card}\, A > \omega_1$. As $\mathtt{card}\, \mathbb{C} = \omega_1$, we can w.l.o.g. assume that $\boldsymbol{b}_p = \boldsymbol{b}_q$ for all $p, q \in A$. It follows by the Δ-system lemma that there is a set $A' \subseteq A$ of the same cardinality $\mathtt{card}\, A' = \mathtt{card}\, A > \omega_1$, and a finite set $d \subseteq \mathcal{I}^+$, such that $|p|^+ = d$ for all $p \in A'$. Then we have $S_p \neq S_q$ for all $p \neq q$ in A', easily leading to a contradiction, as in the proof of Lemma 2. □

3.3. Outline of Product Extensions

We consider sets of the form $\mathbf{Q}[U]$, U being a system in **L**, as forcing notions over **L**. Accordingly, we'll study $\mathbf{Q}[U]$-generic extensions $\mathbf{L}[G]$ of the ground universe **L**. Define some elements of these extensions. Suppose that $G \subseteq \mathbf{Q}^*$. Put $|G| = \bigcup_{p \in G} |p|$; $|G| \subseteq \mathcal{I}$. Let

$$\boldsymbol{b}_G = \bigcup_{p \in G} \boldsymbol{b}_p, \quad \text{and} \quad S_G(\nu) = S_{G(\nu)} = \bigcup_{p \in G} S_p(\nu)$$

for any $\nu \in |G|$, where $G(\nu) = \{p(\nu) : p \in G\} \subseteq Q^*$.

Thus, $S_G(\nu) \subseteq \mathbf{SEQ}^{\mathbf{L}}$, and $S_G(\nu) = \varnothing$ for any $\nu \notin |G|$.

By the way, this defines a sequence $\vec{S}_G = \{S_G(\nu)\}_{\nu \in \mathcal{I}}$ of subsets of $\mathbf{SEQ}$.

If $c \subseteq \mathcal{I}^+$ then let $G \restriction c = \{p \in G : |p|^+ \subseteq c\}$. It will typically happen that $G \restriction c = \{p \restriction c : p \in G\}$. Put $G \restriction_{\neq\nu} = \{p \in G : \nu \notin |p|^+\} = G \restriction (\mathcal{I}^+ \smallsetminus \{\nu\})$.

If U is a system in **L**, then any $\mathbf{Q}[U]$-generic set $G \subseteq \mathbf{Q}[U]$ splits into the family of sets $G(\nu)$, $\nu \in \mathcal{I}$, and a separate map $\boldsymbol{b}_G : \omega \xrightarrow{\text{onto}} \mathscr{P}(\omega) \cap \mathbf{L}$. It will follow from (ii) of the next lemma that $\mathbf{Q}[U]$-generic extensions of **L** satisfy $\omega_1 = \omega_2^{\mathbf{L}}$.

Lemma 9. *Let U be a system in* **L**, *and $G \subseteq \mathbf{Q}[U]$ be a set $\mathbf{Q}[U]$-generic over* **L**. *Then*:

(i) $\boldsymbol{b}_G$ *is a $\mathbb{C}$-generic map from ω onto $\mathscr{P}(\omega) \cap \mathbf{L}$*;

(ii) *if $\nu \in \mathcal{I}$ then $\mathbf{L}[G(\nu)] = \mathbf{L}[S_G(\nu)]$ and* $\omega_1^{\mathbf{L}[\boldsymbol{b}_G]} = \omega_1^{\mathbf{L}[G(\nu)]} = \omega_2^{\mathbf{L}} = \omega_1^{\mathbf{L}[G]}$;

(iii) $\mathbf{L}[G] = \mathbf{L}[\vec{S}_G]$ *and* $|G|^+ = \mathcal{I}$;

(iv) *if $\nu \in \mathcal{I}$ and $c \in \mathbf{L}[G \restriction_{\neq\nu}]$, $\nu \notin c \subseteq \mathcal{I}^+$, then* $\mathbf{L}[G \restriction c] \subseteq \mathbf{L}[G \restriction_{\neq\nu}]$;

(v) *if $\nu \in \mathcal{I}$ then* $S_G(\nu) \notin \mathbf{L}[G \restriction_{\neq\nu}]$;

(vi) *if* $\nu \in \mathcal{I}$ *then the set* $G(\nu) = \{p(\nu) : p \in G\} \in \mathbf{L}[G]$ *is* $P[U(\nu)]$*-generic over* $\mathbf{L}$*, hence if* $f \in \mathbf{FUN}^{\mathbf{L}}$ *then* $f \in U(\nu)$ *iff* $S_G(\nu)$ *does not cover* f *.*

Proof. Proofs of (i) and (iii)–(vi) are similar to ([7] (Lemma 9)). To prove $\omega_1^{\mathbf{L}[G(\nu)]} = \omega_2^{\mathbf{L}}$ in (ii) apply Lemma 3 (v). Finally, to see that $\omega_2^{\mathbf{L}}$ remains a cardinal in $\mathbf{L}[G]$ apply Lemma 8. □

3.4. Names for Sets in Product Extensions

The next definition introduces *names* for elements of product-generic extensions of $\mathbf{L}$ considered.

Assume that in $\mathbf{L}$, $K \subseteq \mathbf{Q}^*$, e.g., $K = \mathbf{Q}[U]$, where U is a system, and X is any set. By $\mathbf{N}_X(K)$ (K*-names* for subsets of X) we denote the set of all sets $\tau \subseteq K \times X$ in $\mathbf{L}$. Furthermore, $\mathbf{SN}_X(K)$ (small names) consist of all ω_1-size names $\tau \in \mathbf{SN}_X(K)$; in other words, it is required that $\mathtt{card}\,\tau \leq \omega_1$.

Suppose that $\tau \in \mathbf{N}_X(\mathbf{Q}^*)$. We put

$$\mathtt{dom}\,\tau = \{p : \exists x\,(\langle p, x\rangle \in \tau\}, \quad |\tau|^+ = \bigcup_{p \in \mathtt{dom}\,\tau} |p|^+, \quad |\tau| = \bigcup_{p \in \mathtt{dom}\,\tau} |p|.$$

If $G \subseteq \mathbf{Q}^*$ then define

$$\tau[G] = \{x \in X : (\tau''x) \cap G \neq \varnothing\}, \quad \text{where} \quad \tau''x = \{p : \langle p, x\rangle \in \tau\},$$

so that $\tau[G] \subseteq X$. If φ is a formula in which some names $\tau \in \mathbf{SN}^{\omega}_{\omega}(\mathbf{Q}^*)$ occur, and $G \subseteq \mathbf{Q}^*$, then accordingly $\varphi[G]$ is the result of substitution of $\tau[G]$ for each name τ in φ.

Lemma 10. *Suppose that* $X \in \mathbf{L}$, $\mathtt{card}\,X \leq \omega_1$ *in* $\mathbf{L}$*,* U *is a system in* $\mathbf{L}$*, and* $G \subseteq \mathbf{Q}[U]$ *is a set* $\mathbf{Q}[U]$*-generic over* $\mathbf{L}$*. Then for any set* $Y \in \mathbf{L}[G]$, $Y \subseteq X$*, there is a name* $\tau \in \mathbf{SN}_X(\mathbf{Q}[U])$ *in* $\mathbf{L}$ *such that* $Y = \tau[G]$*. If in addition* $c \in \mathbf{L}$, $c \subseteq \mathcal{I}^+$*, and* $Y \in \mathbf{L}[G \restriction c]$*, then there is a name* $\tau \in \mathbf{SN}_X(\mathbf{Q}[U] \restriction c)$ *in* $\mathbf{L}$ *such that* $Y = \tau[G]$*.*

Proof. It follows from general forcing theory that there is a name $\sigma \in \mathbf{N}_X(\mathbf{Q}[U])$, not necessarily an ω_1-size name, such that $X = \sigma[G]$. Let $Q_x = \sigma''x$ for all $x \in X$. Arguing in $\mathbf{L}$, put

$$\tau = \{\langle p, x\rangle \in \sigma : x \in X \wedge p \in A_x\},$$

where $A_x \subseteq Q_x$ is a maximal antichain for any x. We observe that $\mathtt{card}\,A_x \leq \omega_1$ in $\mathbf{L}$ for all x by Lemma 8, hence $\tau \in \mathbf{SN}_X(\mathbf{Q}[U])$. And on the other hand, we have $\tau[G] = \sigma[G] = Y$.

To prove the additional claim, note that by the product forcing theorem if $Y \in \mathbf{L}[G \restriction c]$ then the original name σ can be chosen in $\mathbf{N}_X(\mathbf{Q}[U] \restriction c)$, and repeat the argument. □

3.5. Names for Reals in Product Extensions

Now we introduce *names for reals* (elements of ω^ω) in generic extensions of $\mathbf{L}$ considered. This is an important particular case of the content of Section 3.4.

Assume that in $\mathbf{L}$, $K \subseteq \mathbf{Q}^*$, e.g., $K = \mathbf{Q}[U]$, where U is a system. By $\mathbf{N}^{\omega}_{\omega}(K)$ (K*-names* for reals in ω^ω) we denote the set of all $\tau \subseteq K \times (\omega \times \omega)$ such that the sets $\tau''\langle j, k\rangle = \{p : \langle p, \langle j, k\rangle\rangle \in \tau\}$ satisfy the following requirement:

$$\text{if } k \neq k',\ p \in \tau''\langle j, k\rangle,\ p' \in \tau''\langle j, k'\rangle, \text{ then conditions } p, p' \text{ are incompatible.}$$

We let $\tau''j = \bigcup_k \tau''\langle j, k\rangle$, $\mathtt{dom}\,\tau = \bigcup_{j,k<\omega} \tau''\langle j, k\rangle$, $|\tau|^+ = \bigcup\{|p|^+ : p \in \mathtt{dom}\,\tau\}$.

Let $\mathbf{SN}^{\omega}_{\omega}(K)$ (small names) consist of all ω_1-size names $\tau \in \mathbf{N}^{\omega}_{\omega}(K)$; in other words, it is required that $\mathtt{card}\,(\tau''\langle j, k\rangle) \leq \omega_1$ for all $j, k < \omega$.

Define the restrictions $\mathbf{SN}^{\omega}_{\omega}(K) \restriction c = \{\tau \in \mathbf{SN}^{\omega}_{\omega}(K) : |\tau|^+ \subseteq c\}$.

A name $\tau \in \mathbf{SN}^\omega_\omega(K)$ is *K-full* iff the set $\tau''j$ is pre-dense in K for any $j < \omega$. A name $\tau \in \mathbf{SN}^\omega_\omega(K)$ is K-full *below* some $p_0 \in K$, iff all sets $\tau''j$ are pre-dense in K below p_0, i.e., any condition $q \in K$, $q \leqslant p_0$, is compatible with some $r \in \tau_j$ (and this holds for all $j < \omega$).

Suppose that $\tau \in \mathbf{SN}^\omega_\omega(\mathbf{Q}^*)$. A set $G \subseteq K$ is *minimally τ-generic* iff it is compatible in itself (if $p, q \in G$ then there is $r \in G$ with $r \leqslant p$, $r \leqslant q$), and intersects each set $\tau''x$, $x \in X$. In this case, put

$$\tau[G] = \{\langle j,k\rangle \in \omega^\omega \times \omega^\omega : (\tau''\langle j,k\rangle) \cap G \neq \varnothing\},$$

so that $\tau[G] \in \omega^\omega$ and $\tau[G](j) = k \iff \tau''\langle j,k\rangle \cap G \neq \varnothing$. If φ is a formula in which some names $\tau \in \mathbf{SN}^\omega_\omega(\mathbf{Q}^*)$ occur, and a set $G \subseteq \mathbf{Q}^*$ is minimally τ-generic for any name τ in φ, then accordingly $\varphi[G]$ is the result of substitution of $\tau[G]$ for each name τ in φ.

Lemma 11. *Suppose that U is a system in* $\mathbf{L}$, *and $G \subseteq \mathbf{Q}[U]$ is $\mathbf{Q}[U]$-generic over* $\mathbf{L}$. *Then for any real $x \in \mathbf{L}[G] \cap \omega^\omega$ there is a $\mathbf{Q}[U]$-full name $\tau \in \mathbf{SN}^\omega_\omega(\mathbf{Q}[U])$ in* $\mathbf{L}$ *such that $x = \tau[G]$. If in addition $c \in \mathbf{L}$, $c \subseteq \mathcal{I}^+$, and $x \in \mathbf{L}[G \restriction c]$, then there is a $\mathbf{Q}[U]$-full name $\tau \in \mathbf{SN}^\omega_\omega(\mathbf{Q}[U] \restriction c)$ in* $\mathbf{L}$ *such that $x = \tau[G]$.* □

Proof. It follows from general forcing theory that there is a $\mathbf{Q}[U]$-full name $\sigma \in \mathbf{N}^\omega_\omega(\mathbf{Q}[U])$, not necessarily an ω_1-size name, such that $f = \sigma[G]$. Then all sets $Q_j = \sigma''j$, $j < \omega$, are pre-dense in $\mathbf{Q}[U]$. Arguing in $\mathbf{L}$, put $\tau = \{\langle p, \langle j,k\rangle\rangle \in \sigma : j,k < \omega \wedge p \in A_j\}$, where $A_j \subseteq Q_j$ is a maximal antichain for any $j < \omega$. We conclude by Lemma 8 that $\mathtt{card}\, A_j \leq \omega_1$ in $\mathbf{L}$ for all j, hence in fact $\tau \in \mathbf{SN}^\omega_\omega(\mathbf{Q}[U])$. And on the other hand, we have $\tau[G] = \sigma[G] = f$. □

Equivalent names. Names $\tau, \mu \in \mathbf{SN}^\omega_\omega(\mathbf{Q}^*)$ are *equivalent* iff conditions q, r are incompatible whenever $q \in \tau''\langle j,k\rangle$ and $r \in \mu''\langle j,k'\rangle$ for some j and $k' \neq k$. Names τ, μ are equivalent *below* some $p \in \mathbf{Q}^*$ iff the triple of conditions p, q, r is incompatible (that is, no common strengthening) whenever $q \in \tau''\langle j,k\rangle$ and $r \in \mu''\langle j,k'\rangle$ for some j and $k' \neq k$.

Lemma 12. *Suppose that in* $\mathbf{L}$, *$p \in \mathbf{Q}^*$, and names $\mu, \tau \in \mathbf{SN}^\omega_\omega(\mathbf{Q}^*)$ are equivalent (resp., equivalent below p). If $G \subseteq \mathbf{Q}^*$ is minimally μ-generic and minimally τ-generic (resp., and containing p), then $\mu[G] = \tau[G]$.*

Proof. Suppose that this is not the case. Then by definition there exist numbers j and $k' \neq k$ and conditions $q \in G \cap (\tau''\langle j,k\rangle)$ and $r \in G \cap (\mu''\langle j,k'\rangle)$. Then p, q, r are compatible (as elements of the same generic set), contradiction. □

The next lemma provides a useful transformation of names. Recall that $p' \wedge p$ is defined in Section 3.1.

Lemma 13 (in $\mathbf{L}$). *If $p \in \mathbf{Q}^*$ and $\tau \in \mathbf{SN}^\omega_\omega(\mathbf{Q}^*)$, then*

$$\tau_{\leqslant p} = \{\langle p' \wedge p, \langle j,k\rangle\rangle : \langle p', \langle j,k\rangle\rangle \in \tau \text{ and } p' \text{ is compatible with } p\}$$

is still a name in $\mathbf{SN}^\omega_\omega(\mathbf{Q}^*)$, equivalent to τ below p, and $|\tau_{\leqslant p}|^+ \subseteq |\tau|^+ \cup |p|^+$.

If U is a system and $p \in \mathbf{Q}[U]$, $\tau \in \mathbf{SN}^\omega_\omega(\mathbf{Q}[U])$, then $\tau_{\leqslant p} \in \mathbf{SN}^\omega_\omega(\mathbf{Q}[U])$.

Moreover, if τ is $\mathbf{Q}[U]$-full below p then $\tau_{\leqslant p}$ is $\mathbf{Q}[U]$-full below p, too.

Proof. Routine. □

3.6. Permutations

We continue to argue in $\mathbf{L}$. There are three important families of transformations of the whole system of objects related to product forcing, considered in this Subsection and the two following ones.

We begin with *permutations*, the first family. Let $\mathbf{B}\mathrm{IJ}$ be the set of all bijections $\pi : \mathcal{I} \xrightarrow{\text{onto}} \mathcal{I}$, i.e., permutations of the set $\mathcal{I}$, such that the set $|\pi| = \{\nu \in \mathcal{I} : \pi(\nu) \neq \nu\}$ (the *essential domain*) satisfies $\mathtt{card}\,|\pi| \leq \omega_1$. Please note that π is the identity outside of $|\pi|$. Any permutation $\pi \in \mathbf{B}\mathrm{IJ}$ acts onto:

- sets $e \subseteq \mathcal{I}$: by $\pi \cdot e := \{\pi(\nu) : \nu \in e\}$;
- systems U: by $(\pi \cdot U)(\pi(\nu)) := U(\nu)$ for all $\nu \in |U|$—then $|\pi \cdot U| = \pi \cdot |U|$;
- conditions $p \in \mathbf{Q}^*$: if $-1 \in |p|^+$ then $-1 \in |\pi \cdot p|^+$ and $\boldsymbol{b}_{\pi \cdot p} = \boldsymbol{b}_p$, and if $\nu \in |p|$ then $(\pi \cdot p)(\pi(\nu)) := p(\nu)$, so $|\pi \cdot p| = \pi \cdot |p|$;
- sets $G \subseteq \mathbf{Q}^*$: by $\pi \cdot G := \{\pi \cdot p : p \in G\}$—then $\pi \cdot G \subseteq \mathbf{Q}^*$;
- names $\tau \in \mathbf{SN}^{\omega}_{\omega}(\mathbf{Q}^*)$: by $\pi \cdot \tau := \{\langle \pi \cdot p, \langle \ell, k \rangle \rangle : \langle p, \langle \ell, k \rangle \rangle \in \tau\} \in \mathbf{SN}^{\omega}_{\omega}(\mathbf{Q}^*)$.

Lemma 14 (routine). *If $\pi \in \mathbf{B}\mathrm{IJ}$ then $p \longmapsto \pi \cdot p$ is an order-preserving bijection of $\mathbf{Q}^*$ onto $\mathbf{Q}^*$, and if U is a system then $p \in \mathbf{Q}[U] \iff \pi \cdot p \in \mathbf{Q}[\pi \cdot U]$.* □

3.7. Multi-Lipschitz Transformations

Still arguing in $\mathbf{L}$, we let $\mathbf{L}\mathrm{IP}^{\mathcal{I}}$ be the $\mathcal{I}$-product of the group $\mathbf{L}\mathrm{IP}$ (see Section 2.3), this will be our *second family* of transformations, called *multi-Lipschitz*. Thus, a typical element $\boldsymbol{\lambda} \in \mathbf{L}\mathrm{IP}^{\mathcal{I}}$ is $\boldsymbol{\lambda} = \{\lambda_\nu\}_{\nu \in |\boldsymbol{\lambda}|}$, where $|\boldsymbol{\lambda}| = \mathtt{dom}\,\boldsymbol{\lambda} \subseteq \mathcal{I}^+$ has ω_1-size, $\lambda_\nu \in \mathbf{L}\mathrm{IP}$, $\forall \nu$. Define the action of any $\boldsymbol{\lambda} \in \mathbf{L}\mathrm{IP}^{\mathcal{I}}$ on:

- systems U: $|\boldsymbol{\lambda} \cdot U| := |U|$, and $(\boldsymbol{\lambda} \cdot U)(\nu) := \lambda_\nu \cdot U(\nu)$ for all elements $\nu \in |\boldsymbol{\lambda}| \cap |U|$, but $(\boldsymbol{\lambda} \cdot U)(\nu) := U(\nu)$ for all $\nu \in |U| \smallsetminus |\boldsymbol{\lambda}|$;
- conditions $p \in \mathbf{Q}^*$: $|\boldsymbol{\lambda} \cdot p|^+ = |p|^+$, if $-1 \in |p|^+$ then $\boldsymbol{b}_{\boldsymbol{\lambda} \cdot p} = \boldsymbol{b}_p$, if $\nu \in |p| \cap |\boldsymbol{\lambda}|$ then $(\boldsymbol{\lambda} \cdot p)(\nu) = \lambda_\nu \cdot p(\nu)$, but if $\nu \in |p| \smallsetminus |\boldsymbol{\lambda}|$, then $(\boldsymbol{\lambda} \cdot p)(\nu) = p(\nu)$;
- sets $G \subseteq \mathbf{Q}^*$: $\boldsymbol{\lambda} \cdot G := \{\boldsymbol{\lambda} \cdot p : p \in G\}$;
- names $\tau \in \mathbf{SN}^{\omega}_{\omega}(\mathbf{Q}^*)$: $\boldsymbol{\lambda} \cdot \tau := \{\langle \boldsymbol{\lambda} \cdot p, \langle n, k \rangle \rangle : \langle p, \langle n, k \rangle \rangle \in \tau\}$;

In the first two items, we refer to the action of $\lambda_\nu \in \mathbf{L}\mathrm{IP}$ on sets $u \subseteq \mathbf{F}\mathrm{UN}$ and on forcing conditions, as defined in Section 2.3.

Lemma 15 (routine). *If $\boldsymbol{\lambda} \in \mathbf{L}\mathrm{IP}^{\mathcal{I}}$ then $p \longmapsto \pi \cdot p$ is an order-preserving bijection of $\mathbf{Q}^*$ onto $\mathbf{Q}^*$, and if U is a system then $p \in \mathbf{Q}[U] \iff \boldsymbol{\lambda} \cdot p \in \mathbf{Q}[\boldsymbol{\lambda} \cdot U]$.* □

Lemma 16. *Suppose that U, V are systems, $|U| = |V|$, $p \in \mathbf{Q}[U]$, $q \in \mathbf{Q}[V]$, $|p| = |q|$, and sets $F^{\vee}_p(\nu)$, $F^{\vee}_q(\nu)$ are i-similar for all $\nu \in |p| = |q|$. Then there is $\boldsymbol{\lambda} \in \mathbf{L}\mathrm{IP}^{\mathcal{I}}$ such that $|\boldsymbol{\lambda}| = |U| = |V|$, $\boldsymbol{\lambda} \cdot U = V$, and $F^{\vee}_q(\nu) = F^{\vee}_{\boldsymbol{\lambda} \cdot p}(\nu)$ for all $\nu \in |p| = |q|$.*

Proof. Apply Lemma 5 componentwise for every $\nu \in \mathcal{I}$. □

3.8. Multi-Substitutions

Assume that conditions $p, q \in \mathbf{Q}^*$ satisfy the following:

$$\left.\begin{array}{ll} \text{(6i)} & -1 \in |p|^+ = |q|^+ \quad \text{and} \quad \mathtt{lh}\,\boldsymbol{b}_p = \mathtt{lh}\,\boldsymbol{b}_q, \quad \text{and} \\ \text{(6ii)} & \text{if } \nu \in |p| \text{ then } F_p(\nu) = F_q(\nu) \text{ and } S_p(\nu) \cup S_q(\nu) \subseteq F^{\vee}_p(\nu) = F^{\vee}_q(\nu). \end{array}\right\} \qquad (6)$$

In particular, (4) of Section 2.4 holds for all ν. We define a transformation H_{pq} acting as follows. First, we let $\mathbf{D}_{pq}$, the domain of H_{pq}, contain all conditions $r \in \mathbf{Q}^*$ such that

(a) if $-1 \in |r|^+$ and $\boldsymbol{b}_p \neq \boldsymbol{b}_q$, then $\boldsymbol{b}_p \subseteq \boldsymbol{b}_r$ or $\boldsymbol{b}_q \subseteq \boldsymbol{b}_r$;

(b) if $\nu \in |r| \cap |p|$ and $p(\nu) \neq q(\nu)$, then $r(\nu) \leqslant p(\nu)$ or $r(\nu) \leqslant q(\nu)$, thus, in other words, $r(\nu) \in d_{p(\nu)p(\nu)}$ in the sense of Section 2.4.

Please note that all conditions $r \leqslant p$ and all $r \leqslant p$ belong to $\mathbf{D}_{pq}$. On the other hand, if $r \in \mathbf{Q}^*$ satisfies $|r| \cap |p| = \varnothing$ and (a), then r belongs to $\mathbf{D}_{pq}$ as well. In particular, $\odot \in \mathbf{D}_{pq}$.

If $r \in \mathbf{D}_{pq}$, then define $r' = H_{pq}(r) \in \mathbf{Q}^*$ so that $|r'|^+ = |r|^+$ and:

(a1) if $-1 \in |r|^+$ and $\boldsymbol{b}_p = \boldsymbol{b}_q$ then simply $\boldsymbol{b}_{r'} = \boldsymbol{b}_r$,

(a2) if $-1 \in |r|^+$ and $\boldsymbol{b}_p \neq \boldsymbol{b}_q$, then by (a) either $\boldsymbol{b}_r = \boldsymbol{b}_p{}^\frown s$ or $\boldsymbol{b}_r = \boldsymbol{b}_q{}^\frown s$, where $s \in \mathscr{P}(\omega)^{<\omega}$—we put $\boldsymbol{b}_{r'} = \boldsymbol{b}_q{}^\frown s$ in the first case, and $\boldsymbol{b}_{r'} = \boldsymbol{b}_p{}^\frown s$ in the second case;

(b1) if either $\nu \in |r| \smallsetminus |p|$, or $\nu \in |r| \cap |p| \wedge p(\nu) = q(\nu)$, then put $r'(\nu) = r(\nu)$,

(b2) if $\nu \in |p| = |q|$ and $p(\nu) \neq q(\nu)$, then we put $r'(\nu) = h_{p(\nu)q(\nu)}(r(\nu))$, where $h_{p(\nu)q(\nu)}$ is defined in Section 2.4.

Transformations of the form H_{pq} will be called *multi-substitutions*.

Lemma 17 (in **L**). (i) *If U is a system and $p_0, q_0 \in \mathbf{Q}[U]$ then there exist conditions $p, q \in \mathbf{Q}[U]$ with $p \leqslant p_0$, $q \leqslant q_0$, satisfying (6).*

(ii) *If conditions $p, q \in \mathbf{Q}^*$ satisfy (6), then H_{pq} is an order automorphism of $\mathbf{D}_{pq} = \mathbf{D}_{qp}$, and we have $H_{pq} = (H_{pq})^{-1} = H_{qp}$ and $H_{pq}(p) = q$.*

(iii) *If U is a system, and $p, q \in \mathbf{Q}[U]$ satisfy (6), then H_{pq} maps the set $\mathbf{Q}[U] \cap \mathbf{D}_{pq}$ onto itself order-preserving.*

Proof. Apply Lemma 6 componentwise. □

Corollary 1 (of Lemma 17). *If U is a system then $\mathbf{Q}[U]$ is* **homogeneous** *in the following sense: if $p_0, q_0 \in \mathbf{Q}[U]$ then there exist stronger conditions $p \leqslant p_0$ and $q \leqslant q_0$ in $\mathbf{Q}[U]$, such that the according lower cones $\{p' \in \mathbf{Q}[U] : p' \leqslant p\}$ and $\{q' \in \mathbf{Q}[U] : q' \leqslant q\}$ are order-isomorphic.* □

Action of H_{pq} on names. Assume that conditions $p, q \in \mathbf{Q}^*$ satisfy (6). Let $\mathbf{SN}^{\omega}_{\omega}(\mathbf{Q}^*)_{pq}$ contain all names $\tau \in \mathbf{SN}^{\omega}_{\omega}(\mathbf{Q}^*)$ such that $\operatorname{dom}\tau \subseteq \mathbf{D}_{pq}$. If $\tau \in \mathbf{SN}^{\omega}_{\omega}(\mathbf{Q}^*)_{pq}$ then put

$$H_{pq} \cdot \tau = \{\langle H_{pq}(p'), \langle n, k\rangle\rangle : \langle p', \langle n, k\rangle\rangle \in \tau\}.$$

Then obviously $H_{pq} \cdot \tau \in \mathbf{SN}^{\omega}_{\omega}(\mathbf{Q}^*)_{qp}$.

4. The Basic Forcing Notion and the Model

In this paper, we let $\mathbf{ZFC}^-$ be **ZFC** minus the Power Set axiom, with the schema of Collection instead of Replacement, with **AC** is assumed in the form of well-orderability of every set, and with the axiom: "ω_1 exists". See [8] on versions of **ZFC** sans the Power Set axiom in detail.

Let $\mathbf{ZFC}^-_2$ be $\mathbf{ZFC}^-$ plus the axioms: $\mathbf{V} = \mathbf{L}$, and the axiom "every set x satisfies $\operatorname{card} x \leq \omega_1$".

4.1. Jensen—Solovay Sequences

Arguing in **L**, let U, V be systems. Suppose that M is any transitive model of $\mathbf{ZFC}^-_2$. Define $U \preccurlyeq_M U'$ iff $U \preccurlyeq U'$ and the following holds:

(a) the set $\Delta(U, U') = \bigcup_{\nu \in |U|}(U'(\nu) \smallsetminus U(\nu))$ is *multiply* **Seq**-*generic* over M, in the sense that every sequence $\langle f_1, \ldots f_m\rangle$ of pairwise different functions $f_\ell \in \Delta(U, U')$ is generic over M in the sense of $\mathbf{Seq} = \omega_1{}^{<\omega_1}$ as the forcing notion in **L**, and

(b) if $\nu \in |U|$ then $U'(\nu) \smallsetminus U(\nu)$ is dense in **Fun**, therefore uncountable.

Let **JS**, *Jensen—Solovay pairs*, be the set of all pairs $\langle M, U\rangle$ of:

- a transitive model $M \models \mathbf{ZFC}^-_2$, and a system U,

– such that the sets ω_1 and U belong to M—then sets $\mathbf{SEQ}$, $\mathbf{Q}[U]$ also belong to M.

Let **sJS**, *small Jensen—Solovay pairs*, be the set of all pairs $\langle M, U\rangle \in \mathbf{JS}$ such that both U and M have cardinality $\le \omega_1$. We define:

$\langle M, U\rangle \preccurlyeq \langle M', U'\rangle$ ($\langle M', U'\rangle$ extends $\langle M, U\rangle$) iff $M \subseteq M'$ and $U \preccurlyeq_M U'$;
$\langle M, U\rangle \prec \langle M', U'\rangle$ (strict extension) iff $\langle M, U\rangle \preccurlyeq \langle M', U'\rangle$ and $\forall\, \nu \in \mathcal{I}\,(U(\nu) \subsetneqq U'(\nu))$.

Lemma 18 (in **L**). *If $\langle M, U\rangle \in \mathbf{sJS}$ and $z \subseteq \mathcal{I}$, $\mathtt{card}\, z \le \omega_1$, then there is a pair $\langle M', U'\rangle \in \mathbf{sJS}$, such that $\langle M, U\rangle \prec \langle M', U'\rangle$ and $z \subseteq |U'|$.*

Proof. Let $d = |U| \cup z$. By definition $\mathbf{SEQ}$ is *ω-closed* as a forcing: any $\subseteq$-increasing sequence $\{s_n\}_{n<\omega}$ of $s_n \in \mathbf{SEQ}$ has the least upper bound in $\mathbf{SEQ}$, equal to the union of all s_n. It follows that the countable-support product $\mathbf{SEQ}^{(d\times\omega_1)}$ is ω-closed, too. Therefore, as $\mathtt{card}\, M \le \omega_1$, there exists a system $\vec{f} = \{f_{\nu\xi}\}_{\nu\in d, \xi<\omega_1} \in (\mathbf{Fun})^{d\times\omega_1}$, $\mathbf{SEQ}^{(d\times\omega_1)}$-generic over M. Now define $U'(\nu) = U(\nu) \cup \{f_{\nu\xi} : \xi < \omega_1\}$ for each $\nu \in d$ (assuming that $U(\nu) = \varnothing$ in case $\nu \notin |U|$), and let $M' \models \mathbf{ZFC}^-_1$ be any transitive model of cardinality ω_1, satisfying $M \subseteq M'$ and containing U'. □

Lemma 19 (in **L**). *Suppose that pairs $\langle M, U\rangle \preccurlyeq \langle M', U'\rangle \preccurlyeq \langle M'', U''\rangle$ belong to* **JS**. *Then $\langle M, U\rangle \preccurlyeq \langle M'', U''\rangle$. Thus $\preccurlyeq$ is a partial order on* **JS**.

Proof. We claim that $F = \bigcup_{\nu\in|U|}(U''(\nu) \smallsetminus U(\nu))$ is multiply $\mathbf{SEQ}$-generic over M. Suppose, for the sake of brevity, that $F = \{f, g\}$, where $f \in U'(\nu) \smallsetminus U(\nu)$—then $f \in M'$, $g \in U''(\mu) \smallsetminus U'(\mu)$, and $\nu, \mu \in |U|$. (The general case does not differ much.) By definition, f is Cohen generic over M and g is Cohen generic over M'. Therefore, g is Cohen generic over $M[f]$, because $M[f] \subseteq M'$ (as $f \in M'$). It remains to apply the product forcing theorem. □

Now, still in **L**, a *Jensen—Solovay sequence* of length $\lambda \le \omega_2$ is any strictly $\prec$-increasing λ-sequence $\{\langle M_\xi, U_\xi\rangle\}_{\xi<\lambda}$ of pairs $\langle M_\xi, U_\xi\rangle \in \mathbf{sJS}$, satisfying $U_\eta = \bigvee_{\xi<\eta} U_\xi$ on limit steps. Let $\overrightarrow{\mathbf{JS}}_\lambda$ be the set of all such sequences.

Lemma 20 (in **L**). *Let λ be a limit ordinal, and $\{\langle M_\xi, U_\xi\rangle\}_{\xi<\lambda} \in \overrightarrow{\mathbf{JS}}_\lambda$. Put $U = \bigvee_{\xi<\lambda} U_\xi$. Then*

(i) $U_\xi \preccurlyeq_{M_\xi} U$ *for every* ξ.

(ii) *If moreover $\lambda < \omega_2$ and $M \models \mathbf{ZFC}^-_2$ is a transitive model containing $\{\langle M_\xi, U_\xi\rangle\}_{\xi<\lambda}$ then $\langle M, U\rangle \in \mathbf{sJS}$ and $\langle M_\xi, U_\xi\rangle \prec \langle M, U\rangle$, $\forall\,\xi$.*

(iii) *The same is true in case $\lambda = \omega_2$, but then the model M is not necessarily a ω_1-sizemodel, and we require $\langle M, U\rangle \in \mathbf{JS}$ rather than* **sJS**, *of course.*

Proof. The same arguments work as in the proof of Lemma 19. □

4.2. Stability of Dense Sets

If U is a system, D is a pre-dense subset of $\mathbf{P}[U]$, and U' is another system extending U, then in principle D does not necessarily remain maximal in $\mathbf{P}[U']$, a bigger set. This is where the genericity requirement (a) in Section 4.1 plays its role to *seal* the pre-density of sets in M w. r. t. further extensions. This is the content of the following key theorem. Moreover, the product forcing arguments will allow us to extend the stability result in pre-dense sets not necessarily in M, as in items (ii), (iii) of the theorem.

Theorem 2 (stability of dense sets). *Assume that, in **L**, $\langle M, U\rangle \in \mathbf{sJS}$, U' is a system, and $U \preccurlyeq_M U'$. If D is a pre-dense subset of $\mathbf{Q}[U]$ (resp., pre-dense below some $p \in \mathbf{Q}[U]$) then D remains pre-dense in $\mathbf{Q}[U']$ (resp., pre-dense below p) in each of the following three cases:*

(i) $D \in M$;

(ii) $D \in M[G]$, *where* $G \subseteq P$ *is* P*-generic over* $\mathbf{L}$, *and* $P \in M$ *is a PO set*;

(iii) $D \in M[H]$, *where* $H \subseteq U'(\nu_1)$ *is finite*, $\nu_1 \in \mathcal{I}$ *is fixed, and* $D \subseteq \mathbf{Q}[U]\restriction_{\neq\nu_1} = \{q \in \mathbf{Q}[U] : \nu_1 \notin |q|\}$.

Proof. Arguing in $\mathbf{L}$, we consider only the case of sets D pre-dense in $\mathbf{Q}[U]$ itself; the case of pre-density below some $p \in \mathbf{Q}[U]$ is treated similarly.

(i) Suppose, towards the contrary, that a condition $p \in \mathbf{Q}[U']$ is incompatible with each $q \in D$. As $D \subseteq \mathbf{P}[U]$, we can w.l.o.g. assume that $|p| \subseteq |U|$.

We are going to define a condition $p' \in \mathbf{Q}[U]$, also incompatible with each $q \in D$, contrary to the pre-density. To maintain the construction, consider the finite sequence $\vec{f} = \langle f_1, \ldots, f_m \rangle$ of all elements $f \in \mathbf{FUN}$ occurring in $\bigcup_{\nu \in |p|} F_p(\nu)$ but not in U. It follows from $U \preccurlyeq_M U'$ that $\vec{f}$ is $\mathbf{SEQ}^m$-generic over M. Moreover, p being incompatible with D is implied by the fact that $\vec{f}$ meets a certain family of dense sets in $\mathbf{SEQ}^m$, of cardinality $\leq \omega_1$ in M. Therefore, we will be able to simulate this in M, getting a sequence $\vec{g} \in M$ which meets the same dense sets, and hence yields a condition $p' \in \mathbf{Q}[U]$, also incompatible with each $q \in D$.

To present the key idea in sufficient detail in a rather simplified subcase, we assume that

$$|p| - \{\nu_0\} \text{ is a singleton; } \nu_0 \in |U|. \tag{7}$$

Then $p(\nu_0) = \langle S_p(\nu_0) ; F_p(\nu_0) \rangle \in Q[U'(\nu_0)]$, where $S_p(\nu_0) \subseteq \mathbf{SEQ}$ and $F_p(\nu_0) \subseteq U'(\nu_0)$ are finite sets. The (finite) set $X = F_p(\nu_0) \smallsetminus U(\nu_0)$ is multiply $\mathbf{SEQ}$-generic over M since $U \preccurlyeq_M U'$. To make the argument even more transparent, we suppose that

$$X = \{f, g\}, \text{ where } f \neq g \text{ and the pair } \langle f, g \rangle \text{ is } \mathbf{SEQ}^2\text{-generic over } M. \tag{8}$$

(The general case follows the same idea and can be found in [4]; we leave it to the reader.)

Thus, $F_p(\nu_0) = F \cup \{f, g\}$, where $F = F_p(\nu_0) \cap U(\nu_0) \in M$ is by definition a finite set.

The plan is to replace the functions f, g by some functions $f', g' \in U(\nu_0)$ so that the incompatibility of p with conditions in D will be preserved.

It holds by the choice of p and Lemma 1 that $D = D_1(f, g) \cup D_2$, where

$$\begin{aligned} D_1(f,g) &= \{q \in D : A_q \cap F_p^{\vee}(\nu_0) \neq \varnothing\}, \text{ where } A_q = S_q(\nu_0) \smallsetminus S_p(\nu_0) \subseteq \mathbf{SEQ}; \\ D_2 &= \{q \in D : (S_p(\nu_0) \smallsetminus S_q(\nu)) \cap F_q^{\vee}(\nu_0) \neq \varnothing\} \in M; \end{aligned}$$

and D_1 depends on f, g via $F_p(\nu_0)$. The equality $D = D_1(f, g) \cup D_2$ can be rewritten as $\Delta \subseteq D_1(f, g)$, where $\Delta = D \smallsetminus D_2 \in M$. Furthermore, $\Delta \subseteq D_1(f, g)$ is equivalent to

$$\forall A \in \mathscr{A}\, (A \cap F_p^{\vee}(\nu) \neq \varnothing), \text{ where } \mathscr{A} = \{A_q : q \in D\} \in M, \tag{9}$$

and each $A_q = S_q(\nu_0) \smallsetminus S_p(\nu_0) \subseteq \mathbf{SEQ}$ is finite. Recall that $F_p(\nu_0) = F \cup \{f, g\}$, therefore $F_p^{\vee}(\nu_0) = Z \cup S(f, g)$, where $Z = \{h\restriction\mu : 1 \leq \mu < \omega_1 \wedge h \in F\} \in M$ and $S(f, g) = \bigcup_{1 \leq \mu < \omega_1} \{f\restriction\mu, g\restriction\mu\}$. Thus, (9) is equivalent to

$$\forall A' \in \mathscr{A}'\, (A' \cap S(f, g) \neq \varnothing), \text{ where } \mathscr{A}' = \{A_q \smallsetminus Z : q \in D\} \in M. \tag{10}$$

Please note that each $A' \in \mathscr{A}'$ is a finite subset of $\mathbf{SEQ}$, so we can re-enumerate $\mathscr{A}' = \{A'_\kappa : \kappa < \omega_1\}$ in M and rewrite (10) as follows:

$$\forall \kappa < \omega_1\, (A'_\kappa \cap S(f, g) \neq \varnothing), \text{ where each } A'_\kappa \subseteq \mathbf{SEQ} \text{ is finite.} \tag{11}$$

As the pair $\langle f,g\rangle$ is $\mathbf{SEQ}$-generic, there is an index $\mu_0 < \omega_1$ such that (11) is forced over M by $\langle \sigma_0, \tau_0\rangle$, where $\sigma_0 = f\restriction \mu_0$ and $\tau_0 = g\restriction \mu_0$. In other words, $A'_\kappa \cap S(f',g') \neq \varnothing$ holds for all $\kappa < \omega_1$ whenever $\langle f',g'\rangle$ is $\mathbf{SEQ}$-generic over M and $\sigma_0 \subset f'$, $\tau_0 \subset g'$. It follows that for any $\kappa < \omega_1$ and sequences $\sigma, \tau \in \mathbf{SEQ}$ extending resp. σ_0, τ_0 there are sequences $\sigma', \tau' \in \mathbf{SEQ}$ extending resp. σ, τ, at least one of which extends one of sequences $w \in A'_\kappa$. This allows us to define, in M, a pair of sequences $f', g' \in \mathbf{FUN}$, such that $\sigma_0 \subset f'$, $\tau_0 \subset g'$, and for any $\kappa < \omega_1$ at least one of f', g' extends one of $w \in A'_k$. In other words, we have

$$\forall \kappa < \omega_1 (A'_\kappa \cap S(f',g') \neq \varnothing) \quad \text{and} \quad \forall A' \in \mathscr{A}' \, (A' \cap S(f',g') \neq \varnothing).$$

It follows that the condition p' defined by $|p'| = \{\nu_0\}$, $S_{p'}(\nu_0) = S_p(\nu_0)$, $F_{p'}(\nu_0) = F \cup \{f', g'\}$, still satisfies $\forall A \in \mathscr{A} \, (A \cap F^{\vee}_{p'}(\nu_0) \neq \varnothing)$ (compare with (9)), and further $D = D_1(f',g') \cup D_2$, thus p' is incompatible with each $q \in D$. Yet $p' \in M$ since $f', g' \in M$, which contradicts the pre-density of D.

(ii) The above proof works with $M[G]$ instead of M since the set X as in the proof is multiple $\mathbf{SEQ}$-generic over $M[G]$ by the product forcing theorem.

(iii) Assuming w.l.o.g. that $H \subseteq U'(\nu_1) \setminus U(\nu_1)$, we conclude that $M[H]$ is a $\mathbf{SEQ}$-generic extension of M. Now, if $p \in \mathbf{Q}[U']\restriction_{\neq \nu_1}$, then, following the above argument, let $\nu_0 \in |p|$, $\nu_0 \neq \nu_1$. By the definition of $\preccurlyeq$ the set $F = F_p(\nu_0) \setminus U(\nu_0)$ is multiply $\mathbf{SEQ}$-generic not only over M but also over $M[H]$. This allows the carrying out of the same argument as above. □

Corollary 2. *Under the assumptions of Theorem 2, if a set $G \subseteq \mathbf{Q}[U']$ is $\mathbf{Q}[U']$-generic over a transitive model $M' \models \mathbf{ZFC}^-_2$ containing M and U' (including the case $M' = \mathbf{L}$), then the intersection $G \cap \mathbf{Q}[U]$ is $\mathbf{Q}[U]$-generic over M.*

Proof. If a set $D \in M$, $D \subseteq \mathbf{Q}[U]$, is pre-dense in $\mathbf{Q}[U]$, then it is pre-dense in $\mathbf{Q}[U']$ by Theorem 2, and hence $G \cap D \neq \varnothing$ by the genericity. □

Corollary 3 (in **L**). *Under the assumptions of Theorem 2, if $\tau \in M \cap \mathbf{SN}^{\omega}_{\omega}(\mathbf{Q}[U])$ is a $\mathbf{Q}[U]$-full name then τ remains $\mathbf{Q}[U']$-full, and if $p \in \mathbf{Q}[U]$ and τ is $\mathbf{Q}[U]$-full below p, then τ remains $\mathbf{Q}[U']$-full below p.* □

4.3. Complete Sequences and the Basic Forcing Notion

In **L**, we say that a pair $\langle M, U\rangle \in \mathbf{sJS}$ *solves* a set $D \subseteq \mathbf{sJS}$ iff either $\langle M, U\rangle \in D$ or there is no pair $\langle M', U'\rangle \in D$ that extends $\langle M, U\rangle$. Let $D^{\mathtt{solv}}$ be the set of all pairs $\langle M, U\rangle \in \mathbf{sJS}$ which solve a given set $D \subseteq \mathbf{sJS}$. A sequence $\{\langle M_\xi, U_\xi\rangle\}_{\xi<\omega_2} \in \overrightarrow{\mathbf{JS}}_{\omega_2}$ is called *n-complete* ($n \geq 3$) iff it intersects every set of the form $D^{\mathtt{solv}}$, where $D \subseteq \mathbf{sJS}$ is a $\Sigma^{\mathrm{H}\omega_2}_{n-2}(\mathrm{H}\omega_2)$ set.

Recall that $\mathrm{H}\omega_2$ is the collection of all sets x whose transitive closure $\mathrm{TC}(x)$ has cardinality $\mathtt{card}\,(\mathrm{TC}(x)) < \omega_2$. Furthermore, $\Sigma^{\mathrm{H}\omega_2}_{n-2}(\mathrm{H}\omega_2)$ means definability by a Σ_{n-2} formula of the $\in$-language, in which any definability parameters in $\mathrm{H}\omega_2$ are allowed, while $\Sigma^{\mathrm{H}\omega_2}_{n-2}$ means parameter-free definability. Similarly, $\Delta^{\mathrm{H}\omega_2}_{n-1}(\{\omega_1\})$ in the next theorem means that ω_1 is allowed as a sole parameter. It is a simple exercise that sets $\{\mathbf{SEQ}\}$ and $\mathbf{SEQ}$ are $\Delta^{\mathrm{H}\omega_2}_1(\{\omega_1\})$ under $\mathbf{V} = \mathbf{L}$.

Generally, we refer to e.g., ([9] (Part B, 5.4)), or ([10] (Chapter 13)) on the Lévy hierarchy of $\in$-formulas and definability classes Σ^H_n, Π^H_n, Δ^H_n for any transitive set H.

Theorem 3 (in **L**). *Let $n \geq 2$. There is a sequence $\{\langle M_\xi, U_\xi\rangle\}_{\xi<\omega_2} \in \overrightarrow{\mathbf{JS}}_{\omega_2}$ of class $\Delta^{\mathrm{H}\omega_2}_{n-1}(\{\omega_1\})$, hence, $\Delta^{\mathrm{H}\omega_2}_{n-1}$ in case $n \geq 3$, n-complete in case $n \geq 3$, and such that $\xi \in |U_{\xi+1}|$ for all $\xi < \omega_2$.*

Proof. To account forω_1 as a parameter, note that the set ω_1 is $\Sigma^{\mathrm{H}\omega_2}_1$, and hence the singleton $\{\omega_1\}$ is $\Delta^{\mathrm{H}\omega_2}_2$. Indeed "being ω_1" is equivalent to the conjunction of "being uncountable"—which is $\Pi^{\mathrm{H}\omega_2}_1$, and "every smaller ordinal is countable"—which is $\Sigma_1^{\mathrm{H}\omega_2}$ since the quantifier "for all smaller ordinals" is bounded, hence, it does not increase the complexity.

It follows that $\Delta_{n-1}^{\mathrm{H}\omega_2}(\{\omega_1\}) = \Delta_{n-1}^{\mathrm{H}\omega_2}$ in case $n \geq 3$, supporting the "hence" claim of the theorem.

Then, it can be verified that the sets Q^*, $\mathbf{Q}^*$, **sJS** are $\Delta_1^{\mathrm{H}\omega_2}(\{\omega_1\})$. (Indeed "being finite" and "being countable" are $\Delta_1^{\mathrm{H}\omega_2}$ relations, while "being of cardinality ω_1" is $\Delta_1^{\mathrm{H}\omega_2}(\{\omega_1\})$; the Π_1 definition says that there is no injection from ω_1 into a given set.)

Define pairs $\langle M_\xi, U_\xi\rangle$, $\xi < \omega_2$, by induction. Let U_0 be the null system with $|U_0| = \varnothing$, and M_0 be the least CTM of $\mathbf{ZFC}_2^-$. If $\lambda < \omega_1$ is a limit, then put $U_\lambda = \bigvee_{\xi<\lambda} U_\xi$ and let M_λ be the least CTM of $\mathbf{ZFC}_2^-$ containing the sequence $\{\langle M_\xi, U_\xi\rangle\}_{\xi<\lambda}$. If $\langle M_\xi, U_\xi\rangle \in \mathbf{sJS}$ is defined, then by Lemma 18 there is a pair $\langle M', U'\rangle \in \mathbf{sJS}$ with $\langle M_\xi, U_\xi\rangle \prec \langle M', U'\rangle$ and $\xi \in |U'|$. Further let $\Theta \subseteq \omega_1 \times \mathrm{H}\omega_2$ be a universal $\Sigma_{n-2}^{\mathrm{H}\omega_2}$ set, and if $\xi < \omega_2$ then $D_\xi = \{z \in \mathbf{sJS} : \langle \xi, z\rangle \in \Theta\}$. Let $\langle M_{\xi+1}, U_{\xi+1}\rangle$ be the $<_{\mathbf{L}}$-least pair $\langle M, U\rangle \in D_\xi{}^{\mathtt{solv}}$ satisfying $\langle M', U'\rangle \preccurlyeq \langle M, U\rangle$, where $<_{\mathbf{L}}$ is the Gödel wellordering of $\mathbf{L}$, the constructible universe. This completes the inductive construction of $\langle M_\xi, U_\xi\rangle \in \mathbf{sJS}$, $\xi < \omega_2$.

To check the definability property, make use of the well-known fact that the restriction $<_{\mathbf{L}}\restriction \mathrm{H}\omega_2$ is a $\Delta_1^{\mathrm{H}\omega_2}$ relation, and if $n \geq 1$, $p \in \omega^\omega$ is any parameter, and $R(x, y, z, \dots)$ is a finitary $\Delta_n^{\mathrm{H}\omega_2}(p)$ relation on HC then the relations $\exists x <_{\mathbf{L}} y\, R(x, y, z, \dots)$ and $\forall x <_{\mathbf{L}} y\, R(x, y, z, \dots)$ (with arguments $y, z, \dots$) are $\Delta_n^{\mathrm{H}\omega_2}(p)$ as well. □

Definition 2 (in **L**). ***Fix a number*** $\mathtt{n} \geq 2$ ***during the proof of Theorem 1.***

- *Let* $\vec{\mathbb{js}} = \{\langle \mathbb{M}_\xi, \mathbb{U}_\xi\rangle\}_{\xi<\omega_2} \in \overrightarrow{\mathbf{JS}}_{\omega_2}$ *be any* $\mathtt{n}$*-complete Jensen–Solovay sequence of class* $\Delta_{\mathtt{n}-1}^{\mathrm{H}\omega_2}$ *as in Theorem 3—in case* $\mathtt{n} \geq 3$*, or just any Jensen–Solovay sequence of class* $\Delta_1^{\mathrm{H}\omega_2}(\{\omega_1\})$*—in case* $\mathtt{n} = 2$*, as in Theorem 3, including* $\xi \in |\mathbb{U}_{\xi+1}|$ *for all* ξ *in both cases.*
- *Put* $\mathbb{U} = \bigvee_{\xi<\omega_1} \mathbb{U}_\xi$*, so* $\mathbb{U}(\nu) = \bigcup_{\xi<\omega_2, \nu\in|\mathbb{U}_\xi|} \mathbb{U}_\xi(\nu)$ *for all* $\nu \in \mathcal{I}$*. Thus,* $\mathbb{U} \in \mathbf{L}$ *is a system and* $|\mathbb{U}| = \mathcal{I}$ *since* $\xi \in |\mathbb{U}_{\xi+1}|$ *for all* ξ*.*

We define $\mathbb{Q} = \mathbf{Q}[\mathbb{U}]$ *(the* ***basic forcing notion****), and* $\mathbb{Q}_\xi = \mathbf{Q}[\mathbb{U}_\xi]$ *for* $\xi < \omega_2$*. Thus,* $\mathbb{Q}$ *is the finite-support product of the set* $\mathbb{C}$ *and sets* $\mathbb{Q}(\nu) = Q[\mathbb{U}(\nu)]$*,* $i \in \mathcal{I}$*; so that* $\mathbb{Q} \in \mathbf{L}$*.* □

Corollary 4. *Suppose that in* **L**, $\xi < \omega_2$ *and* M *is a TM of* $\mathbf{ZFC}_2^-$ *containing the sequence* $\vec{\mathbb{js}}$*. Then*

(i) $\langle M, \mathbb{U}\rangle \in \mathbf{JS}$, $\langle \mathbb{M}_\xi, \mathbb{U}_\xi\rangle \prec \langle M, \mathbb{U}\rangle$*, and if* $\nu \in \mathcal{I}$ *then* $\mathtt{card}(\mathbb{U}_\xi(\nu)) = \omega_1 < \omega_2 = \mathtt{card}(\mathbb{U}(\nu))$ *in* **L**.

(ii) *If* $G \subseteq \mathbb{Q}$ *is a set* $\mathbb{Q}$*-generic over* **L** *then the set* $G^\xi = G \cap \mathbb{Q}_\xi$ *is* $\mathbb{Q}_\xi$*-generic over* $\mathbb{M}_\xi$*.*

Proof. Make use of Lemma 20 and Corollary 2 in Section 4.2. □

Lemma 21 (in **L**). *The binary relation* $f \in \mathbb{U}(\nu)$*, the sets* $\mathbb{Q}$ *and* $\mathbf{SN}_\omega^\omega(\mathbb{Q})$ *(*$\mathbb{Q}$*-names for reals in* ω^ω*), and the set of all* $\mathbb{Q}$*-full names in* $\mathbf{SN}_\omega^\omega(\mathbb{Q})$ *are* $\Delta_{\mathtt{n}-1}^{\mathrm{H}\omega_2}(\{\omega_1\})$*, and even* $\Delta_{\mathtt{n}-1}^{\mathrm{H}\omega_2}$ *in case* $\mathtt{n} \geq 3$*.*

Proof. The sequence $\{\langle \mathbb{M}_\xi, \mathbb{U}_\xi\rangle\}_{\xi<\omega_1}$ is $\Delta_{\mathtt{n}-1}^{\mathrm{H}\omega_2}$ by definition, hence the relation $f \in \mathbb{U}(\nu)$ is $\Sigma_{\mathtt{n}-1}^{\mathrm{H}\omega_2}$. On the other hand, if $f \in \mathbf{Fun}$ belongs to some $\mathbb{M}_\xi$ then $f \in \mathbb{U}(\nu)$ obviously implies $f \in \mathbb{U}_\xi(\nu)$, leading to a $\Pi_{\mathtt{n}-1}^{\mathrm{HC}}$ definition of the relation $f \in \mathbb{U}(\nu)$. To prove the last claim, note that by Corollary 3 if a name $\tau \in \mathbf{SN}_\omega^\omega(\mathbb{P}_\xi) \cap \mathbb{M}_\xi$ is $\mathbb{P}_\xi$-full then it remains $\mathbb{P}$-full. □

4.4. Basic Generic Extension

The proof of Theorem 1 makes use of a generic extension of the form $\mathbf{L}[G \restriction z]$, where $G \subseteq \mathbb{Q}$ is a set $\mathbb{Q}$-generic over **L**, and $z \subseteq \mathcal{I}^+$, $z \notin \mathbf{L}$. The following two theorems will play the key role in the proof. Define formulas Γ_ν ($\nu \in \mathcal{I}$) as follows:

$$\Gamma_\nu(S) :=_{\text{def}} S \subseteq \mathbf{SEQ}^{\mathbf{L}} \wedge \forall f \in \mathbf{FUN}^{\mathbf{L}}\, (f \in \mathbb{U}(\nu) \iff S \text{ does not cover } f).$$

Lemma 22. *Suppose that a set* $G \subseteq \mathbb{Q}$ *is* $\mathbb{Q}$*-generic over* **L**, *and* $\nu \in \mathcal{I}$, $c \in \mathbf{L}[G]$, $\varnothing \neq c \subseteq \mathcal{I}^+$*. Then*

(i) $\omega_1^{\mathbf{L}[G \restriction c]} = \omega_2^{\mathbf{L}}$,

(ii) *if* $-1 \in c$ *then* $\boldsymbol{b}_G \in \mathbf{L}[G \restriction c]$, *and if* $\nu \in c$ *then* $S_G(\nu) \in \mathbf{L}[G \restriction c]$,

(iii) $\mathbb{F}_\nu(S_G(\nu))$ *holds,*

(iv) $S_G(\nu) \notin \mathbf{L}[G \restriction_{\neq\nu}]$, *and generally, there are no sets* $S \subseteq \mathbf{SEQ}^{\mathbf{L}}$ *in* $\mathbf{L}[G \restriction_{\neq\nu}]$ *satisfying* $\mathbb{F}_\nu(S)$.

Proof. To prove (i) apply Lemma 9 (ii); (ii) is easy. Furthermore, Lemma 9 (vi) immediately implies (iii).

To prove (iv), we need more work. Let $X = \mathbf{SEQ}^{\mathbf{L}}$. Suppose towards the contrary that some $S \in \mathbf{L}[G \restriction_{\neq\nu}]$, $S \subseteq X = \mathbf{SEQ}^{\mathbf{L}}$ satisfies $\mathbb{F}_\nu(S)$. It follows from Lemma 10 (with $U = \mathbb{U}$ and $c = \mathcal{I}^+ \smallsetminus \{\nu\}$), that there is a name $\tau \in \mathbf{SN}_X(\mathbb{Q}) \restriction_{\neq\nu}$ in $\mathbf{L}$ such that $S = \tau[G \restriction_{\neq\nu}]$. There is an ordinal $\xi < \omega_1$ satisfying $\tau \in \mathbb{M}_\xi$ and $\tau \in \mathbf{SN}_X(\mathbb{Q}_\xi \restriction_{\neq\nu})$. Then $S = \tau[G^\xi \restriction_{\neq\nu}]$, where $G^\xi = G \cap \mathbb{P}_\xi$ is $\mathbb{P}_\xi$-generic over $\mathbb{M}_\xi$ by Corollary 4 (ii), and by the way S belongs to $\mathbb{M}_\xi[G^\xi \restriction_{\neq\nu}]$ by the choice of ξ.

Please note that $F = \mathbb{U}(\nu) \smallsetminus \mathbb{U}_\xi(\nu) \neq \varnothing$ by Corollary 4 (i). Let $f \in F$. Then f is Cohen generic over the model $\mathbb{M}_\xi$ by Corollary 4. On the other hand, $G^\xi \restriction_{\neq\nu}$ is $\mathbb{P}_\xi \restriction_{\neq\nu}$-generic over $\mathbb{M}_\xi[f]$ by Theorem 2 (iii). Therefore f is Cohen generic over $\mathbb{M}_\xi[G^\xi \restriction_{\neq\nu}]$ as well.

Recall that $S \in \mathbb{M}_\xi[G^\xi \restriction_{\neq\nu}]$ and $\mathbb{F}_\nu(S)$ holds, hence S does not cover f. As f is Cohen generic over $\mathbb{M}_\xi[G^\xi \restriction_{\neq\nu}]$, it follows that there is a sequence $s \in \mathbf{SEQ}^{\mathbf{L}}$, $s \subset f$, such that S contains no subsequences of f extending s. Take any $\mu \in \mathcal{I}$, $\mu \neq \nu$. By Corollary 4 (i), there exists a function $g \in \mathbb{U}(\mu) \smallsetminus \mathbb{U}_\xi(\mu)$, $g \notin \mathbb{U}(\nu)$, satisfying $s \subset g$. Then, S covers g by $\mathbb{F}_\nu(S)$. However, this is absurd by the choice of s. □

The proof of the next important *elementary equivalence theorem* will be given below in Section 6.3.

Theorem 4 (elementary equivalence theorem). *Assume that in* $\mathbf{L}$, $-1 \in d \subseteq \mathcal{I}^+$, *sets* $Z', Z \subseteq \mathcal{I} \smallsetminus d$ *satisfy* $\mathtt{card}\,(\mathcal{I} \smallsetminus Z) \le \omega_1$ *and* $\mathtt{card}\,(\mathcal{I} \smallsetminus Z') \le \omega_1$, *the symmetric difference* $Z \,\Delta\, Z'$ *is at most countable, and the complementary set* $\mathcal{I} \smallsetminus (d \cup Z \cup Z')$ *is infinite.*

Let $G \subseteq \mathbb{Q}$ *be* $\mathbb{Q}$*-generic over* $\mathbf{L}$, *and* $x_0 \in \mathbf{L}[G \restriction d]$ *be any real. Then any closed* $\Sigma^1_{\mathfrak{n}}$ *formula* φ, *with real parameters in* $\mathbf{L}[x_0]$, *is simultaneously true in* $\mathbf{L}[x_0, G \restriction Z]$ *and in* $\mathbf{L}[x_0, G \restriction Z']$.

4.5. The Main Theorem Modulo the Elementary Equivalence Theorem: The Model

Here we begin **the proof of Theorem 1 on the base of Theorem 4 of Section 4.4.** We fix a number $\mathfrak{n} \ge 2$ during the proof. The goal is to define a generic extension of $\mathbf{L}$ in which for any set $x \subseteq \omega$ the following is true: $x \in \mathbf{L}$ iff $x \in \Delta^1_{\mathfrak{n}+1}$. The model is a part of the basic generic extension defined in Section 4.4.

In the notation of Definition 2 in Section 4.3, consider a set $G \subseteq \mathbb{Q}$, $\mathbb{Q}$-generic over $\mathbf{L}$. Then $\boldsymbol{b}_G = \bigcup G(-1)$ is a $\mathbb{C}$-generic map from ω onto $\mathscr{P}(\omega) \cap \mathbf{L}$ by Lemma 9 (i). We define

$$w[G] = \{\omega k + 2^j : k < \omega \wedge j \in \boldsymbol{b}_G(k)\} \cup \{\omega k + 3^j : j, k < \omega\} \subseteq \omega^2, \tag{12}$$

and $w^+[G] = \{-1\} \cup w[G]$. We also define, for any $m < \omega$,

$$w_{\ge m}[G] = \{\omega k + \ell \in w[G] : k \ge m\}, \quad w_{<m}[G] = \{\omega k + \ell \in w[G] : k < m\},$$

and accordingly $w^+_{\ge m}[G] = \{-1\} \cup w_{\ge m}[G]$ and $w^+_{<m}[G] = \{-1\} \cup w_{<m}[G]$.

With these definitions, each kth slice

$$w_k[G] = \{\omega k + 2^j : j \in \boldsymbol{b}_G(k)\} \cup \{\omega k + 3^j : j < \omega\} \tag{13}$$

of $w[G]$ is necessarily infinite and coinfinite, and it codes the target set $\boldsymbol{b}_G(k)$ since

$$\boldsymbol{b}_G(k) = \{j < \omega : \omega k + 2^j \in w_k[G]\} = \{j < \omega : \omega k + 2^j \in w^+[G]\}. \tag{14}$$

It will be important below that definition (12) is *monotone w.r.t.* $\boldsymbol{b}_G$, i.e., if $\boldsymbol{b}_G(k) \subseteq \boldsymbol{b}_{G'}(k)$ for all k, then $w[G] \subseteq w[G']$ and $w^+[G] \subseteq w^+[G']$. Non-monotone modifications, like e.g.,

$$w[G] = \{\omega k + 2^j : j \in \boldsymbol{b}_G(k)\} \cup \{\omega k + 3^j : j \notin \boldsymbol{b}_G(k)\}$$

would not work. Finally, let

$$W = [\omega^2, \omega_2) = \{\zeta : \omega^2 \le \zeta < \omega_2\}.$$

Anyway, $w^+[G] \subseteq \omega^2 = \omega \cdot \omega$ (the ordinal product) is a set in the model $\mathbf{L}[\boldsymbol{b}_G] = \mathbf{L}[w^+[G]] = \mathbf{L}[w[G]] = \mathbf{L}[w_{\ge m}[G]]$ for each m, containing -1, while $w_{<m}[G] \in \mathbf{L}$ for all m. We are going to prove the following lemma:

Lemma 23. *The model* $\mathbf{L}[G\restriction(w^+[G] \cup W)]$ *witnesses Theorem 1. That is, let a set* $G \subseteq \mathbb{Q}$ *be* $\mathbb{Q}$*-generic over* $\mathbf{L}$*. Then it holds in* $\mathbf{L}[G\restriction(w^+[G] \cup W)]$ *that*

(i) $w[G]$ *is* $\Sigma^1_{\mathrm{n}+1}$ *and each set* $x \in \mathbf{L}$, $x \subseteq \omega$ *is* $\Delta^1_{\mathrm{n}+1}$;

(ii) *if* $x \subseteq \omega$ *is* $\Delta^1_{\mathrm{n}+1}$ *then* $x \in \mathbf{L}$.

Recall that if $Z \subseteq \mathcal{I}^+$ then $G\restriction Z = \{p \in G : |p|^+ \subseteq Z\}$.

Proof (Claim (i) of the lemma). Consider an arbitrary ordinal $\nu = \omega k + \ell$; $k, \ell < \omega$. We claim that

$$\nu \in w[G] \iff \exists S\, \Gamma_\nu(S) \tag{15}$$

holds in $\mathbf{L}[G\restriction(w^+[G] \cup W)]$. Indeed, assume that $\nu \in w[G]$. Then $S = S_G(\nu) \in \mathbf{L}[G\restriction w^+[G]]$, and we have $\Gamma_\nu(S)$ in $\mathbf{L}[G\restriction(w^+[G] \cup W)]$ by Lemma 22 (ii), (iii). Conversely assume that $\nu \notin w[G]$. Then we have $w^+[G] \in \mathbf{L}[\boldsymbol{b}_G] \subseteq \mathbf{L}[G\restriction w^+[G]] \subseteq \mathbf{L}[G\restriction_{\neq\nu}]$, but $\mathbf{L}[G\restriction_{\neq\nu}]$ contains no S with $\Gamma_\nu(S)$ by Lemma 22 (iv).

However, the right-hand side of (15) defines a $\Sigma^{\mathrm{H}\omega_2}_{\mathrm{n}}(\{\omega_1{}^{\mathbf{L}}, \mathbf{SEQ}^{\mathbf{L}}\})$ relation in $\mathbf{L}[G\restriction(w^+[G] \cup W)]$ by Lemma 21. (Indeed, $(\mathrm{H}\omega_2)^{\mathbf{L}} = \mathbf{L}_{\omega_2{}^{\mathbf{L}}} = \mathbf{L}_{\omega_1}$ in $\mathbf{L}[G\restriction(w^+[G] \cup W)]$, therefore $(\mathrm{H}\omega_2)^{\mathbf{L}}$ is $\Sigma^{\mathrm{H}\omega_2}_1$ in $\mathbf{L}[G\restriction(w^+[G] \cup W)]$.) On the other hand, the sets $\{\omega_1{}^{\mathbf{L}}\}$ and $\{\mathbf{SEQ}^{\mathbf{L}}\}$ remain $\Delta^{\mathrm{H}\omega_2}_2$ singletons in $\mathbf{L}[G\restriction(w^+[G] \cup W)]$, so they can be eliminated since $\mathrm{n} \ge 2$. This yields $w[G] \in \Sigma^{\mathrm{HC}}_{\mathrm{n}}$ in $\mathbf{L}[G\restriction(w^+[G] \cup W)]$. It follows that $w[G] \in \Sigma^1_{\mathrm{n}+1}$ by ([10] (Lemma 25.25)), as required.

Consider an arbitrary set $x \in \mathbf{L}$, $x \subseteq \omega$. By genericity there exists $k < \omega$ such that $\boldsymbol{b}_G(k) = x$. Then $x = \{j : \omega k + 2^j \in w[G]\}$ by (12), therefore x is $\Sigma^1_{\mathrm{n}+1}$ as well. However, $\omega \setminus x \in \Sigma^1_{\mathrm{n}+1}$ by the same argument. Thus, x is $\Delta^1_{\mathrm{n}+1}$ in $\mathbf{L}[G\restriction(w^+[G] \cup W)]$, as required. (Claim (i) of Lemma 23) □

4.6. Proof of the Key Claim of Lemma 23

The **proof of Lemma** 23 (ii) is based on several intermediate lemmas.
Recall that $W = [\omega^2, \omega_2) = \{\xi : \omega^2 \le \xi < \omega_2\}$.

Lemma 24 (compare with Lemma 33 in [7]). *Suppose that* $G \subseteq \mathbb{Q}$ *is* $\mathbb{Q}$*-generic over* $\mathbf{L}$*, and* $m < \omega$*. Let* $c \subseteq w_{<m}[G]$ *be any set in* $\mathbf{L}$*. Then any closed* Σ^1_{n} *formula* Φ*, with reals in* $\mathbf{L}[G\restriction(c \cup w^+_{\ge m}[G] \cup W)]$ *as parameters, is simultaneously true in* $\mathbf{L}[G\restriction(c \cup w^+_{\ge m}[G] \cup W)]$ *and in* $\mathbf{L}[G\restriction(w^+[G] \cup W)]$*.*

It follows that if $c' \subseteq c \subseteq w_{<m}[G]$ *in* $\mathbf{L}$*, then any closed* $\Sigma^1_{\mathrm{n}+1}$ *formula* Ψ*, with parameters in* $\mathbf{L}[G\restriction(c' \cup w^+_{\ge m}[G] \cup W)]$*, true in* $\mathbf{L}[G\restriction(c' \cup w^+_{\ge m}[G] \cup W)]$*, is true in* $\mathbf{L}[G\restriction(c \cup w^+_{\ge m}[G] \cup W)]$ *as well.*

Proof (Lemma 24). There is an ordinal $\xi < \omega_2$ such that all parameters in φ belong to $\mathbf{L}[G\restriction Y]$, where $Y = c \cup w^+_{\ge m}[G] \cup X$ and $X = [\omega^2, \xi) = \{\gamma : \omega^2 \le \gamma < \xi\}$. The set Y belongs to $\mathbf{L}[\boldsymbol{b}_G]$, in fact, $\mathbf{L}[Y] = \mathbf{L}[\boldsymbol{b}_G]$. Therefore $G\restriction Y$ is equi-constructible with the pair $\langle \boldsymbol{b}_G, \{S_G(\nu)\}_{\nu \in X'}\rangle$, where $\boldsymbol{b}_G$ is a map from ω onto, essentially, $\omega_1{}^{\mathbf{L}}$. It follows that there is a real x_0 with $\mathbf{L}[G\restriction Y] = \mathbf{L}[x_0]$. Then all parameters of φ belong to $\mathbf{L}[x_0]$.

To prepare for Theorem 4 of Section 4.4, put $Z' = [\xi, \omega_2)$, $c = w_{<m}[G] \smallsetminus e$, $Z = e \cup Z'$,

$$d = \{-1\} \cup \{\omega k + j : k \geq m \wedge j < \omega\} \cup X.$$

As $w^+_{\geq m}[G] \subseteq \{-1\} \cup \{\omega k + j : k \geq m \wedge j < \omega\}$, we have $Y = c \cup w^+_{\geq m}[G] \cup X \subseteq d$, and hence $x_0 \in \mathbf{L}[G{\restriction}d]$. It follows by Theorem 4 that φ is simultaneously true in $\mathbf{L}[x_0, G{\restriction}Z]$ and in $\mathbf{L}[x_0, G{\restriction}Z']$. However, $\mathbf{L}[x_0, G{\restriction}Z'] = \mathbf{L}[G{\restriction}(Y \cup Z')] = \mathbf{L}[G{\restriction}(c \cup w^+_{\geq m}[G] \cup W)]$ by construction, while $\mathbf{L}[x_0, G{\restriction}Z] = \mathbf{L}[G{\restriction}(w^+[G] \cup W)]$, and we are done. □

In continuation of the **proof of Lemma** 23 (ii), suppose that

(†) $\varphi(\cdot)$ and $\psi(\cdot)$ are parameter-free Σ^1_{n+1} formulas that provide a Δ^1_{n+1} definition for a set $x \subseteq \omega$, $x \in \mathbf{L}[G{\restriction}(w^+[G] \cup W)]$, i.e., we have

$$x = \{\ell < \omega : \varphi(\ell)\} = \{\ell < \omega : \neg\,\psi(\ell)\}$$

in $\mathbf{L}[G{\restriction}(w^+[G] \cup W)]$. Thus, the equivalence $\forall\,\ell\,(\varphi(\ell) \iff \neg\,\psi(\ell))$ is forced to be true in $\mathbf{L}[\underline{G}{\restriction}(w^+[\underline{G}] \cup \check{W})]$ by a condition $p_0 \in G$.

Here, $\underline{G}$ is the canonical $\mathbb{Q}$-name for the generic set $G \subseteq \mathbb{Q}$, as usual, while $\check{W}$ is a name for $W \in \mathbf{L}$.

Lemma 25. *Assume* (†). *If* $\ell < \omega$ *then the sentence "*$\mathbf{L}[\underline{G}{\restriction}(w^+[\underline{G}] \cup \check{W})] \models \varphi(\ell)$*" is* $\mathbb{Q}$*-decided by* p_0.

Proof. Suppose, for the sake of simplicity, that p_0 is the empty condition $\odot$ (i.e., $|p_0|^+ = \varnothing$); the general case does not differ much. Then $\forall\,\ell\,(\varphi(\ell) \iff \neg\,\psi(\ell))$ holds in $\mathbf{L}[G{\restriction}(w^+[G] \cup W)]$ for **any** generic set $G \subseteq \mathbb{Q}$.

Say that conditions $p, q \in \mathbb{Q} = \mathbf{Q}[\mathbb{U}]$ are *close neighbours* iff $-1 \in |p|^+ \cap |q|^+$ and one of the following holds:

(I) $\boldsymbol{b}_p = \boldsymbol{b}_q$ (recall that $\boldsymbol{b}_p = p(-1)$), or

(II) $p{\restriction}_{\neq -1} = q{\restriction}_{\neq -1}$, $\mathtt{lh}\,\boldsymbol{b}_p = \mathtt{lh}\,\boldsymbol{b}_q$, and either (a) $\boldsymbol{b}_p(k) \subseteq \boldsymbol{b}_q(k)$ for all $k < \mathtt{lh}\,\boldsymbol{b}_p$, or (b) $\boldsymbol{b}_q(k) \subseteq \boldsymbol{b}_p(k)$ for all $k < \mathtt{lh}\,\boldsymbol{b}_p$.

Proposition 1. *If conditions* $p, q \in \mathbb{Q}$ *are close neighbours, satisfying* (6) *in Section* 3.8, $\ell < \omega$, *and* p $\mathbb{Q}$*-forces the sentence "*$\mathbf{L}[\underline{G}{\restriction}(w^+[\underline{G}] \cup \check{W})] \models \varphi(\ell)$*", then so does* q.

Proof (Proposition). Suppose on the contrary that q does not force "$\mathbf{L}[\underline{G}{\restriction}(w^+[\underline{G}] \cup \check{W})] \models \varphi(\ell)$". As p, q satisfy (6), the associated transformation H_{pq} maps the set $\mathbb{Q}_{\leqslant p} = \{p' \in \mathbb{Q} : p' \leqslant p\}$ onto $\mathbb{Q}_{\leqslant q} = \{q' \in \mathbb{Q} : q' \leqslant q\}$ order-preserving by Lemma 17 (with $U = \mathbb{U}$). By the choice of q, there is a set $G_q \subseteq \mathbb{Q}_{\leqslant q}$, generic over $\mathbf{L}$, containing q, and such that $\varphi(\ell)$ is false in $\mathbf{L}[G_q{\restriction}(w^+[G_q] \cup W)]$. Then $\psi(\ell)$ is true in $\mathbf{L}[G_q{\restriction}(w^+[G_q] \cup W)]$ by (†) (and the assumption that $p_0 = \odot$).

The set $G_p = \{(H_{pq})^{-1}(q') : q' \in G_q \wedge q' \leqslant q\} \subseteq \mathbb{Q}_{\leqslant p}$ is $\mathbb{Q}$-generic over $\mathbf{L}$ as well (as H_{pq} is an order isomorphism), and contains p, and hence $\varphi(\ell)$ is true and $\psi(\ell)$ false in $\mathbf{L}[G_p{\restriction}(w^+[G_p] \cup W)]$.

Case 1: (I) holds, i.e., $\boldsymbol{b}_p = \boldsymbol{b}_q$. Then by definition $\boldsymbol{b}_{G_p} = \boldsymbol{b}_{G_q}$, so that $w^+[G_p] = w^+[G_q]$. On the other hand, the sets G_p and G_q are equi-constructible by means of the application of H_{pq}, and hence $G_p{\restriction}(w^+[G_p] \cup W)$ and $G_q{\restriction}(w^+[G_q] \cup W)$ are equi-constructible, that is, the classes $\mathbf{L}[G_p{\restriction}(w^+[G_p] \cup W)]$ and $\mathbf{L}[G_q{\restriction}(w^+[G_q] \cup W)]$ coincide. However, $\varphi(\ell)$ is true in one of them and false in the other one, a contradiction.

Case 2: (II) holds. Let $m = \mathtt{lh}\,\boldsymbol{b}_p = \mathtt{lh}\,\boldsymbol{b}_q$. Then $\boldsymbol{b}_{G_p}(k) = \boldsymbol{b}_{G_q}(k)$ for all $k \geq m$ via H_{pq}. This implies $\mathbf{L}[\boldsymbol{b}_{G_p}] = \mathbf{L}[\boldsymbol{b}_{G_q}]$, and also implies $w^+_{\geq m}[G_p] = w^+_{\geq m}[G_q]$, while the difference between the sets $w_{<m}[G_p]$, $w_{<m}[G_q]$ is that for any $k < m$ and any j,

$$\omega k + 2^j \in w_{<m}[G_q] \iff j \in \boldsymbol{b}_q(k) \quad \text{and} \quad \omega k + 2^j \in w_{<m}[G_p] \iff j \in \boldsymbol{b}_p(k). \tag{16}$$

Moreover, (II) implies $G_p{\restriction}_{\neq -1} = G_q{\restriction}_{\neq -1}$, and hence $S_{G_p}(\nu) = S_{G_q}(\nu)$ for all $\nu \in \mathcal{I}$ via H_{pq}. We conclude that $\mathbf{L}[G_p{\restriction}Z] = \mathbf{L}[G_q{\restriction}Z]$ for any set $Z \in \mathbf{L}[\boldsymbol{b}_{G_p}]$, $Z \subseteq \mathcal{I}^+$, in particular, $\mathbf{L}[G_p{\restriction}(w^+[G_q] \cup W)] = \mathbf{L}[G_q{\restriction}(w^+[G_q] \cup W)]$.

If now (II) (a) holds, then $c' = w_{<m}[G_p] \subsetneq c = w_{<m}[G_q] = c' \cup z$ by (16), where

$$z = \{\omega k + 2^j : k < m \wedge j \in \boldsymbol{b}_q(k) \smallsetminus \boldsymbol{b}_p(k)\} \in \mathbf{L}.$$

However, $\varphi(\ell)$ holds in $\mathbf{L}[G_p{\restriction}(w^+[G_p] \cup W)]$, see above. It follows by Lemma 24 that $\varphi(\ell)$ holds in $\mathbf{L}[G_p{\restriction}(w^+[G_q] \cup W)]$. However, we know that $\mathbf{L}[G_p{\restriction}(w^+[G_q] \cup W)] = \mathbf{L}[G_q{\restriction}(w^+[G_q] \cup W)]$. Thus, $\varphi(\ell)$ holds in $\mathbf{L}[G_q{\restriction}(w^+[G_q] \cup W)]$, which is a contradiction to the above. If (II) (b) holds, then argue similarly using the formula $\psi(\ell)$. (Proposition 1) □

Coming back to Lemma 25, suppose towards the contrary that "$\mathbf{L}[\underline{G}{\restriction}(w^+[\underline{G}] \cup \check{W})] \models \varphi(\ell)$" is not $\mathbb{Q}$-decided by $p_0 = \odot$. There are two conditions $p, q \in \mathbb{Q}$ such that p $\mathbb{Q}$-forces "$\mathbf{L}[\underline{G}{\restriction}(w^+[\underline{G}] \cup \check{W})] \models \varphi(\ell)$" while q $\mathbb{Q}$-forces the negation. We may w.l.o.g. assume, by Lemma 17 (i), that p, q satisfy (6) of Section 3.8. We claim that p, q can be connected by a finite chain of conditions in $\mathbb{Q}$ in which each two consecutive terms are close neighbours in the sense above, satisfying (6) in Section 3.8— then Proposition 1 implies a contradiction and concludes the proof of Lemma 25.

Thus, it remains to prove the connection claim. Let $p' \in \mathbb{Q}$ be defined by $\boldsymbol{b}_{p'} = \boldsymbol{b}_p$ and $p'{\restriction}_{\neq -1} = q{\restriction}_{\neq -1}$. Then p, p' are close neighbours and (6) holds for this pair as it holds for p, q. Let $r \in \mathbb{Q}$ be defined by $\boldsymbol{b}_r(k) = \boldsymbol{b}_p(k) \cup \boldsymbol{b}_q(k)$ for all $k < \ell = \mathtt{lh}\,\boldsymbol{b}_p = \mathtt{lh}\,\boldsymbol{b}_q$ and $p'{\restriction}_{\neq -1} = q{\restriction}_{\neq -1}$. Still r is a close neighbour to both p' and q, and (6) holds for p', r and q, r. Thus, the chain p—p'—r—q proves the connection claim. (Lemma 25) □

Now, to accomplish the **proof of Lemma** 23 (ii), apply Lemma 25.

(Lemma 23 (ii)) □

(Theorem 1 modulo Theorem 4 of Section 4.4) □

5. Forcing Approximation

To prove Theorem 4 of Section 4.4 and thus complete the proof of Theorem 1 in the next Section 6, we define here a forcing-like relation **forc**, and exploit certain symmetries of objects related to **forc**. This similarity will allow us to only outline really analogous issues but concentrate on several things which bear some difference.

We argue under Blanket Assumption 1.

Recall that $\mathbf{ZFC}^-$ is **ZFC** minus the Power Set axiom, with the schema of Collection instead of Replacement, with the axiom "ω_1 exists", and with **AC** in the form of wellorderability of every set, and $\mathbf{ZFC}^-_2$ is $\mathbf{ZFC}^-$ plus the axioms: $\mathbf{V} = \mathbf{L}$, and "every set x satisfies $\mathtt{card}\,x \le \omega_1$".

5.1. Formulas

Here we introduce a language that will help us to study analytic definability in $\mathbf{Q}[U]$-generic extensions, for different systems U, and their submodels.

Let $\mathcal{L}$ be the 2nd order Peano language, with variables of type 1 over ω^ω. If $K \subseteq \mathbf{Q}^*$ then an $\mathcal{L}(K)$ *formula* is any formula of $\mathcal{L}$, with some free variables of types 0, 1 replaced by resp. numbers in ω and names in $\mathbf{SN}^{\omega}_{\omega}(K)$, and some type 1 quantifiers are allowed to have *bounding indices* B (i.e., $\exists^B$, $\forall^B$) such that $B \subseteq \mathcal{I}^+$ satisfies either $\mathtt{card}\,B \le \omega_1$ or $\mathtt{card}(\mathcal{I} \smallsetminus B) \le \omega_1$ (in $\mathbf{L}$). In particular, $\mathcal{I}^+$ itself can serve as an index, and the absence

If φ is a $\mathcal{L}(\mathbf{Q}^*)$ formula, then let

$$\begin{aligned} \operatorname{NAM}\varphi &= \text{the set of all names } \tau \text{ that occur in } \varphi; \\ \operatorname{IND}\varphi &= \text{the set of all quantifier indices } B \text{ which occur in } \varphi; \\ |\varphi|^+ &= \textstyle\bigcup_{\tau\in\operatorname{NAM}\varphi}|\tau|^+ \quad \text{(a set of } \omega_1\text{-size);} \\ \|\varphi\| &= |\varphi|^+\cup\left(\bigcup\operatorname{IND}\varphi\right) \ - \text{ so that } |\varphi|^+\subseteq\|\varphi\|\subseteq\mathcal{I}^+. \end{aligned}$$

If a set $G \subseteq \mathbf{Q}^*$ is *minimally φ-generic* (that is, minimally τ-generic w.r.t. every name $\tau \in \operatorname{NAM}\varphi$, in the sense of Section 3.5), then the *valuation* $\varphi[G]$ is the result of substitution of $\tau[G]$ for any name $\tau \in \operatorname{NAM}\varphi$, and changing each quantifier $\exists^B x$, $\forall^B x$ to resp. $\exists\,(\forall)\,x \in \omega^\omega \cap \mathbf{L}[G\restriction B]$, while index-free type 1 quantifiers are relativized to ω^ω; $\varphi[G]$ is a formula of $\mathcal{L}$ with real parameters, and *some* quantifiers of type 1 relativized to certain submodels of $\mathbf{L}[G]$.

An *arithmetic* formula in $\mathcal{L}(K)$ is a formula with no quantifiers of type 1 (names in $\mathbf{SN}^\omega_\omega(K)$ are allowed). If $n < \omega$ then let a $\mathcal{L}\Sigma^1_n(K)$, resp., $\mathcal{L}\Pi^1_n(K)$ formula be a formula of the form

$$\exists^\circ x_1\,\forall^\circ x_2\,\ldots\,\forall^\circ(\exists^\circ)\,x_{n-1}\,\exists\,(\forall)\,x_n\,\psi\,,\quad \forall^\circ x_1\,\exists^\circ x_2\,\ldots\,\exists^\circ(\forall^\circ)\,x_{n-1}\,\forall\,(\exists)\,x_n\,\psi$$

respectively, where ψ is an arithmetic formula in $\mathcal{L}(K)$, all variables x_i are of type 1 (over ω^ω), the sign $^\circ$ means that this quantifier can have a bounding index as above, and it is required that the rightmost (closest to the kernel ψ) quantifier does not have a bounding index.

If in addition $M \models \mathbf{ZFC}^-$ is a transitive model and $K \subseteq \mathbf{Q}^*$ then define

$\mathcal{L}\Sigma^1_n(K,M) =$ all $\mathcal{L}\Sigma^1_n(K)$ formulas φ such that $\operatorname{NAM}\varphi \subseteq \mathbf{SN}^\omega_\omega(K) \cap M$ and each index $B \in \operatorname{IND}\varphi$ satisfies the requirement: either $B \in M$ or $\mathcal{I} \smallsetminus B \in M$.

Define $\mathcal{L}\Pi^1_n(K,M)$ similarly.

5.2. Forcing Approximation

We introduce a convenient forcing-type relation $p\ \mathbf{forc}^M_U\ \varphi$ for pairs $\langle M, U\rangle$ in $\mathbf{sJS}$ and formulas φ in $\mathcal{L}(K)$, associated with the truth in K-generic extensions of $\mathbf{L}$, where $K = \mathbf{Q}[U] \subseteq \mathbf{Q}^*$ and $U \in \mathbf{L}$ is a system.

(F1) First, writing $p\ \mathbf{forc}^M_U\ \varphi$, it is assumed that:

- (a) $\langle M, U\rangle \in \mathbf{sJS}$ and p belongs to $\mathbf{Q}[U]$,
- (b) φ is a closed formula in $\mathcal{L}\Pi^1_k(\mathbf{Q}[U], M) \cup \mathcal{L}\Sigma^1_{k+1}(\mathbf{Q}[U], M)$ for some $k \geq 1$, and each name $\tau \in \operatorname{NAM}\varphi$ is $\mathbf{Q}[U]$-full below p.

Under these assumptions, the sets U, $\mathbf{Q}[U]$, p, $\operatorname{NAM}\varphi$ belong to M.

The definition of **forc** goes on by induction on the complexity of formulas.

(F2) If $\langle M, U\rangle \in \mathbf{sJS}$, $p \in \mathbf{Q}[U]$, and φ is a closed formula in $\mathcal{L}\Pi^1_1(\mathbf{Q}[U], M)$ (then by definition it has no quantifier indices), then: $p\ \mathbf{forc}^M_U\ \varphi$ iff (F1) holds and p $\mathbf{Q}[U]$-forces $\varphi[\underline{G}]$ over M in the usual sense. Please note that the forcing notion $\mathbf{Q}[U]$ belongs to M in this case by (F1).

(F3) If $\varphi(x) \in \mathcal{L}\Pi^1_k(\mathbf{Q}[U], M)$, $k \geq 1$, then:

- (a) $p\ \mathbf{forc}^M_U\ \exists^B x\,\varphi(x)$ iff there is a name $\tau \in M \cap \mathbf{SN}^\omega_\omega(\mathbf{Q}[U])\restriction B$, $\mathbf{Q}[U]$-full below p (by (F1)b) and such that $p\ \mathbf{forc}^M_U\ \varphi(\tau)$.
- (b) $p\ \mathbf{forc}^M_U\ \exists\,x\,\varphi(x)$ iff there is a name $\tau \in M \cap \mathbf{SN}^\omega_\omega(\mathbf{Q}[U])$, $\mathbf{Q}[U]$-full below p (by (F1)b) and such that $p\ \mathbf{forc}^M_U\ \varphi(\tau)$.

(F4) If $k \geq 2$, φ is a closed $\mathcal{L}\Pi^1_k(\mathbf{Q}[U], M)$ formula, $p \in \mathbf{Q}[U]$, and (F1) holds, then p $\mathbf{forc}^M_U$ φ iff we have $\neg\, q$ $\mathbf{forc}^{M'}_{U'}$ $\varphi^{\neg}$ for every pair $\langle M', U'\rangle \in \mathbf{sJS}$ extending $\langle M, U\rangle$, and every condition $q \in \mathbf{Q}[U']$, $q \leqslant p$, where $\varphi^{\neg}$ is the result of canonical conversion of $\neg\,\varphi$ to $\mathcal{L}\Sigma^1_k(\mathbf{Q}[U], M)$.

The next theorem classifies the complexity of **forc** in terms of projective hierarchy. Please note that if $\langle M, U\rangle \in \mathbf{sJS}$ and $k \geq 1$ then any formula φ in $\mathcal{L}\Pi^1_k(\mathbf{Q}[U], M) \cup \mathcal{L}\Sigma^1_{k+1}(\mathbf{Q}[U], M)$ belongs to M if we somehow "label" any large index $B \in \text{IND}\,\varphi$ (such that $\mathtt{card}(\mathcal{I} \smallsetminus B) \leq \omega_1$) by its small complement $\mathcal{I} \smallsetminus B \in M$. Therefore, the sets

$$\begin{aligned}\mathbf{Forc}(\Pi^1_k) &= \{\langle M, U, p, \varphi\rangle : \langle M, U\rangle \in \mathbf{sJS} \wedge p \in \mathbf{Q}[U] \wedge \\ &\qquad \wedge\ \varphi \text{ is a closed formula in } \mathcal{L}\Pi^1_k(\mathbf{Q}[U], M) \wedge p\ \mathbf{forc}^M_U\ \varphi\},\end{aligned}$$

and $\mathbf{Forc}(\Sigma^1_k)$ similarly defined, are subsets of $\mathrm{H}\omega_2$ (in **L**).

Lemma 26 (in **L**). *The sets* $\mathbf{Forc}(\Pi^1_1)$ *and* $\mathbf{Forc}(\Sigma^1_2)$ *belong to* $\Delta^{\mathrm{H}\omega_2}_1$.
If $k \geq 2$ *then the sets* $\mathbf{Forc}(\Pi^1_k)$ *and* $\mathbf{Forc}(\Sigma^1_{k+1})$ *belong to* $\Pi^{\mathrm{H}\omega_2}_{k-1}$.

Proof (sketch). Suppose that φ is $\mathcal{L}\Pi^1_1$. Under the assumptions of the theorem, items (F1)a, (F1)b of (F1) are $\Delta^{\mathrm{H}\omega_2}_1$ relations, while (F2) is reducible to a forcing relation over M that we can relativize to M. The inductive step goes on straightforwardly using (F3), (F4). Please note that the quantifier over names in (F3) is a bounded quantifier (bounded by M), hence it does not add any extra complexity. □

5.3. Further Properties of Forcing Approximations

The notion of names $\nu, \tau \in \mathbf{SN}^{\omega}_{\omega}(\mathbf{Q}^*)$ being equivalent *below* some $p \in \mathbf{Q}^*$, is introduced in Subsection 3.5. We continue with a couple of routine lemmas.

Lemma 27. *Suppose that* $\langle M, U\rangle$, p, φ *satisfy* (F1) *of Section* 5.2, *and* $\text{NAM}\,\varphi = \{\tau_1, \ldots, \tau_m\}$. *Let* $\mu_1, \ldots, \mu_m$ *be another list of names in* $\mathbf{SN}^{\omega}_{\omega}(\mathbf{Q}[U])$, $\mathbf{Q}[U]$*-full below* p, *and such that* τ_j *and* μ_j *are equivalent below* p *for each* $j = 1, \ldots, m$. *Then* p $\mathbf{forc}^M_U$ $\varphi(\tau_1, \ldots, \tau_m)$ *iff* p $\mathbf{forc}^M_U$ $\varphi(\mu_1, \ldots, \mu_m)$.

Proof. Suppose that φ is $\mathcal{L}\Pi^1_1$. Let $G \subseteq \mathbf{Q}[U]$ be a set $\mathbf{Q}[U]$-generic over M, and containing p. Then $\tau_\ell[G] = \mu_\ell[G]$ for all ℓ by Lemma 12. This implies the result required, by (F2) of Section 5.2.

The induction steps $\mathcal{L}\Pi^1_k \to \mathcal{L}\Sigma^1_{k+1}$ and $\mathcal{L}\Sigma^1_k \to \mathcal{L}\Pi^1_k$ are carried out by an easy reduction to items (F3), (F4) of Section 5.2. □

Lemma 28 (in **L**). *Let* $\langle M, U\rangle$, p, φ *satisfy* (F1) *of Section* 5.2. *Then*:

(i) *if* $k \geq 2$, φ *is* $\mathcal{L}\Pi^1_k(\mathbf{Q}[U], M)$, *and* p $\mathbf{forc}^M_U$ φ, *then* p $\mathbf{forc}^M_U$ $\varphi^{\neg}$ *fails;*

(ii) *if* p $\mathbf{forc}^M_U$ φ, $\langle M, U\rangle \preccurlyeq \langle M', U'\rangle \in \mathbf{sJS}$, *and* $q \in \mathbf{Q}[U']$, $q \leqslant p$, *then* q $\mathbf{forc}^{M'}_{U'}$ φ.

Proof. Claim (i) immediately follows from (F4) of Section 5.2.

To prove (ii) let $\varphi = \varphi(\tau_1, \ldots, \tau_m)$ be a closed formula in $\mathcal{L}\Pi^1_1(\mathbf{Q}[U], M)$, where all $\mathbf{Q}[U]$-names τ_j belong to M and are $\mathbf{Q}[U]$-full below p. Then all names τ_j remain $\mathbf{Q}[U']$-full below p by Corollary 3 in Section 4.2, therefore below q as well since $q \leqslant p$. Consider a set $G' \subseteq \mathbf{Q}[U']$, $\mathbf{Q}[U']$-generic over M' and containing q. We have to prove that $\varphi[G']$ is true in $M'[G']$. Please note that the set $G = G' \cap \mathbf{Q}[U]$ is $\mathbf{Q}[U]$-generic over M by Corollary 2 in Section 4.2, and we have $p \in G$. Moreover, the valuation $\varphi[G']$ coincides with $\varphi[G]$ since all names in φ belong to $\mathbf{SN}^{\omega}_{\omega}(\mathbf{Q}[U])$. And $\varphi[G]$ is true in $M[G]$ as p $\mathbf{forc}^M_U$ φ. It remains to apply Mostowski's absoluteness (see [10] (p. 484) or [11]) between the models $M[G] \subseteq M'[G']$.

The induction steps related to (F3), (F4) of Section 5.2 are easy. □

5.4. Transformations and Invariance

To prove Theorem 4 of Section 4.4, we make use of the transformations considered in Sections 3.6–3.8. In addition to the definitions given there, define, in **L**, the action of any transformation $\pi \in \mathbf{BIJ}$ (permutation), $\lambda \in \mathbf{LIP}^{\mathcal{I}}$ (multi-Lipschitz), or one of the form H_{pq} (multisubstitution), on $\mathcal{L}$-formulas with quantifier indices and names in $\mathbf{SN}^{\omega}_{\omega}(\mathbf{Q}^*)$ as parameters.

(I) Assume that $\pi \in \mathbf{BIJ}$. To get $\pi\varphi$ replace each quantifier index B (in $\exists^B$ or $\forall^B$) by $\pi \cdot B$ and each name $\tau \in \mathbf{SN}^{\omega}_{\omega}(\mathbf{Q}^*)$ by $\pi \cdot \tau$.

(II) Assume that $\lambda \in \mathbf{LIP}^{\mathcal{I}}$. To get $\lambda\varphi$ replace each name $\tau \in \mathbf{SN}^{\omega}_{\omega}(\mathbf{Q}^*)$ in φ by $\boldsymbol{\alpha} \cdot \tau$, but do not change quantifier indices.

(III) Assume that $p, q \in \mathbf{Q}^*$ satisfy (6) of Section 3.8, and all names τ occurring in φ belong to $\mathbf{SN}^{\omega}_{\omega}(\mathbf{Q}^*)_{pq}$. To get $H_{pq}\varphi$ replace each name $\tau \in \mathbf{SN}^{\omega}_{\omega}(\mathbf{Q}^*)_{pq}$ in φ by $H_{pq} \cdot \tau \in \mathbf{SN}^{\omega}_{\omega}(\mathbf{Q}^*)_{qp}$, but do not change quantifier indices.

Lemma 29 (in **L**). *Suppose that* $\langle M, U \rangle \in \mathbf{sJS}$, $p \in \mathbf{Q}[U]$, $k \geq 1$, φ *is a formula in* $\mathcal{L}\Sigma^1_{k+1}(\mathbf{Q}[U], M) \cup \mathcal{L}\Pi^1_k(\mathbf{Q}[U], M)$, *and* $\pi \in \mathbf{BIJ}$ *is* **coded in** M *in the sense that* $|\pi| \in M$ *and* $\pi \restriction |\pi| \in M$. *Then:* p $\mathbf{forc}^M_U$ φ *iff* $(\pi \cdot p)$ $\mathbf{forc}^M_{\pi \cdot U}$ $\pi\varphi$. □

Proof. Under the conditions of the lemma, π acts as an isomorphism on all relevant domains and preserves all relevant relations between the objects involved. Thus, $\langle M, \pi \cdot U \rangle$, $\pi \cdot p$, $\pi\varphi$ still satisfy (F1) in Section 5.2. This allows proof of the lemma by induction on the complexity of φ.

Base. Suppose that φ is a closed formula in $\mathcal{L}\Pi^1_1(\mathbf{Q}[U], M)$. Then $\pi\varphi$ is a closed formula in $\mathcal{L}\Pi^1_1(\mathbf{Q}[\pi \cdot U], M)$. Moreover, the map $p \longmapsto \pi \cdot p$ is an order isomorphism (in M) $\mathbf{Q}[U] \stackrel{\text{onto}}{\longrightarrow} \mathbf{Q}[\pi \cdot U]$ by Lemma 14. We conclude that a set $G \subseteq P$ is $\mathbf{Q}[U]$-generic over M iff $\pi \cdot G$ is, accordingly, $\mathbf{Q}[\pi \cdot U]$-generic over M, and the valuated formulas $\varphi[G]$ and $(\pi\varphi)[\pi \cdot G]$ coincide. Now the result for Π^1_1 formulas follows from (F2) in Section 5.2.

Step $\Pi^1_n \to \Sigma^1_{n+1}$, $n \geq 1$. Let $\psi(x)$ be a $\mathcal{L}\Pi^1_k(\mathbf{Q}[U], M)$ formula, and φ be $\exists\, x\, \psi(x)$. Assume p $\mathbf{forc}^M_U$ φ. By definition there is a name $\tau \in \mathbf{SN}^{\omega}_{\omega}(\mathbf{Q}[U]) \cap M$, $\mathbf{Q}[U]$-full below the given $p \in \mathbf{Q}[U]$, such that p $\mathbf{forc}^M_U$ $\psi(\tau)$. Then, by the inductive hypothesis, we have $\pi \cdot p$ $\mathbf{forc}^M_{\pi \cdot U}$ $(\pi\psi)(\pi \cdot \tau)$, and hence by definition $\pi \cdot p$ $\mathbf{forc}^M_{\pi \cdot U}$ $\pi\varphi$.

The case of φ being $\exists^B x\, \psi(x)$ is similar.

Step $\Sigma^1_n \to \Pi^1_n$, $n \geq 2$. This is somewhat less trivial. Assume that φ is a closed $\mathcal{L}\Pi^1_k(\mathbf{Q}[U], M)$ formula; all names in φ belong to $\mathbf{SN}^{\omega}_{\omega}(\mathbf{Q}[U]) \cap M$ and are $\mathbf{Q}[U]$-full below p. Then $\pi\varphi$ is a closed $\mathcal{L}\Pi^1_k(\mathbf{Q}[\pi \cdot U], M)$ formula, whose all names belong to $\mathbf{SN}^{\omega}_{\omega}(\mathbf{Q}[\pi \cdot U]) \cap M$ and are $\mathbf{Q}[\pi \cdot U]$-full below $\pi \cdot p$. Suppose that p $\mathbf{forc}^M_U$ φ **fails**.

By definition there exist a pair $\langle M_1, U_1 \rangle \in \mathbf{sJS}$ with $\langle M, U \rangle \preccurlyeq \langle M_1, U_1 \rangle$, and a condition $q \in \mathbf{Q}[U_1]$, $q \leqslant p$, such that q $\mathbf{forc}^{M_1}_{U_1}$ $\varphi^{\neg}$. Then $(\pi \cdot q)$ $\mathbf{forc}^{M_1}_{\pi \cdot U_1}$ $\pi\varphi^{\neg}$ by the inductive hypothesis. Yet the pair $\langle M_1, \pi \cdot U_1 \rangle$ belongs to $\mathbf{sJS}$ and extends $\langle M, \pi \cdot U \rangle$. (Recall that $U \in M$ and π is coded in M.) In addition, $\pi \cdot q \in \mathbf{Q}[\pi \cdot U_1]$, and $\pi \cdot q \leqslant \pi \cdot p$. Therefore, the statement $(\pi \cdot p)$ $\mathbf{forc}^M_{\pi \cdot U}$ $\pi\varphi$ fails, as required. □

Lemma 30 (in **L**). *Suppose that* $\langle M, U \rangle \in \mathbf{sJS}$, $p \in \mathbf{Q}[U]$, $k \geq 1$, φ *is a formula in* $\mathcal{L}\Pi^1_k(\mathbf{Q}[U], M) \cup \mathcal{L}\Sigma^1_{k+1}(\mathbf{Q}[U], M)$, *and* $\boldsymbol{\alpha} \in \mathbf{LIP}^{\mathcal{I}} \cap M$. *Then:* p $\mathbf{forc}^M_U$ φ *iff* $(\boldsymbol{\alpha} \cdot p)$ $\mathbf{forc}^M_{\boldsymbol{\alpha} \cdot U}$ $\boldsymbol{\alpha}\varphi$.

Proof. Similar to the previous one, but with a reference to Lemma 15 rather than Lemma 14. □

Lemma 31 (in **L**). *Assume that* $\langle M, U \rangle \in \mathbf{sJS}$, *conditions* $p, q \in \mathbf{Q}[U]$ *satisfy* (6) *of Section* 3.8, $k \geq 1$, φ *is a closed formula in* $\mathcal{L}\Pi^1_k(\mathbf{Q}[U], M) \cup \mathcal{L}\Sigma^1_{k+1}(\mathbf{Q}[U], M)$ *with all names in* $\mathbf{SN}^{\omega}_{\omega}(\mathbf{Q}^*)_{pq}$ (see Section 3.8), *and* $r \in \mathbf{Q}[U]$, $r \leqslant p$. *Then:* r $\mathbf{forc}^M_U$ φ *iff* $H_{pq} \cdot r$ $\mathbf{forc}^M_U$ $H_{pq}\, \varphi$.

Proof. Similar to the proof of Lemma 29, except for the step $\Pi^1_k \to \Sigma^1_{k+1}$, $k \geq 1$, where we need to take additional care to keep the names involved in $\mathbf{SN}^\omega_\omega(\mathbf{Q}[U])_{pq}$. Thus, let $\psi(x)$ be a $\mathcal{L}\Pi^1_k(\mathbf{Q}[U], M)$ formula, with names in $\mathbf{SN}^\omega_\omega(\mathbf{Q}[U])_{pq}$, and let φ be $\exists\, x\, \psi(x)$. Assume that r $\mathbf{forc}^M_U$ φ.

By definition there is a name $\tau \in \mathbf{SN}^\omega_\omega(\mathbf{Q}[U]) \cap M$, $\mathbf{Q}[U]$-full below r, such that r $\mathbf{forc}^M_U$ $\psi(\tau)$. Please note that τ does not necessarily belong to $\mathbf{SN}^\omega_\omega(\mathbf{Q}[U])_{pq}$. However, the restricted name $\tau' = \tau_{\leqslant r}$ (see Lemma 13 in Section 3.8) is still a name in $\mathbf{SN}^\omega_\omega(\mathbf{Q}[U])$ because $r \in \mathbf{Q}[U]$, and we have $r' \in \mathtt{dom}\,\tau' \Longrightarrow r' \leqslant r \leqslant p$, so that $\tau' \in \mathbf{SN}^\omega_\omega(\mathbf{Q}[U])_{pq}$. Moreover, τ' is equivalent to τ below r by Lemma 13. We conclude that r $\mathbf{forc}^M_U$ $\psi(\tau')$, by Lemma 27.

Then, by the inductive hypothesis, we have $H_{pq} \cdot r$ $\mathbf{forc}^M_U$ $(H_{pq}\,\psi)(H_{pq} \cdot \tau')$, and hence by definition $H_{pq} \cdot r$ $\mathbf{forc}^M_U$ $H_{pq}\,\varphi$ via $H_{pq} \cdot \tau'$. □

6. Elementary Equivalence Theorem

The goal of this section is to prove Theorem 4 of Section 4.4, and accomplish the proof of Theorem 1. We make use of the relation **forc** defined above, and exploit certain symmetries in **forc** studied in Section 5.4.

6.1. *Hidden Invariance*

To explain the idea, one may note first that elementary equivalence of subextensions of a given generic extension is usually a corollary of the fact that the forcing notion considered is enough homogeneous, or in different words, invariant w. r. t. a sufficiently large system of order-preserving transformations. The forcing notion $\mathbb{Q} = \mathbf{Q}[\mathbb{U}]$ we consider, as well as basically any $\mathbf{Q}[U]$, is invariant w. r. t. multi-substitutions by Lemma 17. However, for a straightaway proof of Theorem 4 we would naturally need the invariance under permutations of Section 3.6—to interchange the domains Z and Z', whereas $\mathbb{Q}$ is definitely not invariant w. r. t. permutations.

On the other hand, the relation **forc** is invariant w.r.t. both permutations (Lemma 29) and multi-Lipschitz (Lemma 30), as well as still w. r. t. multi-substitutions by Lemma 31. To bridge the gap between **forc** (not explicitly connected with $\mathbb{Q}$ in any way) and $\mathbb{Q}$-generic extensions, we prove Lemma 33, which ensures that **forc** admits a forcing-style association with the truth in $\mathbb{Q}$-generic extensions, bounded to formulas of type $\Sigma^1_\mathfrak{n}$ and below. This key result will be based on the $\mathfrak{n}$-completeness property (Definition 2 in Section 4.3). Speaking loosely, one may say that some transformations, i.e., permutations and multi-Lipschitz, are *hidden* in construction of $\mathbb{Q}$, so that they do not act per se, but their influence up to $\mathfrak{n}$th level, is preserved.

This method of *hidden invariance*, i.e., invariance properties (of an auxiliary forcing-type relationship like **forc**) hidden in $\mathbb{Q}$ by a suitable generic-style construction of $\mathbb{Q}$, was introduced in Harrington's notes [3] in a somewhat different terminology. We may note that the hidden invariance technique is well known in some other fields of mathematics, including more applied fields, see e.g., [12,13].

6.2. *Approximations of the $\mathfrak{n}$-Complete Forcing Notion*

We return to the forcing notion $\mathbb{Q} = \mathbf{Q}[\mathbb{U}]$ defined in $\mathbf{L}$ as in Definition 2 in Section 4.3 for a given number $\mathfrak{n} \geq 2$ of Theorem 1. **Arguing in L**, we let the pairs $\langle \mathbb{M}_\xi, \mathbb{U}_\xi \rangle$, $\xi < \omega_2$, also be as in Definition 2. Let $\mathbf{forc}_\xi$ denote the relation $\mathbf{forc}^{\mathbb{M}_\xi}_{\mathbb{U}_\xi}$, and let p $\mathbf{forc}_\infty$ φ mean: $\exists\, \xi < \omega_2\, (p\ \mathbf{forc}_\xi\ \varphi)$.

Claims (i), (ii) of Lemma 28 take the form:

(I) p $\mathbf{forc}_\xi$ φ and p $\mathbf{forc}_\eta$ $\varphi^\neg$ $(\xi, \eta < \omega_2)$ contradict to each other.

(II) If p $\mathbf{forc}_\xi$ φ and $\xi \leq \zeta < \omega_2$, $q \in \mathbf{Q}[\mathbb{U}_\zeta]$, $q \leqslant p$, then q $\mathbf{forc}_\zeta$ φ.

The next lemma shows that $\mathbf{forc}_\infty$ satisfies a key property of forcing relations up to the level of $\Pi^1_{\mathfrak{n}-1}$ formulas.

Lemma 32. *If φ is a closed formula in $\mathcal{L}\Pi^1_k(\mathbb{Q})$, $2 \le k < \mathfrak{n}$, $p \in \mathbb{Q}$, and all names in φ are $\mathbb{Q}$-full below p, then there is a condition $q \in \mathbb{Q}$, $q \leqslant p$, such that either q $\mathbf{forc}_\infty$ φ, or q $\mathbf{forc}_\infty$ $\varphi^\neg$.*

Proof. As the names considered are ω_1-sizeobjects, there is an ordinal $\eta < \omega_2$ such that $p \in \mathbb{Q}_\eta$, and all names in φ belong to $\mathbb{M}_\eta \cap \mathbf{SN}^\omega_\omega(\mathbb{Q}_\eta)$; then all names in φ are $\mathbb{Q}_\eta$-full below p, of course. As $k < \mathfrak{n}$, the set D of all pairs $\langle M, U\rangle \in \mathbf{sJS}$ that extend $\langle \mathbb{M}_\eta, \mathbb{U}_\eta\rangle$ and there is a condition $q \in \mathbf{Q}[U]$, $q \leqslant p$, satisfying q $\mathbf{forc}^M_U$ $\varphi^\neg$, belongs to $\mathbf{\Sigma}^{\mathrm{HC}}_{\mathfrak{n}-2}$ by Lemma 26. Therefore, by the $\mathfrak{n}$-completeness of the sequence $\{\langle \mathbb{M}_\xi, \mathbb{U}_\xi\rangle\}_{\xi<\omega_1}$, there is an ordinal ζ, $\eta \le \zeta < \omega_1$, such that $\langle \mathbb{M}_\zeta, \mathbb{U}_\zeta\rangle \in D^{\mathtt{solv}}$.

We have two cases.

Case 1: $\langle \mathbb{M}_\zeta, \mathbb{U}_\zeta\rangle \in D$. Then there is a condition $q \in K[\mathbb{U}_\zeta]$, $q \leqslant p$, satisfying q $\mathbf{forc}^{\mathbb{M}_\zeta}_{\mathbb{U}_\zeta}$ $\varphi^\neg$, that is, q $\mathbf{forc}_\infty$ $\varphi^\neg$. However, obviously $q \in \mathbb{Q}$.

Case 2: there is no pair $\langle M, U\rangle \in D$ extending $\langle \mathbb{M}_\zeta, \mathbb{U}_\zeta\rangle$. Prove p $\mathbf{forc}_\zeta$ φ. Suppose otherwise. Then by the choice of η and (F4) in Section 5.2, there exist: a pair $\langle M, U\rangle \in \mathbf{sJS}$ extending $\langle \mathbb{M}_\zeta, \mathbb{U}_\zeta\rangle$, and a condition $q \in \mathbf{Q}[U]$, $q \leqslant p$, such that q $\mathbf{forc}^M_U$ $\varphi^\neg$. Then $\langle M, U\rangle \in D$, a contradiction. □

Now we prove another key lemma which connects, in a forcing-style way, the relation $\mathbf{forc}_\infty$ and the truth in $\mathbb{Q}$-generic extensions of $\mathbf{L}$, up to the level of $\Sigma^1_\mathfrak{n}$ formulas.

Lemma 33. *Suppose that φ is a formula in $\mathcal{L}\Pi^1_k(\mathbb{Q}) \cup \mathcal{L}\Sigma^1_{k+1}(\mathbb{Q})$, $1 \le k < \mathfrak{n}$, and all names in φ are $\mathbb{Q}$-full. Let $G \subseteq \mathbb{Q}$ be $\mathbb{Q}$-generic over $\mathbf{L}$. Then $\varphi[G]$ is true in $\mathbf{L}[G]$ iff there is a condition $p \in G$ such that p $\mathbf{forc}_\infty$ φ.*

Proof. We proceed by induction and begin with **the case of $\mathcal{L}\Pi^1_1$ formulas**. Consider a closed formula φ in $\mathcal{L}\Pi^1_1(\mathbb{Q})$. As names in the formulas considered are ω_1-sizenames in $\mathbf{SN}^\omega_\omega(\mathbb{Q})$, there is an ordinal $\xi < \omega_2$ such that φ is a $\mathcal{L}\Pi^1_1(\mathbb{Q}_\xi)$ formula. Please note that since $G \subseteq \mathbb{P}$ is $\mathbb{Q}$-generic over $\mathbf{L}$, the smaller set $G_\xi = G \cap \mathbb{Q}_\xi$ is $\mathbb{Q}_\xi$-generic over $\mathbb{M}_\xi$ by Corollary 2 in Section 4.2, and the formulas $\varphi[G]$, $\varphi[G_\xi]$ coincide by the choice of ξ. Therefore

$\varphi[G]$ holds in $\mathbf{L}[G]$:

iff $\varphi[G_\xi]$ holds in $\mathbb{M}_\xi[G_\xi]$ by the Mostowski absoluteness [10] (p. 484),

iff there is $p \in G_\xi$ which $\mathbb{Q}_\xi$-forces φ over $\mathbb{M}_\xi$,

iff $\exists\, p \in G_\xi$ (p $\mathbf{forc}_\xi$ φ) by (F2) in Section 5.2,

easily getting the result required since ξ is arbitrary.

The step from $\mathcal{L}\Sigma^1_k$ to $\mathcal{L}\Pi^1_k$, $k \ge 2$. Prove the theorem for a $\mathcal{L}\Pi^1_k(\mathbb{Q})$ formula φ, assuming that the result holds for $\varphi^\neg$. Suppose that $\varphi[G]$ is false in $\mathbf{L}[G]$. Then $\varphi^\neg[G]$ is true, and hence by the inductive hypothesis, there is a condition $p \in G \restriction c$ such that p $\mathbf{forc}_\infty$ $\varphi^\neg$. Then it follows from (I) and (II) above that q $\mathbf{forc}_\infty$ φ fails for all $q \in G$.

Conversely let p $\mathbf{forc}_\infty$ φ fail for all $p \in G$. Then by Lemma 32 there exists $q \in G$ satisfying q $\mathbf{forc}_\infty$ $\varphi^\neg$. It follows that $\varphi^\neg[G]$ is true by the inductive hypothesis, therefore $\varphi[G]$ is false.

The step from $\mathcal{L}\Pi^1_k$ to $\mathcal{L}\Sigma^1_{k+1}$. Let $\varphi(x)$ be a $\mathcal{L}\Pi^1_k(\mathbb{Q})$ formula; prove the result for a formula $\exists^B x\, \varphi(x)$. If $p \in G$ and p $\mathbf{forc}_\xi$ $\exists^B x\, \varphi(x)$ then by definition there is a name $\tau \in \mathbb{M}_\xi \cap \mathbf{SN}^\omega_\omega(\mathbb{Q}_\xi) \restriction B$, $\mathbb{Q}_\xi$-full below p, and such that p $\mathbf{forc}_\xi$ $\varphi(\tau)$. Then $\varphi(\tau)[G]$ holds by the inductive hypothesis, and this implies $(\exists^B x\, \varphi(x))[G]$ since obviously $\tau[G] \in \mathbf{L}[G \restriction B]$.

If conversely $(\exists^B x\, \varphi(x))[G]$ is true, then by Lemma 11 there is a $\mathbb{Q}$-full name $\tau \in \mathbf{SN}^\omega_\omega(\mathbb{Q}) \restriction B$ such that $\varphi(\tau)[G]$ is true. Then, by the inductive hypothesis, there is a condition $p \in G$ such that p $\mathbf{forc}_\infty$ $\varphi(\tau)$. Therefore p $\mathbf{forc}_\infty$ $\exists^B x\, \varphi(x)$ by the choice of τ.

The case of $\exists\, x\, \varphi(x)$ is treated similarly. □

6.3. The Elementary Equivalence Theorem

We begin **the proof of Theorem 4 of Section 4.4**, so let d, Z, Z', x_0 be as in the theorem.

Step 1. We assume w.l.o.g. that x_0 itself is the only parameter in the $\Sigma^1_{\mathfrak{n}}$ formula Φ of Theorem 4. By Lemma 11, there exists, in $\mathbf{L}$, a $\mathbb{Q}$-full name $\tau \in \mathbf{SN}^{\omega}_{\omega}(\mathbb{Q})$ such that $x_0 = \tau[G]$ and $|\tau|^+ \subseteq d$. Thus, Φ is $\varphi(\tau[G])$, where $\varphi(\cdot)$ is a parameter-free $\Sigma^1_{\mathfrak{n}}$ formula with a single free variable. Then $|\varphi(\tau)|^+ = |\tau|^+ \subseteq d$.

We also assume w.l.o.g. that the sets Z, Z' satisfy the requirement that $Z \smallsetminus Z'$ *and* $Z' \smallsetminus Z$ *are infinite (countable) sets.* Indeed, otherwise, under the assumptions of Theorem 4, one easily defines a third set Z'' such that each of the pairs Z, Z'' and Z', Z'' still satisfies the assumptions of the theorem, and in addition, all four sets $Z \smallsetminus Z''$, $Z'' \smallsetminus Z$, $Z'' \smallsetminus Z'$ and $Z' \smallsetminus Z''$ are infinite. Please note that this argument necessarily requires that the complementary set $\mathcal{I} \smallsetminus (d \cup Z \cup Z')$ is infinite.

Step 2. We are going to reorganize the quantifier prefix of φ, in particular, by assigning the indices Z and Z' to certain quantifiers, to reflect the relativization to classes $\mathbf{L}[x_0, G \restriction Z]$ and $\mathbf{L}[x_0, G \restriction Z']$. This is not an easy task because generally speaking there is no set $Z_0 \subseteq \mathcal{I}$ in $\mathbf{L}$ satisfying $\mathbf{L}[x_0] = \mathbf{L}[G \restriction Z_0]$. However, nevertheless we will define an $\mathcal{L}\Sigma^1_{\mathfrak{n}}$ formula, say $\psi^Z(v)$, and then $\psi^{Z'}(v)$ by the substitution of Z' for Z, such that the following will hold:

(A) For any set $G \subseteq \mathbb{Q}$, $\mathbb{Q}$-generic over $\mathbf{L}$:

$$\varphi(\tau[G]) \text{ is true in } \mathbf{L}[\tau[G], G \restriction Z] \quad \text{iff} \quad \psi^Z(\tau)[G] \text{ is true in } \mathbf{L}[G], \quad \text{and}$$
$$\varphi(\tau[G]) \text{ is true in } \mathbf{L}[\tau[G], G \restriction Z'] \quad \text{iff} \quad \psi^{Z'}(\tau)[G] \text{ is true in } \mathbf{L}[G].$$

(See Section 5.1 on the interpretation $\psi[G]$ for any $\mathcal{L}$-formula ψ.)

To explain this transformation, assume that $\mathfrak{n} = 4$ for the sake of brevity, and hence $\varphi(v)$ has the form $\exists\, x\, \forall\, y\, \vartheta(v, x, y)$, where ϑ is a Σ^1_2 formula. To begin with, we define

$$\psi^Z_1(v) \;:=\; \exists^Z x'\, \exists\, x \in \mathbf{L}[x', v]\; \forall^Z y'\, \forall\, y \in \mathbf{L}[v, y']\; \vartheta(v, x, y)\,, \tag{17}$$

and define $\psi^{Z'}_1(v)$ accordingly.

Lemma 34. *The formulas* ψ^Z_1, $\psi^{Z'}_1$ *satisfy* (A).

Proof. To prove the implication $\Longrightarrow$, suppose that $\varphi(\tau[G])$ holds in $\mathbf{L}[\tau[G], G \restriction Z]$, so that there is a real $x_1 \in \omega^\omega \cap \mathbf{L}[\tau[G], G \restriction Z]$ satisfying $\forall\, y\, \vartheta(\tau[G], x_1, y)$ in $\mathbf{L}[\tau[G], G \restriction Z]$. By a standard argument there is a real $x' \in \omega^\omega \cap \mathbf{L}[G \restriction Z]$ with $x_1 \in \omega^\omega \cap \mathbf{L}[\tau[G], x']$. We claim that these reals x' and x_1 witness that $\psi^Z_1(\tau)[G]$ holds in $\mathbf{L}[G]$, that is, we have $\forall^Z y'\, \forall\, y \in \mathbf{L}[\tau[G], y']\; \vartheta(\tau[G], x_1, y)$ in $\mathbf{L}[G]$.

Indeed, suppose that $y' \in \omega^\omega \cap \mathbf{L}[G \restriction Z]$ and $y \in \omega^\omega \cap \mathbf{L}[\tau[G], y']$. Then $y \in \mathbf{L}[\tau[G], G \restriction Z]$, of course. Therefore $\vartheta(\tau[G], x_1, y)$ is true in $\mathbf{L}[\tau[G], G \restriction Z]$ by the choice of x_1. We conclude that $\vartheta(\tau[G], x_1, y)$ is true in $\mathbf{L}[G]$ as well by the Shoenfield absoluteness theorem, as ϑ is a Σ^1_2 formula.

The inverse implication is proved similarly. (Lemma) □

Thus, the formulas ψ^Z_1, $\psi^{Z'}_1$ do satisfy (A), but they are not $\mathcal{L}\Sigma^1_{\mathfrak{n}}$ formulas as defined in Section 5.1, of course. It will take some effort to convert them to a $\mathcal{L}\Sigma^1_{\mathfrak{n}}$ form. We must recall some instrumentarium known in Gödel's theory of constructability of reals.

- If $x \in \omega^\omega$ then define reals $(x)_{\rm ev}$ and $(x)_{\rm odd}$ in ω^ω by $(x)_{\rm ev}(k) = x(2k)$ and $(x)_{\rm odd}(k) = x(2k+1)$ for all k. If $y, z \in \omega^\omega$ then define $x{*}y \in \omega^\omega$ such that $(x{*}y)_{\rm ev} = x$, $(x{*}y)_{\rm odd} = y$.
- There is a Π^1_1 set $\mathbf{WO} = \{w \in \omega^\omega : \mathbf{wo}(x)\}$ of *codes of countable ordinals*, defined by a Π^1_1 formula $\mathbf{wo}$, so that $|w|$ is the ordinal coded by $w \in \mathbf{WO}$, and $\omega_1 = \{|w| : w \in \mathbf{WO}\}$, see ([14] (1E))).

As a one more pre-requisite, we make use of a system of maps $f_\xi : \omega^\omega \to \omega^\omega$, $\xi < \omega_1$, such that:

(a) if $x \in \omega^\omega$ then $\mathbf{L}[x] \cap \omega^\omega = \{f_\xi(x) : \xi < \omega_1\}$, and

(b) there exist a Σ^1_1 formula $S(x,y,w)$ and a Π^1_1 formula $P(x,y,w)$ such that if $w \in \mathbf{WO}$ then $f_{|w|}(x) = y \iff S(x,y,w) \iff P(x,y,w)$ for all $x, y \in \omega^\omega$,

see e.g., ([14] (Theorem 2.6)). Recall that $\omega_1^{\mathbf{L}[G]} = \omega_1^{\mathbf{L}[G\restriction Z]} = \omega_2^{\mathbf{L}}$ by Lemma 22.

Now consider the formula

$$\psi_2^Z(v) := \exists^Z x \left(\mathbf{wo}((x)_{\mathrm{ev}}) \wedge \forall^Z y \, [\mathbf{wo}((y)_{\mathrm{ev}}) \Longrightarrow \right. \\ \left. \Longrightarrow \vartheta(v, f_{|(x)_{\mathrm{ev}}|}(v*(x)_{\mathrm{odd}}), f_{|(y)_{\mathrm{ev}}|}(v*(y)_{\mathrm{odd}}))] \right), \tag{18}$$

and define $\psi_2^{Z'}(v)$ similarly.

We keep the global understanding that the quantifiers $\exists^Z, \forall^Z$ are relativized to $\mathbf{L}[G\restriction Z] \cap \omega^\omega$.

Lemma 35. *The formulas $\psi_1^Z(\tau[G])$ and $\psi_2^Z(\tau[G])$ are equivalent in $\mathbf{L}[G]$, and the same for $\psi_1^{Z'}$ and $\psi_2^{Z'}$.*

Proof (Lemma). To prove the implication $\Longrightarrow$, assume that $\psi_1^Z(\tau[G])$ holds in $\mathbf{L}[G]$, and this is witnessed by reals $x' \in \omega^\omega \cap \mathbf{L}[G\restriction Z]$ and $x_1 \in \omega^\omega \cap \mathbf{L}[\tau[G], x'] = \omega^\omega \cap \mathbf{L}[\tau[G]*x']$ satisfying $\forall^Z y' \, \forall y \in \mathbf{L}[\tau[G], y'] \; \vartheta(\tau[G], x_1, y)$ in $\mathbf{L}[G]$. Please note that $\omega_1^{\mathbf{L}[G\restriction Z]} = \omega_1^{\mathbf{L}[G]} = \omega_2^{\mathbf{L}}$ by Lemma 9 (ii). It follows by (a) that there is an ordinal $\xi < \omega_1^{\mathbf{L}[G\restriction Z]}$ with $x_1 = f_\xi(\tau[G]*x')$, and then there is a real $w \in \mathbf{WO} \cap \mathbf{L}[G\restriction Z]$ with $\xi = |w|$.

Now let $\tilde{x} = w*x'$, so that $w = (\tilde{x})_{\mathrm{ev}}$, $x' = (\tilde{x})_{\mathrm{odd}}$, and $x_1 = f_{|(\tilde{x})_{\mathrm{ev}}|}(\tau[G]*(\tilde{x})_{\mathrm{odd}})$. We claim that $\tilde{x}$ witnesses $\psi_2^Z(\tau[G])$ in $\mathbf{L}[G]$. Indeed, assume that $\tilde{y} \in \omega^\omega \cap \mathbf{L}[G\restriction Z]$ and $w = (\tilde{y})_{\mathrm{ev}} \in \mathbf{WO}$, $\eta = |(\tilde{y})_{\mathrm{ev}}|$, and $y_1 = f_\eta(\tau[G]*(\tilde{y})_{\mathrm{odd}})$; we must prove that $\vartheta(\tau[G], x_1, y_1)$ is true in $\mathbf{L}[G]$.

However, we have $y_1 \in \mathbf{L}[\tau[G], y']$ by construction, where $y' = (\tilde{y})_{\mathrm{odd}} \in \mathbf{L}[G\restriction Z]$ by the choice of $\tilde{y}$. Now it follows by the choice of x_1 that $\vartheta(\tau[G], x_1, y_1)$ indeed holds, as required.

The proof of the inverse implication is similar. (Lemma) □

Please note that the formula $\psi_2^Z(v)$ can be converted to the following logically equivalent form:

$$\psi_3^Z(v) := \exists^Z x \, \forall^Z y \left[\mathbf{wo}((x)_{\mathrm{ev}}) \wedge \left(\mathbf{wo}((y)_{\mathrm{ev}}) \Longrightarrow \right. \right. \\ \left. \left. \Longrightarrow \vartheta(v, f_{|(x)_{\mathrm{ev}}|}(v*(x)_{\mathrm{odd}}), f_{|(y)_{\mathrm{ev}}|}(v*(y)_{\mathrm{odd}})) \right) \right]. \tag{19}$$

And here the kernel $[\ldots]$ can be converted to a true Σ^1_2 form, say $\chi(v,x,y)$, with the help of the formulas S and P of (b), and because $\mathbf{wo}(\cdot)$ is Π^1_1 and ϑ is Σ^1_2. This yields a $\mathcal{L}\Sigma^1_4$ formula $\psi^Z(v) := \exists^Z x \, \forall^Z y \, \chi(v,x,y)$, equivalent to ψ_1^Z, and hence satisfying (A) by Lemmas 34 and 35, as required.

Step 3. Assuming that the formula $\Phi := \varphi(\tau[G])$ is true in $\mathbf{L}[x_0, G\restriction Z]$, the transformed formula $\psi^Z(\tau)[G]$ holds in $\mathbf{L}[G]$ by (A). By Lemma 33 there is a condition $p \in G$ such that p $\mathbf{forc}_\infty$ $\psi^Z(\tau)$ that is, there is an ordinal $\xi < \omega_2$ such that p $\mathbf{forc}_\xi$ $\psi^Z(\tau)$—then by definition $p \in \mathbf{Q}[\mathbb{U}_\xi]$. We w.l.o.g. assume that p and ζ satisfy the following two requirements:

(B) $\mathtt{card}\,(|p| \cap (Z \smallsetminus Z')) = \mathtt{card}\,(|p| \cap (Z' \smallsetminus Z))$ (recall that $Z' \smallsetminus Z$, $Z \smallsetminus Z'$ are infinite, Step 1).

(C) $-1 \in |p|^+$, and if $\nu, \nu' \in |p|$ then $S_p(\nu) \subseteq F_p^\vee(\nu)$ and the sets $F_p^\vee(\nu)$ and $F_p^\vee(\nu')$ are i-similar (see Section 2.3).

Please note that if $\xi < \eta < \omega_2$ then still p $\mathbf{forc}_\eta$ $\psi^Z(\tau)$ by Lemma 28. Therefore, we can increase ξ below ω_2 so that the following holds:

(D) the sets d, $Z \smallsetminus Z'$, $Z' \smallsetminus Z$ belong to $\mathbb{M}_\xi$ and are subsets of $|\mathbb{U}_\xi|$.

Step 4. Now, to finalize the proof of Theorem 4, it suffices (by Lemma 33) to prove:

Lemma 36. *We have p $\mathbf{forc}_\infty$ $\psi^{Z'}(\tau)$ as well.*

Proof (Lemma). Let $\delta = d \cup (Z \Delta Z')$; then $\delta \in \mathbb{M}_\xi$ by (D), and $\delta \subseteq |\mathbb{U}_\xi|$. There is a bijection $f \in M$, $f : \delta \overset{\text{onto}}{\longrightarrow} \delta$, such that

(E) $f \restriction d$ is the identity, f maps $Z \smallsetminus Z'$ onto $Z' \smallsetminus Z$ and vice versa.

Then, by (B), f maps $|p|$ onto $|p|$. Let π be the trivial extension of f onto $\mathcal{I}$: $\pi(\nu) = \nu$ for $\nu \notin \delta$. Thus, π is coded in $\mathbb{M}_\xi$ in the sense of Lemma 29, and $|\pi| \subseteq \delta \subseteq |\mathbb{U}_\xi|$. We have p $\mathbf{forc}_{\mathbb{U}_\xi}^{\mathbb{M}_\xi}$ $\psi^Z(\tau)$ by the choice of ξ, hence $\mathbb{U}_\xi \in \mathbb{M}_\xi$ and $p \in \mathbb{P}_\xi = \mathbf{Q}[\mathbb{U}_\xi] \in \mathbb{M}_\xi$. Moreover, $\pi \cdot \tau = \tau$ because $|\tau|^+ \subseteq d$ and $\pi \restriction d$ is the identity by (E). It follows that p' $\mathbf{forc}_{U'}^{\mathbb{M}_\xi}$ $\varphi^{Z'}(\tau)$ by Lemma 29, where $U' = \pi \cdot \mathbb{U}_\xi$, $p' = \pi \cdot p$. Please note that $p' \in \mathbf{Q}[U']$, $|p'|^+ = |p|^+$, $|U'| = |\mathbb{U}_\xi|$, $U' \restriction d = \mathbb{U}_\xi \restriction d$, $p' \restriction d = p \restriction d$. Also note that

(F) if $\nu \in |p'| = |p|$ then the sets $F_p^\vee(\nu)$, $F_{p'}^\vee(\nu)$ are i-similar by (C), (E).

We conclude, by Lemma 16, that there is a transformation $\boldsymbol{\lambda} = \{\lambda_\nu\}_{\nu \in |\mathbb{U}_\xi|} \in \mathbf{LIP}^{\mathcal{I}} \cap \mathbb{M}_\xi$, such that $\boldsymbol{\lambda} \cdot U' = \mathbb{U}_\xi$, $\lambda_\nu =$ the identity for all $\nu \in d$, and $F_p^\vee(\nu) = F_q^\vee(\nu)$ for all $\nu \in |p| = |p'| = |q|$, where $q = \boldsymbol{\lambda} \cdot p' \in \mathbf{Q}[\mathbb{U}_\xi]$. Then we have q $\mathbf{forc}_{\mathbb{U}_\xi}^{\mathbb{M}_\xi}$ $\psi^{Z'}(\tau)$ by Lemma 30. Here $\boldsymbol{\lambda} \cdot \psi^{Z'}(\tau) = \psi^{Z'}(\tau)$ by the choice of $\boldsymbol{\lambda}$, because $|\tau|^+ \subseteq d$. And $q \restriction d = p \restriction d$ holds by the same reason.

It remains to derive p $\mathbf{forc}_{\mathbb{U}_\xi}^{\mathbb{M}_\xi}$ $\psi^{Z'}(\tau)$ from q $\mathbf{forc}_{\mathbb{U}_\xi}^{\mathbb{M}_\xi}$ $\psi^{Z'}(\tau)$. Please note that p, q satisfy (6) of Section 3.8 by construction, hence the transformation H_{qp} is defined. Moreover, the only name τ occurring in $\psi^{Z'}(\tau)$ satisfies $|\tau|^+ \subseteq d$, and $\pi \restriction d$ is the identity by (E). It follows that $\tau \in \mathbf{SN}_\omega^\omega(\mathbf{Q}^*)_{qp}$, and $\pi \cdot \tau = \tau$. We conclude that Lemma 31 is applicable. This yields p $\mathbf{forc}_{\mathbb{U}_\xi}^{\mathbb{M}_\xi}$ $\psi^{Z'}(\tau)$, as required. (Lemma 36) □

(Theorem 4 of Section 4.4) □

(Theorem 1, see Section 4.5) □

7. Conclusions and Discussion

In this study, the method of almost-disjoint forcing was employed to the problem of getting a model of **ZFC** in which the constructible reals are precisely the Δ^1_n reals, for different values $n > 2$. The problem appeared under no 87 in Harvey Friedman's treatise *One hundred and two problems in mathematical logic* [1], and was generally known in the early years of forcing, see, e.g., problems 3110, 3111, 3112 in an early survey [2] by A. Mathias. The problem was solved by Leo Harrington, as mentioned in [1,2] and a sketch of the proof mainly related to the case $n = 3$ in Harrington's own handwritten notes [3].

From this study, it is concluded that the hidden invariance technique (as outlined in Section 6.1) allows the solution of the general case of the problem (an arbitrary $n \geq 3$), by providing a generic extension of **L** in which the constructible reals are precisely the Δ^1_n reals, for a chosen value $n \geq 3$, as sketched by Harrington. The hidden invariance technique has been applied in recent papers [7,15–17] for the problem of getting a set theoretic structure of this or another kind at a pre-selected projective level. We may note here that the hidden invariance technique, as a true mathematical technique, also has multiple applications both in the physical and engineering fields. In this regard, we cite works [18,19] that have exploited this technique (albeit simplified) for engineering applications.

We continue with a brief discussion with a few possible future research lines.

1. Harvey Friedman completes [1] with a modified version of the above problem, defined as Problem 87′: find a model of

$$\mathbf{ZFC} + \text{“for any reals } x, y, \text{ we have: } x \in \mathbf{L}[y] \implies x \text{ is } \Delta^1_3 \text{ in } y\text{”.} \tag{20}$$

This problem was also known in the early years of forcing, see, e.g., problem 3111 in [2]. Problem (20) was solved in the positive by René David [20], where the question is attributed to Harrington. So far it is unknown whether this result generalizes to higher classes Δ^1_n, $n \geq 4$, or Δ^1_∞ and whether it can be strengthened towards $\iff$ instead of $\Longrightarrow$. This is a very interesting and perhaps difficult question.

2. Another question to be mentioned here is the following. Please note that in any extension of $\mathbf{L}$ satisfying Theorem 1, it is true that every universal $\Sigma^1_{\mathfrak{n}+1}$ set $u \subseteq \omega \times \omega$ is by necessity $\Sigma^1_{\mathfrak{n}+1}$ but non-$\Delta^1_{\mathfrak{n}+1}$, and hence nonconstructible. This gives another proof of Theorem 3 in [7]. (It claims, for any $\mathfrak{n} \geq 2$, the existence of a generic extension of $\mathbf{L}$ in which there is a nonconstructible $\Sigma^1_{\mathfrak{n}+1}$ set $a \subseteq \omega$ whereas all $\Delta^1_{\mathfrak{n}+1}$ sets are constructible.) And the problem is, given $\mathfrak{n} \geq 2$, to find a model in which

all $\Delta^1_{\mathfrak{n}+1}$ reals are constructible, but there exists a $\Sigma^1_{\mathfrak{n}+1}$ nonconstructible real $u \subseteq \omega$, *which satisfies* $\mathbf{V} = \mathbf{L}[u]$.

Neither the model considered in Section 4.5 above, nor the model for ([7] (Theorem 3)), suffice to solve the problem, because these models in principle are incompatible with $\mathbf{V} = \mathbf{L}[u]$ for a real u.

3. For any $n < \omega$, let D_{1n} be the set of all reals (here subsets of $\omega = \{0, 1, 2, \dots\}$), definable by a type-theoretic parameter-free formula whose quantifiers have types bounded by n from above. In particular, D_{10}= arithmetically definable reals and D_{11} = analytically definable reals. Alfred Tarski asked in [6] whether it is true that for a given $n \geq 1$, the set D_{1n} belongs to D_{2n}, that is, is itself definable by a type-theoretic parameter-free formula whose quantifiers have types bounded by n. The axiom of constructibility $\mathbf{V} = \mathbf{L}$ implies that $D_{1n} \notin D_{2n}$, so the problem is to find a generic model in which $D_{1n} \in D_{2n}$ holds, and moreso the equality $D_{1n} = \mathbf{L} \cap \mathscr{P}(\omega)$ holds. We believe that such a model can be constructed by an appropriate modification of the methods developed in this paper.

4. It will be interesting to apply the hidden invariance technique to some other forcing notions and coding systems (those not of the almost-disjoint type), such as in [21,22].

5. This is a rather technical question. One may want to consider a smaller extension $\mathbf{L}[w^+[G]]$ instead of $\mathbf{L}[w^+[G], G \restriction W]$ in Lemma 23. Claim (i) of Lemma 23 then holds for such a smaller model in virtue of the same argument as above. However, the proof of Claim (ii) of Lemma 23, as given above for $\mathbf{L}[w^+[G], G \restriction W]$, does not go through for $\mathbf{L}[w^+[G]]$. The obstacle is that if we try to carry out the proof of Lemma 24 for $\mathbf{L}[w^+[G]]$, then it may well happen that say $Z' = \varnothing$, and then Theorem 4 is not applicable. It is an interesting problem to figure out whether in fact Claim (ii) of Lemma 23 holds in $\mathbf{L}[w^+[G]]$.

Supplementary Materials: Table of contents and Index are available online at http://www.mdpi.com/2227-7390/8/9/1477/s1.

Author Contributions: Conceptualization, V.K. and V.L.; methodology, V.K. and V.L.; validation, V.K. and V.L.; formal analysis, V.K. and V.L.; investigation, V.K. and V.L.; writing original draft preparation, V.K.; writing review and editing, V.K.; project administration, V.L.; funding acquisition, V.L. All authors have read and agreed to the published version of the manuscript.

Funding: This research was funded by Russian Foundation for Basic Research RFBR grant number 18-29-13037.

Acknowledgments: We thank the anonymous reviewers for their thorough review and highly appreciate the comments and suggestions, which significantly contributed to improving the quality of the publication.

Conflicts of Interest: The authors declare no conflict of interest. The funders had no role in the design of the study; in the collection, analyses, or interpretation of data; in the writing of the manuscript, or in the decision to publish the results.

References

1. Friedman, H. One hundred and two problems in mathematical logic. *J. Symb. Log.* **1975**, *40*, 113–129. [CrossRef]
2. Mathias, A.R.D. Surrealist landscape with figures (a survey of recent results in set theory). *Period. Math. Hung.* **1979**, *10*, 109–175. The original preprint of this paper is known in typescript since 1968 under the title "A survey of recent results in set theory". [CrossRef]

3. Harrington, L. The Constructible Reals Can Be (Almost) Anything. Preprint Dated May 1974 with the Following Addenda Dated up to October 1975: (A) Models Where Separation Principles Fail, May 74; (B) Separation without Reduction, April 75; (C) The Constructible Reals Can Be (Almost) Anything, Part II, May 75. Available online: http://logic-library.berkeley.edu/catalog/detail/2135 (accessed on 25 August 2020).
4. Jensen, R.B.; Solovay, R.M. Some applications of almost disjoint sets. In *Studies in Logic and the Foundations of Mathematics*; Bar-Hillel, Y., Ed.; North-Holland: Amsterdam, The Netherlands; London, UK, 1970; Volime 59, pp. 84–104.
5. Hinman, P.G. *Recursion-Theoretic Hierarchies*; Perspectives in Mathematical Logic; Springer: Berlin/Heidelberg, Germany; New York, NY, USA, 1978; p. 480.
6. Tarski, A. A problem concerning the notion of definability. *J. Symb. Log.* **1948**, *13*, 107–111. [CrossRef]
7. Kanovei, V.; Lyubetsky, V. Models of set theory in which nonconstructible reals first appear at a given projective level. *Mathematics* **2020**, *8*, 910. [CrossRef]
8. Gitman, V.; Hamkins, J.D.; Johnstone, T.A. What is the theory ZFC without power set? *Math. Log. Q.* **2016**, *62*, 391–406. [CrossRef]
9. Barwise, J. (Ed.) Handbook of mathematical logic. In *Studies in Logic and the Foundations of Mathematics*; North-Holland: Amsterdam, The Netherlands, 1977; Volume 90, p. 375.
10. Jech, T. *Set Theory*, 3rd ed.; Springer: Berlin/Heidelberg, Germany; New York, NY, USA, 2003; p. xiii + 769 .
11. Kanovei, V. An Ulm-type classification theorem for equivalence relations in Solovay model. *J. Symb. Log.* **1997**, *62*, 1333–1351. [CrossRef]
12. Kanovei, V.G. *On Some Questions of Definability in the Third Order Arithmetic and a Generalization of Jensen Minimal Δ^1_3 Real Theorem*; VINITI Deposited Preprint 839/75; VINITI RAS: Moscow, Russia, 1975; pp. 1–48.
13. Kanovei, V. On the nonemptiness of classes in axiomatic set theory. *Math. USSR Izv.* **1978**, *12*, 507–535. [CrossRef]
14. Kanovei, V.G.; Lyubetsky, V.A. On some classical problems in descriptive set theory. *Russ. Math. Surv.* **2003**, *58*, 839–927. [CrossRef]
15. Kanovei, V.; Lyubetsky, V. Definable E_0 classes at arbitrary projective levels. *Ann. Pure Appl. Log.* **2018**, *169*, 851–871. [CrossRef]
16. Kanovei, V.; Lyubetsky, V. Definable minimal collapse functions at arbitrary projective levels. *J. Symb. Log.* **2019**, *84*, 266–289. [CrossRef]
17. Kanovei, V.; Lyubetsky, V. Non-uniformizable sets with countable cross-sections on a given level of the projective hierarchy. *Fundam. Math.* **2019**, *245*, 175–215. [CrossRef]
18. Angiulli, G.; Jannelli, A.; Morabito, F.C.; Versaci, M. Reconstructing the membrane detection of a 1D electrostatic-driven MEMS device by the shooting method: Convergence analysis and ghost solutions identification. *Comput. Appl. Math.* **2018**, *37*, 4484–4498. [CrossRef]
19. Di Barba, P.; Fattorusso, L.; Versaci, M. Electrostatic field in terms of geometric curvature in membrane MEMS devices. *Commun. Appl. Ind. Math.* **2017**, *8*, 165–184. [CrossRef]
20. David, R. Δ^1_3 reals. *Ann. Math. Log.* **1982**, *23*, 121–125. [CrossRef]
21. Friedman, S.D.; Gitman, V.; Kanovei, V. A model of second-order arithmetic satisfying AC but not DC. *J. Math. Log.* **2019**, *19*, 1–39. [CrossRef]
22. Karagila, A. The Bristol model: An abyss called a Cohen reals. *J. Math. Log.* **2018**, *18*, 1–37. [CrossRef]

mathematics
MDPI

Article

On the 'Definability of Definable' Problem of Alfred Tarski †

Vladimir Kanovei *,‡ **and Vassily Lyubetsky** *,‡

Institute for Information Transmission Problems of the Russian Academy of Sciences (Kharkevich Institute), 127051 Moscow, Russia
* Correspondence: kanovei@iitp.ru (V.K.); lyubetsk@iitp.ru (V.L.)
† The supplementary materials to this paper, published separately, include the Index.
‡ These authors contributed equally to this work.

Received: 14 November 2020; Accepted: 7 December 2020; Published: 14 December 2020

check for updates

Abstract: In this paper we prove that for any $m \geq 1$ there exists a generic extension of $\mathbf{L}$, the constructible universe, in which it is true that the set of all constructible reals (here subsets of ω) is equal to the set $\mathbf{D}_{1m}$ of all reals definable by a parameter free type-theoretic formula with types bounded by m, and hence the Tarski 'definability of definable' sentence $\mathbf{D}_{1m} \in \mathbf{D}_{2m}$ (even in the form $\mathbf{D}_{1m} \in \mathbf{D}_{21}$) holds for this particular m. This solves an old problem of Alfred Tarski (1948). Our methods, based on the almost-disjoint forcing of Jensen and Solovay, are significant modifications and further development of the methods presented in our two previous papers in this Journal.

Keywords: definability of definable; tarski problem; type theoretic hierarchy; generic models; almost disjoint forcing

MSC: 03E15; 03E35

Dedicated to the 70-th Anniversary of A. L. Semenov.

Contents

1. Introduction

This paper continues our research project on the issues of definability in models of set theory, that was started in [1–3] among other papers, and most recently in [4,5] in this Journal. Questions of definability of mathematical objects were raised in the course of discussions on the foundations of mathematics, set theory, and the axiom of choice in the early twentieth century, such as, for instance, the famous discussion between Baire, Borel, Hadamard, and Lebesgue published in *Sinq lettres* [6]. Various aspects of definability in models of set theory have since remained the focus of work on the foundations of mathematics, see, for example, [7–13] among many important recent studies.

The topic of this paper goes back to the profound research by Alfred Tarski, who demonstrated in [14] that 'being definable' (in most general, unrestricted sense) is not a mathematically well-defined notion (see Murawski [15] on the history of this discovery and the role of Gödel, and Addison [16] on the modern perspective of the Tarski definability theory). More specificly, restricted notions of definability, in particular, type-theoretic definability, were considered by Tarski in [17] and later work in [18].

Definition 1 (Tarski). *If $m, k < \omega$ then $\mathbf{D}_{km}$ is the set of all elements of order k, definable by a parameter free type-theoretic formula of order m.*

Here elements of order 0 are just natural numbers (members of the set $\omega = \{0, 1, 2, \ldots\}$), elements of order 1 are sets of natural numbers (commonly called *reals* in modern set theory), and generally, elements of order $k+1$ ($k < \omega$) are arbitrary sets of elements of order k (see details in

Section 2.1 below). The order of a type-theoretic formula is the largest order of all its quantified and free variables. The notion of definability is taken in the form:

$$x^k = \{y^{k-1} \text{ of order } k-1 : \varphi(y^{k-1})\}\,, \tag{1}$$

where the upper index routinely denotes the order of a variable or element.

1.1. The Problem

Investigating the definability properties of sets $\mathbf{D}_{km}$, Tarski notes in [18] that $\mathbf{D}_{km} \in \mathbf{D}_{k+1,m+1}$. To prove this result, one can exploit the fact that the truth of all formulas of order m can be suitably expressed by a single formula of order $m+1$. Using such a formula, one easily gets $\mathbf{D}_{km} \in \mathbf{D}_{k+1,m+1}$. Then Tarski turns to the question whether a stronger sentence $\mathbf{D}_{km} \in \mathbf{D}_{k+1,m}$ holds. Tarski comes to the following conclusion (verbatim):

> *the solution of the problem is (trivially) positive if $k = 0$; the solution is negative if $k \geq 2$; in the (perhaps most interesting) case $k = 1$ the problem remains open.*

The negative result for $k \geq 2$ (and $m \geq k-1$, to avoid trivialities) is obtained in [18] (page 110) essentially by virtue of the fact that countable ordinals admit a definable embedding into the set of all elements of order 2. This leaves:

$$\mathbf{D}_{1m} \in \mathbf{D}_{2m} \quad (m \geq 1) \tag{2}$$

as a major open problem in [18].

Tarski notes in [18], with a reference to Gödel's work on constructibility [19], that it seems:

> *very unlikely that an affirmative solution of the problem is possible.*

Tarski does not elaborate on this point, but it is quite clear that the axiom of constructibility $\mathbf{V} = \mathbf{L}$ (and even a weaker hypothesis, see Lemma 2 below) implies $\mathbf{D}_{1m} \notin \mathbf{D}_{2m}$ for all $m \geq 1$, and hence no proof of $\mathbf{D}_{1m} \in \mathbf{D}_{2m}$ for even one single $m \geq 1$ (the "affirmative solution" in Tarski's words), can be maintained in **ZFC**. In other words, the hypothesis:

$$\mathbf{D}_{1m} \notin \mathbf{D}_{2m} \text{ holds for all } m \geq 1$$

(the negative solution of (2) for all $m \geq 1$ simultaneously) does not contradict the **ZFC** axioms. The problem of consistency of the affirmative sentences $\mathbf{D}_{1m} \in \mathbf{D}_{2m}$ was left open in [18].

This paper is devoted to this problem of Alfred Tarski.

1.2. Further Reformulations and Harrington's Statement

The problem emerged once again in the early years of forcing, especially in the case $m = 1$ corresponding to analytic definability in second-order arithmetic. The early survey [20] by A. R. D. Mathias (the original typescript has been known to set theorists since 1968) contains Problem 3112, that requires finding a model of **ZFC** in which it is true that:

> *the set of analytically definable reals is analytically definable*

that is, $\mathbf{D}_{11} \in \mathbf{D}_{21}$. Recall that *reals* in this context mean subsets of ω. Another problem there, P 3110, suggests a sharper form of this statement, namely; find a model in which it is true that

> *analytically definable reals are precisely the constructible reals*

that is, $\mathbf{D}_{11} = \mathscr{P}(\omega) \cap \mathbf{L}$. The set $\mathscr{P}(\omega) \cap \mathbf{L}$ of all constructible reals is (lightface) Σ^1_2, and hence $\mathbf{D}_{21}$, so that the equality $\mathbf{D}_{11} = \mathscr{P}(\omega) \cap \mathbf{L}$ implies $\mathbf{D}_{11} \in \mathbf{D}_{21}$, that is the case $m = 1$ of the sentence (2).

Somewhat later, Problem 87 in Harvey Friedman's survey *One hundred and two problems in mathematical logic* [21] requires to prove that for each n in the domain $2 < n \leq \omega$ there is a model of:

$$\text{ZFC} + \text{"the constructible reals are precisely the } \Delta^1_n \text{ reals"}. \qquad (3)$$

For $n \leq 2$ this is definitely impossible by the Shoenfield absoluteness theorem. As Δ^1_ω is the same as $\mathbf{D}_{11}$ = all analytically definable reals, the case $n = \omega$ in (3) is just a reformulation of $\mathbf{D}_{11} = \mathscr{P}(\omega) \cap \mathbf{L}$.

At the very end of [21], it is noted that Leo Harrington had solved problem (3) affirmatively. A similar remark, see in [20] (p. 166), a comment to P 3110. And indeed, Harrington's handwritten notes [22] present the following major result quoted here verbatim:

Theorem 1 (Harrington [22] (p. 1)). *There are models of* **ZFC** *in which the set of constructible reals is, respectively, exactly the following set of reals*:

$$\Delta^1_3, \Delta^1_4, \ldots, \Delta^1_\omega = projective, \Delta^m_n, 1 \leq n \leq \omega, 2 \leq m \leq \omega.$$

We may note that $\Delta^1_\omega = \mathbf{D}_{11}$ and generally $\Delta^m_\omega = \mathbf{D}_{1m}$ for any $m \geq 2$ in the context of Theorem 1. On the other hand the set $\mathscr{P}(\omega) \cap \mathbf{L}$ of constructible reals is Σ^1_2, and hence $\mathbf{D}_{21}$. Therefore Theorem 1 implies the consistency of the affirmative sentences $\mathbf{D}_1 \in \mathbf{D}_2$ and $\mathbf{D}_{1m} \in \mathbf{D}_{2m}$ for any particular value $m \geq 1$, and hence shows that the Tarski problems considered are independent of **ZFC**.

Based on the almost-disjoint forcing tool of Jensen and Solovay [23], a sketch of a generic extension of **L**, in which it is true that $\omega^\omega \cap \mathbf{L} = \Delta^1_3$, follows in [22] (pp. 2–4). Then a few sentences are added on page 5 of [22], which explain, without much going into details, as how Harrington planned to get some other models claimed by the theorem, in particular, a model in which $\omega^\omega \cap \mathbf{L} = \Delta^1_n$ holds for a given (arbitrary) natural index $n > 3$, and a model in which $\omega^\omega \cap \mathbf{L} = \Delta^1_\omega$, where $\Delta^1_\omega = \bigcup_n \Delta^1_n = \mathbf{D}_{11}$ (all analytically definable reals). This positively solves Problem 87 of [21], including the case $n = \omega$, of course. Different cases of higher order definability are briefly observed in [22] (p. 5) as well.

Yet, for all we know, no detailed proofs have ever emerged in Harrington's published works. An article by Harrington, entitled "Consistency and independence results in descriptive set theory", which apparently might have contained these results among others, was announced in the References list in Peter Hinman's book [24] (p. 462) to appear in *Ann. of Math.*, 1978, but in fact this or a similar article has never been published in *Annals of Mathematics* or any other journal. Some methods sketched in [22] were later used in [25], but with respect to different questions and only in relation to the definability classes of the 2nd and 3rd projective level.

1.3. The Main Theorem

The goal of this paper is to present a complete proof of the following part of Harrington's statement in Theorem 1, related to the consistency of the Tarski sentence $\mathbf{D}_{1m} \in \mathbf{D}_{2m}$ and the equality $\mathbf{D}_{1m} = \mathscr{P}(\omega) \cap \mathbf{L}$, strengthened by extra claims (ii) and (iii). This is **the main result** of this paper.

Theorem 2. *Let* $\mathsf{M} \geq 1$. *There is a generic extension of* **L** *in which it is true that*

(i) $\mathbf{D}_{1\mathsf{M}} = \mathscr{P}(\omega) \cap \mathbf{L}$, *that is, constructible reals are precisely reals in* $\mathbf{D}_{1\mathsf{M}}$ — *in particular*, $\mathbf{D}_{1\mathsf{M}}$ *is a* Σ^1_2 *set, hence*, $\mathbf{D}_{1\mathsf{M}} \in \mathbf{D}_{21}$, *and even moreso*, $\mathbf{D}_{1\mathsf{M}} \in \mathbf{D}_{2\mathsf{M}}$;

(ii) *if* $n \neq \mathsf{M}$ *then* $\mathbf{D}_{1n} \notin \mathbf{D}_{2n}$;

(iii) *the general continuum hypothesis* GCH *holds*.

Thus, for every particular $\mathsf{M} \geq 1$, there exists a generic extension of **L** in which the Tarski sentence $\mathbf{D}_{1\mathsf{M}} \in \mathbf{D}_{2\mathsf{M}}$ holds whereas $\mathbf{D}_{1n} \notin \mathbf{D}_{2n}$ for all other values $n \neq \mathsf{M}$. We recall that $\mathbf{D}_{1\mathsf{M}} \in \mathbf{D}_{2\mathsf{M}}$ *fails* in **L** itself for all M, see above.

Corollary 1. *If* $\mathsf{M} \geq 1$ *then the sentence* $\mathbf{D}_{1\mathsf{M}} \in \mathbf{D}_{2\mathsf{M}}$ *is undecidable in* **ZFC**, *even in the presence of* $\forall n \neq \mathsf{M} (\mathbf{D}_{1n} \notin \mathbf{D}_{2n})$.

This paper is dedicated to the proof of Theorem 2. This will be another application of the methods sketched by Harrington and developed in detail in our previous papers [4,5] in this Journal, but here modified and further developed for the purpose of a solution to the Tarski problem.

We may note that problems of construction of models of set theory in which this or another effect is obtained at a certain prescribed definability level (not necessarily the least possible one) are considered in modern set theory, see e.g., Problem 9 in [26] (Section 9) or Problem 11 in [27] (page 209). Some results of this type have recently been obtained in set theory, namely:

(A) a model [3] in which, for a given $n \geq 3$, there exists a countable non-empty Π^1_n set of reals, containing no OD element, while every countable Σ^1_n set of reals contains only OD reals;
(B) a model [28] in which, for a given $n \geq 2$, there is a Π^1_n real singleton that effectively codes a cofinal map $\omega \to \omega_1^{\mathbf{L}}$, minimal over $\mathbf{L}$, while every Σ^1_n real is constructible;
(C) a model [29] in which, for a given $n \geq 2$, there exists a planar non-ROD-uniformizable lightface Π^1_n set, all of whose vertical cross-sections are countable, whereas all boldface $\mathbf{\Sigma}^1_n$ sets with countable cross-sections are $\mathbf{\Delta}^1_{n+1}$-uniformizable;
(D) a model [30] in which, for a given $n \geq 3$, the Separation principle fails for $\mathbf{\Pi}^1_n$.

Theorem 2 of this paper naturally extends this research line.

1.4. *Structure of the Proof*

To define a model for Theorem 2, we employ the product of two forcing notions. The first forcing $\mathbb{C}$ is a Cohen-style collapse forcing that adjoins a generic collapse map $\zeta : \omega \xrightarrow{\text{onto}} \Xi = \mathscr{P}(\omega) \cap \mathbf{L}$, Section 2.7. The collapse is necessary since any model for Theorem 2 has to satisfy the inequality $\omega_1^{\mathbf{L}} < \omega_1$.

The second forcing notion has the form of the product $\mathbb{P}^{\Omega} = \prod_{n,i<\omega} \mathbb{P}^{\Omega}(n,i) \in \mathbf{L}$, where each factor $\mathbb{P}^{\Omega}(n,i)$ is an almost-disjoint type forcing determined by a set:

$$\mathbb{U}^{\Omega}(n,i) \in \mathbf{L}, \quad \mathbb{U}^{\Omega}(n,i) \subseteq \mathbf{Fun}_{\Omega} = (\Omega^{\Omega}) \cap \mathbf{L},$$

dense in $\mathbf{Fun}_{\Omega}$, where $\Omega = \omega_{\mathbb{M}}^{\mathbf{L}}$ and $\mathbb{M} \geq 1$ is the number we are dealing with in Theorem 2. This forcing $\mathbb{P}^{\Omega}$ adjoins an according system of generic sets $S(n,i) \subseteq \mathbf{Seq}_{\Omega} = (\Omega^{<\Omega}) \cap \mathbf{L}$, such that:

$(*)$ if $f \in \mathbf{Fun}_{\Omega}$ in $\mathbf{L}$ then $S(n,i)$ covers f (that is, $f \restriction \xi \in S(n,i)$ for unbounded-many $\xi < \Omega$) iff $f \notin \mathbb{U}^{\Omega}(n,i)$ (Lemma 15).

Basically any *system* $U \in \mathbf{L}$ of dense sets $U(n,i) \subseteq \mathbf{Fun}_{\Omega}$ defines a similar product forcing $\mathbf{P}[U] = \prod_{n,i<\omega} P[U(n,i)] \in \mathbf{L}$ (see Section 3.2). Forcing notions of the form $\mathbf{P}[U]$ satisfy certain chain and distributivity conditions in $\mathbf{L}$ (Lemma 14), that imply some general properties of related generic extensions (Lemmas 15 and 16).

The key system $\mathbb{U}^{\Omega}$ is defined in Section 4.4 (Definition 6, on the base of Theorem 6 in Section 4.2), in the form of componentwise union $\mathbb{U}^{\Omega} = \bigvee_{\alpha<\Omega^{\oplus}} \mathbb{U}^{\Omega}_{\alpha}$, i.e., $\mathbb{U}^{\Omega}(n,i) = \bigcup_{\alpha<\Omega^{\oplus}} \mathbb{U}^{\Omega}_{\alpha}(n,i)$ for all $n,i < \omega$, where $\Omega^{\oplus} = \omega_{\mathbb{M}+1}^{\mathbf{L}}$ is the $\mathbf{L}$-cardinal next to Ω, and the systems $\mathbb{U}^{\Omega}_{\alpha} \in \mathbf{L}$ are:

- *Increasing*, i.e., $\mathbb{U}^{\Omega}_{\alpha}(n,i) \subseteq \mathbb{U}^{\Omega}_{\gamma}(n,i)$ for all $\alpha < \gamma$ and $n,i < \omega$,
- *Small*, i.e., $\mathtt{card}\, \mathbb{U}^{\Omega}_{\alpha}(n,i) \leq \Omega$ in $\mathbf{L}$ for all $n,i < \omega$, and,
- *Disjoint*, i.e., the components $\mathbb{U}^{\Omega}_{\alpha}(n,i)$ are pairwise disjoint.

We apply a diamond-based argument in Section 4 to ensure that the resulting system $\mathbb{U}^{\Omega} \in \mathbf{L}$ has its different *slices* $\{\mathbb{U}^{\Omega}(n,i)\}_{i<\omega}$ $(n < \omega)$ satisfying different definability and inner genericity requirements (Theorem 6 in Section 4.2), so that the descriptive complexity and the level of inner genericity (or completeness) of nth 'slice' tends to infinity with $n \to \infty$. This is a major novelty of the construction.

Then we consider the key product forcing notion $\mathbb{P}^{\Omega} = \mathbf{P}[\mathbb{U}^{\Omega}] = \prod_{n,i<\omega} \mathbb{P}^{\Omega}(n,i)$. We extend $\mathbf{L}$ by a collapse-generic map $\boldsymbol{\zeta} : \omega \xrightarrow{\text{onto}} \mathscr{P}(\omega) \cap \mathbf{L}$ to $\mathbf{L}$, as above, and define the partial product $\mathbb{P}^{\Omega} \restriction \boldsymbol{w} = \prod_{\langle n,i\rangle \in \boldsymbol{w}} \mathbb{P}^{\Omega}(n,i) \in \mathbf{L}[\boldsymbol{\zeta}]$ as a forcing notion in $\mathbf{L}[\boldsymbol{\zeta}]$, where:

$$\boldsymbol{w} = \boldsymbol{w}[\boldsymbol{\zeta}] = \{\langle n,i\rangle : n \in \omega \wedge i \in \boldsymbol{\zeta}(n)\}.$$

Adjoining a $(\mathbb{P}^{\Omega} \restriction \boldsymbol{w})$-generic set G to $\mathbf{L}[\boldsymbol{\zeta}]$, we get a model $\mathbf{L}[\boldsymbol{\zeta}, G]$ for Theorem 2. In particular, if $x = \boldsymbol{\zeta}(n) \in \mathscr{P}(\omega) \cap \mathbf{L}$, then x is definable in $\mathbf{L}[\boldsymbol{\zeta}, G]$ by means of the equivalence:

$$i \in x \iff \exists S \subseteq \mathbf{Seq}_{\Omega} \, \forall f \in \mathbf{Fun}_{\Omega} \, (S \text{ covers } f \text{ iff } f \notin \mathbb{U}^{\Omega}(n,i)), \tag{4}$$

in which the implication $\Longrightarrow$ follows from $(*)$ via $S = S(n,i)$ (note that $S(n,i) \in \mathbf{L}[\boldsymbol{\zeta}, G]$ since $\langle n,i\rangle \in \boldsymbol{w}$ in case $i \in x = \boldsymbol{\zeta}(n)$), whereas the inverse implication $\Longleftarrow$ is based on the completeness properties of the system $\mathbb{U}^{\Omega}$. It also takes some effort to check that the right-hand side of (4) really defines a $\mathbf{D}_{1\mathbb{M}}$ relation in $\mathbf{L}[\boldsymbol{\zeta}, G]$; for that purpose Theorem 3 is proved beforehand in Section 2.3.

To prove that, conversely, every $x \in \mathbf{D}_{1\mathbb{M}}$ in $\mathbf{L}[\boldsymbol{\zeta}, G]$ belongs to $\mathbf{L}$, we introduce *forcing approximations* in Section 5, a forcing-like relation used to prove the elementary equivalence theorem. Its key advantage is the invariance under some transformations, including the permutations of the index set $\mathcal{I}$, see Section 6.5. The actual forcing notion $\mathbb{P}^{\Omega} = \mathbf{P}[\mathbb{U}^{\Omega}]$ is absolutely not invariant under permutations of $\mathcal{I}$, but the $\mathbb{M}$-completeness property, maintained through the inductive construction of $\mathbb{U}^{\Omega}$ in $\mathbf{L}$, allows us to prove that the auxiliary forcing is in the same relation to the truth in $\mathbb{P}^{\Omega}$-generic extensions, as the true $\mathbb{P}^{\Omega}$-forcing relation (Theorem 10). We call this construction *hidden invariance* (see Section 6.1), and this is the other major novelty of this paper.

Finally, Section 6 presents the proof of the invariance theorem (Theorem 11), with the help of forcing approximations, and thereby completes the proof of Theorem 2.

The flowchart of the proof can be seen in Figure 1 on page 6.

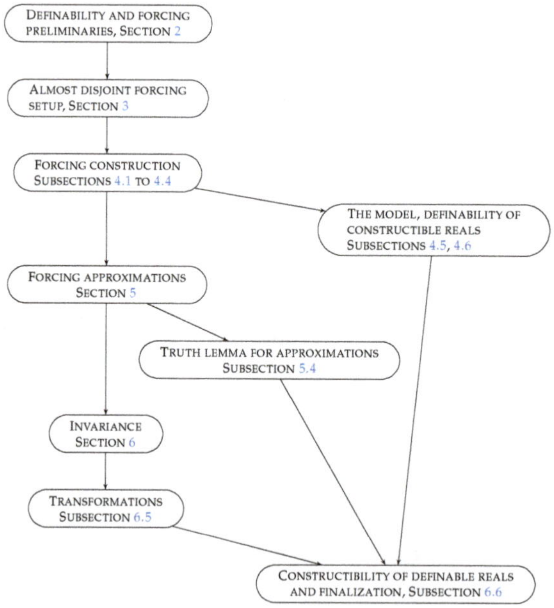

Figure 1. Flowchart of the proof of Theorem 2.

2. Preliminaries

This Section contains several definitions and results that will be very instrumental in the proof of Theorem 2.

2.1. Definability Issues

Beginning with the type-theoretic definability, we recall some details of Tarski's constructions from [18]. The type-theoretic language deals with variables $x^k, y^k, \dots$ of orders $k < \omega$, and includes the Peano arithmetic language for order 0 and the atomic predicate $\in$ of membership used as $x^k \in y^{k+1}$. The *order* of a formula φ is equal to the highest order of all variables in φ. Variables of each order k can be substituted with elements of the corresponding iteration:

$$\mathscr{P}^k(\omega) = \underbrace{\mathscr{P}(\mathscr{P}(\dots\mathscr{P}(\omega)\dots))}_{k \text{ times the powerset operation } \mathscr{P}(\cdot)} \quad , \quad \text{the set of all elements of order } k$$

of the powerset operation. In particular, $\mathscr{P}^0(\omega) = \omega$ (natural numbers), $\mathscr{P}^1(\omega) = \mathscr{P}(\omega)$ (the reals), $\mathscr{P}^2(\omega) = \mathscr{P}(\mathscr{P}(\omega))$ (sets of reals), and so on. Accordingly each quantifier $\exists x^k$, $\forall x^k$ in a type-theoretic formula is naturally relativized to $\mathscr{P}^k(\omega)$, and the truth of a closed type-theoretic formula (with or without parameters) is understood in the sense of such a relativization.

If $k, m < \omega$, $k \geq 1$, then, by Definition 1, $\mathbf{D}_{km}$ is the set of all $x^k \in \mathscr{P}^k(\omega)$, definable in the form:

$$x^k = \{y^{k-1} \in \mathscr{P}^{k-1}(\omega) : \varphi(y^{k-1})\}$$

by a parameter free formula φ of order $\leq m$; thus $\mathbf{D}_{km} \subseteq \mathscr{P}^k(\omega)$.

Remark 1. *We will occasionally extend the definition of $\mathbf{D}_{km}$ to binary relations, especially in the case $k = 1$. Namely a set $X \subseteq \mathscr{P}^{k-1}(\omega) \times \mathscr{P}^{k-1}(\omega)$ belongs to $\mathbf{D}_{km}$ if it is definable by a parameter free formula of order $\leq m$ with two free variables.*

In matters of $\in$-**definability**, we refer to e.g., [31] (Part B, 5.4), or [32] (Chapter 13) on the Lévy hierarchy of $\in$-formulas and definability classes Σ^H_n, Π^H_n, Δ^H_n for any transitive set H. In particular,

Σ^H_n = all sets $X \subseteq H$, definable in H by a parameter-free Σ_n formula;
$\Sigma_n(H)$ = all sets $X \subseteq H$ definable in H by a Σ_n formula with any sets in H as parameters.

Something like $\Sigma^H_n(x)$, $x \in H$, means that only x is admitted as a parameter, while $\Sigma^H_n(P)$, where $P \subseteq H$, means that all $x \in P$ can be parameters. Collections like Π^H_n, $\Pi^H_n(x)$, $\Pi^H_n(P)$ are defined similarly, and $\Delta^H_n = \Sigma^H_n \cap \Pi^H_n$, etc. These definitions usually work with transitive sets of the form:

$$H = \mathbf{H}\varkappa = \{x : \mathtt{card}\,(\mathrm{TC}\,(x)) < \varkappa\}, \quad \text{where } \varkappa \text{ is an infinite cardinal,}$$

and TC is *the transitive closure.* In particular, $\mathbf{HC} = \mathbf{H}\omega_1$, all heredidarily-countable sets.

2.2. Constructibility Issues

As usual, $\mathbf{L}$ is the constructible universe, and $<_{\mathbf{L}}$ will denote the Gödel wellordering of $\mathbf{L}$. Let $\varkappa$ be an infinite regular cardinal. The following are well-known facts in the theory of constructibility, see e.g., [33] and Lemma 6.3 ff in [31] (Section B.5):

1°. The set $\mathbf{H}\varkappa \cap \mathbf{L}$ belongs to $\Sigma^{\mathbf{H}\varkappa}_1$ and is equal to $(\mathbf{H}\varkappa)^{\mathbf{L}} = \mathbf{L}_\varkappa$.

2°. The restriction $<_{\mathbf{L}} \restriction (\mathbf{H}\varkappa)^{\mathbf{L}}$ is a wellordering of $(\mathbf{H}\varkappa)^{\mathbf{L}}$ of length $\varkappa$ and a $\Delta^{(\mathbf{H}\varkappa)^{\mathbf{L}}}_1$ relation.

3°. On the other hand, the set $\mathscr{P}(\omega) \cap \mathbf{L}$ and relation $<_{\mathbf{L}} \restriction (\mathscr{P}(\omega) \cap \mathbf{L})$ belong to Σ^1_2 and to $\mathbf{D}_{21}$.

4°. The map $x \longmapsto \mathrm{pr}\, x = \{y : y <_{\mathbf{L}} x\} : (\mathbf{H}\varkappa)^{\mathbf{L}} \to (\mathbf{H}\varkappa)^{\mathbf{L}}$ is $\Delta_1^{(\mathbf{H}\varkappa)^{\mathbf{L}}}$ as well.

The last statement implies the following useful definability estimation.

5°. Assume that $m \geq 1$ and $P \subseteq (\mathbf{H}\varkappa)^{\mathbf{L}} \times (\mathbf{H}\varkappa)^{\mathbf{L}}$ is $\Delta_m^{(\mathbf{H}\varkappa)^{\mathbf{L}}}$. If $x \in D = \{x \in (\mathbf{H}\varkappa)^{\mathbf{L}} : \exists y\, P(x,y)\}$, then let $y_x \in (\mathbf{H}\varkappa)^{\mathbf{L}}$ be the $<_{\mathbf{L}}$-least witness. Then $P' = \{\langle x, y_x\rangle : x \in D\} \subseteq P$ is $\Delta_m^{(\mathbf{H}\varkappa)^{\mathbf{L}}}$ as well.

Indeed $y = y_x$ is equivalent to $P(x,y) \wedge \forall z \in \mathrm{pr}\, y \neg P(x,z)$, where:

$$\begin{aligned} \forall z \in \mathrm{pr}\, y \neg P(x,z) \quad &\Longleftrightarrow \quad \exists Z\, (Z = \mathrm{pr}\, y \wedge \forall z \in Z \neg P(x,z)) \\ &\Longleftrightarrow \quad \forall Z\, (Z = \mathrm{pr}\, y \Longrightarrow \forall z \in Z \neg P(x,z)), \end{aligned}$$

and the bounded quantifiers $\forall z \in Z$ do not influence the definability class.

We proceed with several easy and rather known lemmas.

Lemma 1. *Assume that $x, y \in \mathscr{P}(\omega) \cap \mathbf{L}$ and $y <_{\mathbf{L}} x$. Then $y \in \Delta^1_2(x)$, and hence if $y \in \mathbf{D}_{1n}$, $n \geq 1$, or $y \in \mathbf{D}_1$, then $x \in \mathbf{D}_{1n}$, resp., $x \in \mathbf{D}_1$ as well.*

Proof. By the Shoenfield absoluteness, it suffices to prove that $y \in \Delta^1_2(x)$ is true in $\mathbf{L}$.

We argue in L. Let $\varkappa = \omega_1$, so that $\mathbf{H}\varkappa = (\mathbf{H}\varkappa)^{\mathbf{L}} = \mathbf{HC}$ (hereditarily countable). The set:

$$P = \{\langle z, f\rangle : z \subseteq \omega \wedge f : \omega \to \mathscr{P}(\omega) \wedge \mathtt{ran}\, f = \mathrm{pr}\, z\}$$

belongs to $\Delta_1^{\mathbf{HC}}$ by 4° since:

$$\begin{aligned} \mathtt{ran}\, f = \mathrm{pr}\, z \quad &\Longleftrightarrow \quad \exists u (u = \mathrm{pr}\, z \wedge \forall n\, (f(n) \in u) \wedge \forall z' \in u\, \exists n\, (f(n) = z')) \\ &\Longleftrightarrow \quad \forall u (u = \mathrm{pr}\, z \Longrightarrow \forall n\, (f(n) \in u) \wedge \forall z' \in u\, \exists n\, (f(n) = z')). \end{aligned}$$

Let f_z be the $<_{\mathbf{L}}$-least f such that $\langle z, f\rangle \in P$; then $P' = \{\langle z, f_z\rangle : z \subseteq \omega\}$ is $\Delta_1^{\mathbf{HC}}$ by 5°. It follows that f_x is $\Delta_1^{\mathbf{HC}}(x)$ (with x as the only parameter). Therefore, as $y <_{\mathbf{L}} x$, we have $y \in \Delta_1^{\mathbf{HC}}(x)$ because $y = f_x(n)$ for some n. It follows that $y \in \Delta^1_2(x)$. (See e.g., [34] (p. 281) on this translation result.) □

Remark 2 (Essentially Tarski [18]). *If $n \geq 1$ and $\omega_1^{\mathbf{L}} = \omega_1$ then $\mathbf{D}_{1n} \notin \mathbf{D}_{2n}$.*

Proof. If $\omega_1^{\mathbf{L}} = \omega_1$ then the set $Y = \mathscr{P}(\omega) \cap \mathbf{L}$ is uncountable. On the other hand $X = \mathbf{D}_{1n}$ is countable, hence $Z = Y \smallsetminus X \neq \varnothing$. Note that $Y \in \mathbf{D}_{21}$ by 3° above. It follows that if $X \in \mathbf{D}_{2n}$ then Z belongs to $\mathbf{D}_{2n}$, too, and then the $<_{\mathbf{L}}$-least element z_0 of the set Z belongs to $\mathbf{D}_{1n}$ because $<_{\mathbf{L}}$ is $\mathbf{D}_{21}$ on Y still by 3°. However $z_0 \notin X = \mathbf{D}_{1n}$ by construction. This is a contradiction. □

Lemma 2. *If $1 \leq n < m < \omega$ and $\mathbf{D}_{1m} \subseteq \mathbf{L}$, then $\mathbf{D}_{1n} \notin \mathbf{D}_{2n}$.*

Proof. We have $\mathbf{D}_{1n} \subsetneqq \mathbf{D}_{1m}$ since $n < m$. Therefore $\mathbf{D}_{1n} \subsetneqq \mathbf{D}_{1m} \subseteq Y = \mathscr{P}(\omega) \cap \mathbf{L}$. If, to the contrary, $\mathbf{D}_{1n} \in \mathbf{D}_{2n}$, then the set $Y \smallsetminus \mathbf{D}_{1n}$ belongs to $\mathbf{D}_{2n}$ as well since $Y \in \mathbf{D}_{21}$ by 3° above. We conclude that the $<_{\mathbf{L}}$-least element $y_0 \in Y \smallsetminus \mathbf{D}_{1n}$ belongs to $\mathbf{D}_{1n}$, because $<_{\mathbf{L}}$ is $\mathbf{D}_{21}$ on Y by 3°. This is a contradiction since $z_0 \notin \mathbf{D}_{1n}$ by construction. □

2.3. Type-Theoretic Definability vs. $\in$-Definability

It occurs that the definability classes in sets of the form $\mathbf{H}\varkappa$ correspond to the Tarski definability classes, in the sense of the following theorem:

Theorem 3. *Assume that the generalized continuum hypothesis $2^\vartheta = \vartheta^+$ holds for all infinite cardinals $\vartheta < \omega_{m-1}$. If $m \geq 1$ and $x \subseteq \omega$, then x is $\mathbf{D}_{1m}$ if x is $\in$-definable in $\mathbf{H}\omega_m$.*

In case $m = 1$ (then $\mathbf{H}\omega_m = \mathbf{H}\omega_1 = \mathbf{HC}$ and the GCH premice is vacuous), this result was explicitly mentioned, in [34] (p. 281), a detailed proof see e.g., [32] (Lemma 25.25).

Proof. The GCH premice of the theorem is equivalent to $\mathscr{P}^m(\omega) \subseteq \mathbf{H}\omega_m$. This implies $\Longrightarrow$: if $x \in \mathbf{D}_{1m}$ then x is surely $\in$-definable in $\mathbf{H}\omega_m$.

The inverse implication takes more effort. We have to somehow *model* the $\in$-structure of $\mathbf{H}\omega_m$ in $\mathbf{D}_{1m}$. For this purpose, if $k < \omega$ and $x, y \in \mathscr{P}^k(\omega)$ then define a quasi-pair $\langle x, y\rangle^k \in \mathscr{P}^k(\omega)$ by induction as follows. If $k = 0$, so that $x, y \in \omega$, then put $\langle x, y\rangle^0 = 2^x \cdot 3^y \in \omega$. If $x, y \in \mathscr{P}^{k+1}(\omega)$ then put $\langle x, y\rangle^{k+1} = \{\langle 0, x'\rangle^k : x' \in x\} \cup \{\langle 1, y'\rangle^k : y' \in y\} \in \mathscr{P}^{k+1}(\omega)$. Note that elements $0 = \varnothing$ and $1 = \{\varnothing\}$ belong to every type-theoretic level $\mathscr{P}^k(\omega)$. It can be easily established by induction that if $x, y, a, b \in \mathscr{P}^k(\omega)$ and $\langle x, y\rangle^k = \langle a, b\rangle^k$ then $x = a$ and $y = b$.

Following [32] (25.13), we associate, with each $r \in \mathscr{P}^m(\omega)$, a binary relation E_r defined so that:

$$x \mathrel{\mathsf{E}_r} y \quad \text{iff} \quad x, y \in M = \mathscr{P}^{m-1}(\omega) \text{ and } \langle x, y\rangle^{m-1} \in r.$$

on the set $M = \mathscr{P}^{m-1}(\omega)$. Let WFE_0 contain all sets $r \in \mathscr{P}^m(\omega)$ such that E_r is an extensional well-founded relation on $|r| = \{0\} \cup \{x \in M : \exists y \in M\,(x \mathrel{\mathsf{E}_r} y \vee y \mathrel{\mathsf{E}_r} x)\}$, with the additional property that 0 is the only *top element* of $|r|$, that is, $0 \mathrel{\mathsf{E}_r} x$ holds for no $x \in |r|$. If $r \in \mathrm{WFE}_0$ then let π_r be the unique 1-1 map defined on $|r|$ and satisfying $\pi_r(x) = \{\pi_r(y) : y \mathrel{\mathsf{E}_r} x\}$ for all $x \in |r|$ — the *transitive collapse*. We put $F(r) = \pi_r(0)$.

Under our assumptions, F is a map from WFE_0 onto $\mathbf{H}\omega_m$, $\in$-definable in $\mathbf{H}\omega_m$.

One easily proves that WFE_0 belongs to $\mathbf{D}_{mm}$, that is, it is type-theoretically definable with quantifiers only over order levels $\le m$. Moreover the binary relations EQ, IN defined on WFE_0 by:

$$r \mathrel{\mathsf{EQ}} q \ \text{ iff } \ F(r) = F(q), \quad \text{and} \quad r \mathrel{\mathsf{IN}} q \ \text{ iff } \ F(r) \in F(q),$$

belong to $\mathbf{D}_{mm}$ as well. Namely, let a *bisimulation for* $r, q \in \mathrm{WFE}_0$ be any binary relation $\mathsf{B} \subseteq |r| \times |q|$ satisfying $0 \mathrel{\mathsf{B}} 0$ and, for all $x \in |r|$ and $y \in |q|$,

$$x \mathrel{\mathsf{B}} y \quad \text{iff} \quad \forall x'\, \exists y'\, (x' \mathrel{\mathsf{E}_r} x \Longrightarrow y' \mathrel{\mathsf{E}_q} y \wedge x' \mathrel{\mathsf{B}} y') \ \wedge\ \forall y'\, \exists x'\, (y' \mathrel{\mathsf{E}_q} y \Longrightarrow x' \mathrel{\mathsf{E}_r} x \wedge x' \mathrel{\mathsf{B}} y').$$

Then, on the one hand, $F(r) = F(q)$ iff there exists a bisimulation for r, q iff there exists $b \in \mathscr{P}^m(\omega)$ such that E_b is a bisimulation for r, q. On the other hand, we can express the property "E_b is a bisimulation for r, q" by a type-theoretic formula with quantifiers only over orders $\le m$, by suitably replacing pairs $\langle\cdot,\cdot\rangle$ with quasipairs $\langle\cdot,\cdot\rangle^{m-1}$.

To treat IN, we have to only change $0 \mathrel{\mathsf{B}} 0$ above to $\exists y_0 \in |q|\,(0 \mathrel{\mathsf{B}} y_0 \wedge y_0 \mathrel{\mathsf{E}_q} 0)$.

Finally if $n < \omega$ then let $r_n = \{\langle i, j\rangle^{m-1} : 1 \le i < j \le n\} \cup \{\langle i, 0\rangle^{m-1} : 1 \le i \le n\}$, so that $r_n \in \mathrm{WFE}_0$ and $F(r_n) = n$.

And now let $x = \{n < \omega : \mathbf{H}\omega_m \models \varphi(n)\} \subseteq \omega$ be $\in$-definable in $\mathbf{H}\omega_m$ by a parameter free formula $\varphi(\cdot)$. Then we have $x = \{n < \omega : \Phi(r_n)\}$, where Φ is obtained from φ by substitution of EQ for $=$ and IN for $\in$ and relativization of all quantifiers to WFE_0. This proves $x \in \mathbf{D}_{1m}$. $\square$

2.4. Reduction to the Powerset Definability

Let $\preccurlyeq$ be the wellordering of $\mathbf{Ord} \times \mathbf{Ord}$ defined so that $\langle \xi, \eta\rangle \preccurlyeq \langle \xi', \eta'\rangle$ iff:

$$\langle \max\{\xi, \eta\}, \xi, \eta\rangle \leqslant_{\mathtt{lex}} \langle \max\{\xi', \eta'\}, \xi', \eta'\rangle$$

lexicographically. Let $\mathbb{p} : \mathbf{Ord} \times \mathbf{Ord} \xrightarrow{\text{onto}} \mathbf{Ord}$ be the order preserving map: $\langle \xi, \eta\rangle \preccurlyeq \langle \xi', \eta'\rangle$ iff $\mathbb{p}(\xi, \eta) \le \mathbb{p}(\xi', \eta')$—the canonical pairing function. Let $\mathbb{p}_1$ and $\mathbb{p}_2$ be the inverse functions, so that $\alpha = \mathbb{p}(\mathbb{p}_1(\alpha), \mathbb{p}_2(\alpha))$ for all α.

Lemma 3 (routine). *If Ω is an infinite cardinal and $\kappa = \Omega^+$, then $\mathbb{p}$ maps $\Omega \times \Omega$ onto Ω bijectively, and the restriction $\mathbb{p} \restriction (\Omega \times \Omega)$ is constructible and $\Delta_1^{\mathbf{H}\kappa}$.*

Now we prove another reduction-type definability theorem.

Theorem 4. *If Ω is a regular cardinal, $\varkappa = \Omega^+$, $X, Y \subseteq \omega$, and X is $\in$-definable in $\mathbf{H}\varkappa$ with Y as the only parameter, then X is $\in$-definable in the structure $\langle \mathscr{P}(\Omega) ; \in, \mathbb{p} \rangle$ with Y as the only parameter.*

Proof (sketch). If $x \subseteq \Omega$ then let $\mathsf{E}'_x = \{\langle \xi, \eta \rangle : \xi, \eta < \Omega \wedge \mathbb{p}(\xi, \eta) \in x\}$ be a binary relation on its domain $|x| = \mathtt{dom}\,\mathsf{E}'_x \cup \mathtt{ran}\,\mathsf{E}'_x$. Following the proof of Theorem 3, let WFE'_0 contain all sets $x \subseteq \Omega$ such that E'_x is an extensional well-founded relation on $|x|$, with the additional property that $0 \in |x|$ and 0 is the only *top element* of $|x|$, that is, $0 \;\mathsf{E}'_x\; \xi$ holds for no $\xi \in |x|$. If $x \in \mathrm{WFE}'_0$ then let φ_x be the unique 1-1 map defined on $|x|$ and satisfying $\varphi_x(\xi) = \{\varphi_x(\eta) : \eta \;\mathsf{E}'_x\; \xi\}$ for all $\xi \in |x|$—the *transitive collapse*. We put $F'(x) = \varphi_x(0)$; F' is a map from WFE'_0 onto $\mathbf{H}\varkappa$, $\in$-definable in $\mathbf{H}\varkappa$.

Both WFE'_0 and the binary relations EQ', IN' defined on WFE'_0 by:

$$x \;\mathsf{EQ}'\; y \text{ iff } F'(x) = F'(y), \quad \text{and} \quad x \;\mathsf{IN}'\; y \text{ iff } F'(x) \in F'(y),$$

are $\in$-definable in $\langle \mathscr{P}(\Omega) ; \in, \mathbb{p} \rangle$ by the same bisimulation argument as in the proof of Theorem 3. Finally if $n < \omega$ then let $x_n = \{\mathbb{p}(i,j) : 1 \le i < j \le n\} \cup \{\mathbb{p}(i,0) : 1 \le i \le n\}$, so that $x_n \in \mathrm{WFE}'_0$ and $F'(x_n) = n$.

Now let $X = \{n < \omega : \mathbf{H}\varkappa \models \Phi(n, Y)\} \subseteq \omega$ be $\in$-definable in $\mathbf{H}\varkappa$ by a formula $\varphi(\cdot, Y)$. Then we have $X = \{n < \omega : \Phi'(x_n)\}$, where Φ' is obtained from Φ by the substitution of EQ' for $=$ and IN' for $\in$ and relativization of all quantifiers to WFE'_0. This proves the theorem. □

2.5. A Useful Result in Forcing Theory

We remind that, by [32] (Chapter 15), if $\varkappa$ is an infinite ordinal, then a forcing notion $P = \langle P ; \le \rangle$:

- Is *$\varkappa$-closed*, if any $\le$-decreasing sequence $\{p_\alpha\}_{\alpha < \lambda}$ in P, of length $\lambda \le \varkappa$, has a lower bound in P;
- Is *$\varkappa$-distributive*, if the intersection of $\varkappa$-many open dense sets is open dense, and a set $D \subseteq P$ is *open*, iff $q \leqslant p \in D \Longrightarrow q \in D$, and *dense*, iff for any $p \in P$ there is $q \in D$, $q \leqslant p$.
- Satisfies *$\varkappa$-chain condition*, or *$\varkappa$-CC*, if every antichain $A \subseteq P$ has cardinality strictly less than $\varkappa$.

We will make use of the following general result in forcing theory.

Lemma 4. *Assume that, in $\mathbf{L}$, $\vartheta < \Omega = \vartheta^+$ are regular infinite cardinals, and $Q, P \in \mathbf{L}$ are forcing notions, Q satisfies Ω-CC in $\mathbf{L}$, and P is ϑ-closed in $\mathbf{L}$. Assume that $\langle F, G \rangle$ is a pair $(Q \times P)$-generic over $\mathbf{L}$. Then,*

(i) *P remains ϑ-distributive in $\mathbf{L}[F]$,*

(ii) *Ω is still a cardinal in $\mathbf{L}[F, G]$,*

(iii) *Every set $X \in \mathbf{L}[F, G]$, $X \subseteq \Omega$, bounded in Ω, belongs to $\mathbf{L}[F]$.*

Proof. (i) Consider any sequence $\{D_\alpha\}_{\alpha < \vartheta}$ in $\mathbf{L}[F]$ of open dense sets $D_\alpha \subseteq P$. Prove that their intersection is dense. Let $\widehat{p} \in P$. Then $D = \{\langle \alpha, p \rangle : \alpha < \vartheta \wedge p \in D_\alpha\}$ belongs to $\mathbf{L}[F]$. Therefore there is a name $t \in \mathbf{L}$, $t \subseteq Q \times (\vartheta \times P)$, satisfying $D = t[F]$. Then $D_\alpha = t_\alpha[F]$ for all α, where $t_\alpha = \{\langle q, p \rangle : \langle q, \langle \alpha, p \rangle \rangle \in t\}$. There exists a condition $q_0 \in F$ which Q-forces

(A) "$t_\alpha[\underline{F}]$ is open dense in P"

over $\mathbf{L}$ for every $\alpha < \vartheta$. We can w.l.o.g. assume that 1_Q forces (A), otherwise replace Q by $Q' = \{q \in Q : q \leqslant q_0\}$. Under this assumption, we have the following:

(B) If $\alpha < \vartheta$, $p \in P$, and $q \in Q$ then there exist $q' \in Q$ and $p' \in P$ such that $q' \leqslant q$, $p' \leqslant p$, and q' Q-forces $p' \in t_\alpha[\underline{F}]$ over $\mathbf{L}$.

Now we prove a stronger fact:

(C) If $\gamma < \vartheta$ and $p \in P$, then there is $p' \in P$, $p' \leqslant p$, such that 1_Q forces $p' \in t_\gamma[\underline{F}]$ over $\mathbf{L}$.

Indeed, arguing in $\mathbf{L}$, and using (B) and the assumption that P is ϑ-closed, we can define a decreasing sequence $\{p_\alpha\}_{\alpha<\eta}$ of conditions in P, where $\eta < \Omega$, and a sequence $\{q_\alpha\}_{\alpha<\eta}$ of conditions in Q, such that $q_0 = q$, q_α is incompatible with q_β whenever $\alpha \neq \beta$, and each q_α Q-forces $p_{\alpha+1} \in t_\gamma[\underline{F}]$. Note that the construction really has to stop at some $\eta < \Omega$ otherwise we have an antichain in Q of cardinality Ω. Thus $A = \{q_\alpha : \alpha < \eta\}$ is a maximal antichain, and on the other hand, as P is ϑ-closed and $\eta < \Omega = \vartheta^+$, there is a condition $p \in P$ satisfying $p \leqslant p_\alpha$ for all $\alpha < \eta$. Then every $q \in A$ Q-forces $p \in t_\gamma[\underline{F}]$ by construction, therefore, as A is a maximal antichain, q witnesses (C).

To accomplish the proof of (i), we define, using (C), a decreasing sequence $\{p_\gamma\}_{\gamma<\vartheta} \in \mathbf{L}$ of conditions in P, such that $p_0 \leqslant \widehat{p}$ and, for any $\gamma < \vartheta$, 1_Q forces $p_{\gamma+1} \in t_\gamma[\underline{F}]$ over $\mathbf{L}$. Once again, there is a condition $p \in P$, $p \leqslant p_\gamma$ for all γ. Then 1_Q forces $p \in t_\gamma[\underline{F}]$ for all γ, hence $p \in \bigcap_\gamma D_\gamma$, as required.

Finally, as Q is Ω-CC in $\mathbf{L}$, Ω remains a cardinal in $\mathbf{L}[F]$. Then, as P is ϑ-distributive in $\mathbf{L}[F]$, we obtain (ii) and (iii) by standard arguments. □

2.6. Definable Names

Let $Q \in \mathbf{L}$ be any forcing notion. It is well known (see, e.g., Lemma 2.5 in Chapter B.4 of [31]) that if $F \subseteq Q$ is a Q-generic filter over $\mathbf{L}$, $X \in \mathbf{L}$, and $Y \in \mathbf{L}[F]$, $Y \subseteq X$, then there is a set $t \in \mathbf{L}$, $t \subseteq Q \times X$, such that:

$$Y = t[Q] := \{x \in X : \exists q \in F\,(\langle q, x\rangle \in t)\}\,;$$

such a t is called a *Q-name* (for Y), whereas $t[G]$ is the *G-valuation*, or *G-interpretation* of t. There is a more comprehensive system of names and valuations, which involves all sets Y in generic extensions, not only those included in the groung model, see e.g., Chapter IV in [35], but it will not be used in this paper. The next theorem claims that in certain cases such a name t as above can be chosen of nearly the same definability level as the set Y itself.

Theorem 5. *Assume that $Q \in \mathbf{L}$ is any forcing, $F \subseteq Q$ is Q-generic over $\mathbf{L}$, $\varkappa > \omega$ is a cardinal in $\mathbf{L}[F]$* (hence, in $\mathbf{L}$, too), *$n \geq 1$, $H = (\mathbf{H}\varkappa)^{\mathbf{L}}$, $H[F] = (\mathbf{H}\varkappa)^{\mathbf{L}[F]}$, and $Y \in \mathbf{L}[F]$, $Y \subseteq H$. Then,*

(i) *If Y belongs to Σ_n^H* (hence to $\mathbf{L}$), *then Y also belongs to $\Sigma_n^{H[F]}$;*

(ii) *If $Q \in H$ and Y belongs to $\Sigma_n(H[F])$* (meaning Σ_n in $H[F]$ with arbitrary definability parameters in $H[F]$ allowed) *then there exists a $\Sigma_n(H)$ name $t \in \mathbf{L}$, $t \subseteq Q \times H$, such that $Y = t[F]$.*

Proof. To prove (i) note that $H = H[F] \cap \mathbf{L}$. But the formula "x is contructible" is Σ_1 [31] (Part B, 5.4). It follows that H is $\Sigma_1^{H[F]}$. Now the result is clear: We formally relativize, to the $\Sigma_1^{H[F]}$ set H, all quantifiers in the Σ_n definition of Y in H, getting a Σ_n definition of Y in $H[F]$.

To prove (ii), assume that $Q \in H$. We utilize a more complex system of representation of sets in $\mathbf{L}[F]$, affecting all these sets, not just subsets of sets in $\mathbf{L}$. We take it from [36]. Inductively on the $\in$-rank $\mathtt{rk}\,(a)$, each set a is mapped to the set $K(a) = \{K(b) : \exists q \in F\,(\langle q, b\rangle \in a)\}$ (depends on F!). The next lemma continues the proof of Theorem 5.

Lemma 5. $H[F] = \{K(a) : a \in H\}$.

Proof. From right to left, an elementary induction argument works. Prove it from left to right. Induction by the $\in$-rank $\mathtt{rk}\,(x)$, for each $x \in H[F]$ we define a set $a_x \in H$ such that $x = K(a_x)$. If $x = \varnothing$, then $a_x = \varnothing$ will do. Assume that $\mathtt{rk}\,(x) > 0$ and a_y is already defined for each $y \in x$.

The set $A = \{a_y : y \in x\} \in L[F]$, $A \subseteq H$ has cardinality $< \varkappa$ in $L[F]$. Moreover, there is a set $B \in \mathbf{L}$, $B \subseteq H$, of cardinality $\leq \varkappa$ in $\mathbf{L}$, such that $A \subseteq B$. (Indeed, $H \in \mathbf{L}$ has cardinality $\varkappa$ in $\mathbf{L}$. Let $\{t_\alpha\}_{\alpha<\varkappa}$ be a constructible enumeration of elements of H. As $\operatorname{card} A < \varkappa$ strictly, there is $\gamma < \varkappa$ such that $A \subseteq B = \{t_\alpha : \alpha < \gamma\}$. The set B is as required.)

According to the above, we have $A = \tau[F]$ for some $\tau \in \mathbf{L}$, $\tau \subseteq Q \times B$. Then $\tau \in H$. On the other hand, it is easy to check that $x = \{K(b) : b \in A\} = K(\tau)$, that is, you can take $a_x = \tau$. This ends the proof of the lemma. □

In continuation of the proof of Theorem 5(ii), we introduce, following [36], the forcing relation $q \Vdash \varphi$ (where $q \in Q$) by induction on the logical complexity of the formula φ (a closed formula with parameters in H); it corresponds to $H[F]$ as a Q-generic extension of H. Below $\leqslant$ is the partial order on Q, and $q \leqslant q'$ means that q is a stronger condition.

(I) $q \Vdash a \in b$ iff $\exists \langle q', c\rangle \in b\,(q \leqslant q' \wedge q \Vdash a = c)$;

(II) $q \Vdash a \neq b$ iff $\exists \langle q', c\rangle \in b\,(q \leqslant q' \wedge q \Vdash c \notin a)$ or $\exists \langle q', c\rangle \in a\,(q \leqslant q' \wedge q \Vdash c \notin b)$;

(III) $q \Vdash \neg \varphi$ iff $\neg \exists q'\,(q' \leqslant q \wedge q' \Vdash \varphi)$;

(IV) $q \Vdash \varphi \vee \psi$ iff $q \Vdash \varphi$ or $q \Vdash \psi$;

(V) $q \Vdash \exists x \in b\, \varphi(x)$ iff $\exists \langle q', c\rangle \in b\,(q \leqslant q' \wedge q \Vdash \varphi(c))$;

(VI) $q \Vdash \exists x\, \varphi(x)$ iff $\exists c \in H\,(q \Vdash \varphi(c))$.

This definition assumes that some logical connectives are expressed in a certain way via other connectives. For each parameter free formula $\varphi(x_1, \dots, x_k)$, define a set:

$$F_\varphi = \{(q, a_1, \dots, a_k) : a_1, \dots, a_k \in H \wedge q \in Q \wedge q \Vdash \varphi(a_1, \dots, a_k)\}.$$

Lemma 6. *If $k > 1$ and φ is a Σ_k formula, then F_φ is $\Sigma_k^H(\{Q\})$* (Q is allowed as a sole parameter).

Proof. All quantifiers of definitions (I)–(V) are bounded either by the set $Q \in H$, or by a set of the form $Q \times a$, where still $a \in H$. Therefore it is not difficult to show that $F_\varphi \in \Sigma_1^H$ for any bounded formula φ. (The sole unbounded quantifier will express the existence of a full description of all subformulas of the form $a \in b$, $a = b$, that appear in accordance with (I)–(III).) Induction on k proves the result. □

The next lemma is similar to the Truth Lemma as in [36], so the proof is omitted.

Lemma 7. *Let Φ be a closed formula with parameters in H, and Φ' obtained from Φ so that each $a \in H$ is replaced by $K(a)$. Then Φ' is true in $H[F]$ iff there exists $q \in F$ such that $q \Vdash \Phi$.*

Let us finish the proof of Theorem 5(ii). Let $Y \in \Sigma_n(H[F])$, $Y \subseteq H$. There is a parameter free Σ_n formula $\varphi(\cdot,\cdot)$, and a parameter $y \in H[F]$, such that $X = \{x \in H : \varphi(x, y) \text{ holds in } H[F]\}$. For each $x \in H$, we define the set $\check{x} \in H$ by induction, so that $\check{\varnothing} = \varnothing$, and if $x \neq \varnothing$ then $\check{x} = \{\langle q, \check{z}\rangle : q \in Q \wedge z \in x\}$. Then $K(\check{x}) = x$ for all x. It follows by Lemma 7 that:

$$X = \{x \in H : \exists q \in F\,(q \Vdash \varphi(\check{x}, b))\} = t[F],$$

where $b \in H$ is such that $y = K(b)$ (exists by Lemma 5), whereas:

$$t = \{\langle q, x\rangle : q \in Q \wedge x \in H \wedge q \Vdash \varphi(\check{x}, b)\}.$$

Finally, note that the function $x \longmapsto \check{x}$ belongs to $\Delta_1^H(\{Q\})$. We conclude that $t \in \Sigma_n(H)$ by Lemma 6, as required. This completes the proof of Theorem 5. □

2.7. Collapse Forcing

We conclude from Lemma 2 that the construction of any generic extension of $\mathbf{L}$, in which $\mathbf{D}_{1n} \in \mathbf{D}_{2n}$ holds for some $n \geq 1$, has to involve a collapse of $\omega_1^{\mathbf{L}}$ down to ω, explicitly or implicitly. To set up such a collapse in a technically convenient form, we let $\Xi = \mathscr{P}(\omega) \cap \mathbf{L}$ be the set of all constructible sets $x \subseteq \omega$, and let $\mathbb{C} = \Xi^{<\omega}$. Thus $\mathbb{C} \in \mathbf{L}$ is the ordinary Cohen-style collapse forcing that makes Ξ (and $\omega_1^{\mathbf{L}}$ as well) countable in $\mathbb{C}$-generic extensions. The choice of Ξ as the collapse domain, instead of $\omega_1^{\mathbf{L}}$, is made by technical reasons that will be clear below. Note that $\mathbb{C}$ adjoins generic maps $\zeta : \omega \xrightarrow{\text{onto}} \Xi$ to $\mathbf{L}$. A map $\zeta \in \Xi^\omega$ is $\mathbb{C}$-generic over $\mathbf{L}$ iff the set $G_\zeta = \{e \in \mathbb{C} : e \subset \zeta\}$ is $\mathbb{C}$-generic in the usual sense.

Lemma 8 (Routine). *If $\zeta \in \Xi^\omega$ is $\mathbb{C}$-generic over $\mathbf{L}$ then $\omega_\xi^{\mathbf{L}[\zeta]} = \omega_{\xi+1}^{\mathbf{L}}$ for all $\xi \in \mathbf{Ord}$.*

The representation result, as in the beginning of Section 2.6, takes the following form: If $\zeta \in \Xi^\omega$ is $\mathbb{C}$-generic over $\mathbf{L}$, $X \in \mathbf{L}$, and $Y \in \mathbf{L}[\zeta]$, $Y \subseteq X$, then there is a set $t \in \mathbf{L}$, $t \subseteq \mathbb{C} \times X$, such that:

$$Y = t[\zeta] := \{x \in X : \exists e \in G_\zeta \, (\langle e, x\rangle \in t)\} ;$$

such a t is called a $\mathbb{C}$*-name* (for Y).

Theorem 5 is applicable for $Q = \mathbb{C}$ and any $\mathbf{L}$-cardinal $\varkappa \geq \omega_2^{\mathbf{L}}$, whereas if $\xi \in \mathbf{Ord}$, $\xi \geq 1$, then Lemma 4 is applicable for $Q = \mathbb{C}$, $\vartheta = \omega_\xi^{\mathbf{L}}$, $\Omega = \omega_{\xi+1}^{\mathbf{L}}$, and any forcing $P \in \mathbf{L}$, ϑ-complete in $\mathbf{L}$.

3. Almost Disjoint Forcing, Uncountable Version

Here we introduce the main coding tool used in the proof of Theorem 2, an uncountable version of almost disjoint forcing of Jensen–Solovay [23].

3.1. Introduction to almost Disjoint Forcing

Definition 2. *Fix an uncountable successor $\mathbf{L}$-cardinal $\Omega = \omega_{\mu+1}^{\mathbf{L}}$. The value of Ω will be specified in Section 4.5 with respect to the integer $\mathbb{M}$ of Theorem 2, namely, $\Omega = \omega_{\mathbb{M}}^{\mathbf{L}}$, but until then we will view Ω as an arbitrary successor $\mathbf{L}$-cardinal.*

We put $\Omega^\ominus = \omega_\mu^{\mathbf{L}}$ and $\Omega^\oplus = \omega_{\mu+2}^{\mathbf{L}}$. Here $^\oplus$, resp., $^\ominus$ mean the next, resp., previous $\mathbf{L}$-cardinal, which may not be true cardinals in generic extensions of $\mathbf{L}$.

We finally put:

$$\mathbb{H} = (\mathbf{H}\Omega^\oplus)^{\mathbf{L}} = \{x \in \mathbf{L} : \mathtt{card}\,(\mathrm{TC}\,(x)) < \Omega^\oplus \text{ in } \mathbf{L}\}. \tag{5}$$

Moreover if $\mathbf{L}[G]$ is a generic extension of $\mathbf{L}$ then we define:

$$\mathbb{H}[G] = (\mathbf{H}\Omega^\oplus)^{\mathbf{L}[G]} = \{x \in \mathbf{L}[G] : \mathtt{card}\,(\mathrm{TC}\,(x)) < \Omega^\oplus \text{ in } \mathbf{L}[G]\}. \tag{6}$$

provided $\Omega^\oplus$ remains a cardinal in $\mathbf{L}[G]$.

- Let $\mathbf{Seq}_\Omega = (\Omega^{<\Omega} \smallsetminus \{\Lambda\}) \cap \mathbf{L}$, the set of all constructible non-empty sequences s of ordinals $< \Omega$, of length $\mathtt{lh}\, s = \mathtt{dom}\, s < \Omega$, called *strings*. We underline that $\mathbf{Seq}_\Omega \in \mathbf{L}$, and Λ, the empty string, does not belong to $\mathbf{Seq}_\Omega$;
- Let $\mathbf{Fun}_\Omega = \Omega^\Omega \cap \mathbf{L}$ = all constructible Ω-sequences of ordinals $< \Omega$; $\mathbf{Fun}_\Omega \in \mathbf{L}$;
- If $X \subseteq \mathbf{Fun}_\Omega$ then put $X^\vee = \{f \restriction \xi : f \in F_p \wedge 1 \leq \xi < \Omega\}$, a tree in $\mathbf{Seq}_\Omega$, without terminal nodes;
- A set $X \subseteq \mathbf{Fun}_\Omega$ is *dense* iff $X^\vee = \mathbf{Seq}_\Omega$, i.e. for any $s \in \mathbf{Seq}_\Omega$ there is $f \in X$ such that $s \subset f$;
- If $S \subseteq \mathbf{Seq}_\Omega$, $f \in \mathbf{Fun}_\Omega$ then let $S/f = \mathtt{sup}\{\xi < \Omega : f \restriction \xi \in S\}$. If S/f is unbounded in Ω then say that S *covers* f, otherwise S *does not cover* f.

Definition 3 (in $\mathbf{L}$). *$^*\mathcal{P}_\Omega$ is the set of all pairs $p = \langle S_p ; F_p \rangle \in \mathbf{L}$ of sets $F_p \subseteq \mathbf{Fun}_\Omega$, $S_p \subseteq \mathbf{Seq}_\Omega$ of cardinality strictly less than Ω in $\mathbf{L}$. Elements of $^*\mathcal{P}_\Omega$ will be called (forcing) conditions.*

If $p, q \in {}^*\mathcal{P}_\Omega$ then $p \wedge q = \langle S_p \cup S_q ; F_p \cup F_q \rangle$; a condition in $^*\mathcal{P}_\Omega$.

Let $p, q \in {}^*\mathcal{P}_\Omega$. Define $q \leqslant p$ (q is stronger as a forcing condition) iff $S_p \subseteq S_q$, $F_p \subseteq F_q$, and the difference $S_q \smallsetminus S_p$ does not intersect $F_p^\vee$, that is, $S_q \cap F_p^\vee = S_p \cap F_p^\vee$. Here $F_p^\vee = (F_p)^\vee$.

Lemma 9 (in $\mathbf{L}$). *The sets $\mathbf{Seq}_\Omega$, $\mathbf{Fun}_\Omega$, $^*\mathcal{P}_\Omega$ belong to $\mathbf{L}$ and $\mathtt{card}\,(\mathbf{Seq}_\Omega) = \Omega$ while $\mathtt{card}\,(\mathbf{Fun}_\Omega) = \mathtt{card}\,{}^*\mathcal{P}_\Omega = \Omega^\oplus$ in $\mathbf{L}$.*

Clearly $q \leqslant p$ iff $S_p \subseteq S_q$, $F_p \subseteq F_q$, and $S_q \cap F_p^\vee = S_p \cap F_p^\vee$.

Lemma 10 (in $\mathbf{L}$). *Conditions $p, q \in {}^*\mathcal{P}_\Omega$ are compatible in $^*\mathcal{P}_\Omega$ iff* 1) *$S_q \smallsetminus S_p$ does not intersect $F_p^\vee$, and* 2) *$S_p \smallsetminus S_q$ does not intersect $F_q^\vee$. Therefore any $p, q \in P^*$ are compatible in P^* iff $p \wedge q \leqslant p$ and $p \wedge q \leqslant q$.*

Proof. If (1), (2) hold then $p \wedge q \leqslant p$ and $p \wedge q \leqslant q$, thus p, q are compatible. □

If $u \subseteq \mathbf{Fun}_\Omega$ then put $P[u] = \{p \in {}^*\mathcal{P}_\Omega : F_p \subseteq u\}$. Thus if $u \in \mathbf{L}$ then $P[u] \in \mathbf{L}$.

Any conditions $p, q \in P[u]$ are compatible in $P[u]$ iff they are compatible in $^*\mathcal{P}_\Omega$ iff $p \wedge q = \langle S_p \cup S_q ; F_p \cup F_q \rangle \in P[u]$ satisfies both $(p \wedge q) \leqslant p$ and $(p \wedge q) \leqslant q$. Thus we say that conditions $p, q \in {}^*\mathcal{P}_\Omega$ are compatible (or incompatible) without an indication which set $P[u]$ containing p, q is considered.

Lemma 11 (in $\mathbf{L}$). *Let $\varnothing \neq u \subseteq \mathbf{Fun}_\Omega$. Then it is true in $\mathbf{L}$ that $\mathtt{card}\, P[u] \leq \Omega^\oplus$, and the forcing notion $P[u]$ satisfies $\Omega^\oplus$-CC, and is $\Omega^\ominus$-closed, hence $\Omega^\ominus$-distributive. Moreover $P[u]$ satisfies $\Omega^\oplus$-CC in any generic extension $\mathbf{L}[H]$ of $\mathbf{L}$, in which $\Omega^\oplus$ remains a cardinal.*

Proof. The closed/distributive claim is obvious on the base of the cardinality restrictions in Definition 3. To prove the $\Omega^\oplus$-CC claim, argue in $\mathbf{L}[H]$. If $p \neq q$ belong to an antichain $A \subseteq P[u]$ then $S_p \neq S_q$ by Lemma 10. Let $M = \{S_p : p \in {}^*\mathcal{P}_\Omega\}$ = all subsets $S \subseteq \mathbf{Seq}_\Omega$, $S \in \mathbf{L}$, with $\mathtt{card}\, S < \Omega$ in $\mathbf{L}$. Then M is a set of cardinality Ω in $\mathbf{L}$, hence in $\mathbf{L}[H]$ as well. □

If $u \subseteq \mathbf{Fun}_\Omega$ in $\mathbf{L}$, and $G \subseteq P[u]$ is a $P[u]$-generic set, then put $S_G = \bigcup_{p \in G} S_p$; thus $S_G \subseteq \mathbf{Seq}_\Omega$. The next lemma witnesses that forcing notions of the form $P[u]$ belong to the type of *almost disjoint* (AD, for brevity) forcing, invented in [23] (§ 5).

Lemma 12. *Suppose that, in $\mathbf{L}$, $u \subseteq \mathbf{Fun}_\Omega$ is dense. Let $G \subseteq P[u]$ be a set $P[u]$-generic over $\mathbf{L}$. Then:*

(i) *If $f \in \mathbf{Fun}_\Omega$ in $\mathbf{L}$ then $f \in u \iff S_G$ does not cover f;*

(ii) *$G = \{p \in P[u] : S_p \subseteq S_G \wedge (S_G \smallsetminus S_p) \cap F_p^\vee = \varnothing\}$, hence $\mathbf{L}[G] = \mathbf{L}[S_G]$.*

Proof. (i) Let $f \in u$. The set $D_f = \{p \in P[u] : f \in F_p\}$ is dense in $P[u]$. (Let $q \in P[u]$. Define $p \in P[u]$ so that $S_p = S_q$ and $F_p = F_q \cup \{f\}$. Then $p \in D_f$ and $p \leqslant q$.) Therefore $D_f \cap G \neq \varnothing$. Pick any $p \in D_f \cap G$. Then $f \in F_p$. Now every $r \in G$ is compatible with p, and hence $S_r / f \subseteq S_p / f$ by Lemma 10. Thus $S_G / f = S_p / f$ is bounded in Ω. Let $f \notin u$. If $\xi < \Omega$ then the set $D_{f\xi} = \{p \in P[u] : \sup(S_p / f) > \xi\}$ is dense in $P[u]$. (If $q \in P[u]$ then $\mathtt{card}\,(F_q^\vee) < \Omega$. As $f \notin u$, there is $\eta > \xi$, $\eta < \Omega$, with $f \restriction \eta \notin F_q^\vee$. Define p so that $F_p = F_q$ and $S_p = S_q \cup \{f \restriction \eta\}$. Then $p \in D_{f\xi}$ and $p \leqslant q$.) Let $p \in D_{f\xi} \cap G$. Then $\sup(S_G / f) > \xi$. As $\xi < \Omega$ is arbitrary, S_G / f is unbounded.

(ii) Consider any $p \in P[u]$. Suppose $p \in G$. Then $S_p \subseteq S_G$. If there exists $s \in (S_G \smallsetminus S_p) \cap F_p^\vee$ then by definition we have $s \in S_q$ for some $q \in G$. However, then p, q are incompatible by Lemma 10, a contradiction. Now suppose $p \notin G$. Then there exists $q \in G$ incompatible with p. By Lemma 10, there are two cases. First, there exists $s \in (S_q \smallsetminus S_p) \cap F_p^\vee$. Then $s \in S_G \smallsetminus S_p$, so p is not compatible with S_G. Second, there exists $s \in (S_p \smallsetminus S_q) \cap F_q^\vee$. Then any condition $r \leqslant q$ satisfies $s \notin S_r$. Therefore $s \notin S_G$, so $S_p \not\subseteq S_G$, and p is not compatible with S_G. □

3.2. Product Almost Disjoint Forcing

Arguing under the assumptions and notation of Definition 2, we consider $\mathcal{I} = \omega \times \omega$, the cartesian product, as the index set for a product forcing.

Definition 4 (in $\mathbf{L}$). *$^*\mathbf{P}_{\Omega}$ (note the boldface upright form) is the $\mathbf{L}$-product of $\mathcal{I}$ copies of $^*\mathcal{P}_{\Omega}$ (Definition 3 in Section 3.1), ordered componentwise: $p \leqslant q$ (p is stronger) iff $p(n,i) \leqslant q(n,i)$ in $^*\mathcal{P}_{\Omega}$ for all $n,i < \omega$.*

That is, $^*\mathbf{P}_{\Omega} \in \mathbf{L}$ and $^*\mathbf{P}_{\Omega}$ consists of all maps $p \in \mathbf{L}$, $p : \mathcal{I} \to {}^*\mathcal{P}_{\Omega}$. If $p \in {}^*\mathbf{P}_{\Omega}$ then put $\mathbf{F}_p(n,i) = F_{p(n,i)}$ and $\mathbf{S}_p(n,i) = S_{p(n,i)}$ for all $n,i < \omega$, so that $p(n,i) = \langle \mathbf{S}_p(n,i) ; \mathbf{F}_p(n,i) \rangle$, where $\mathbf{S}_p : \mathcal{I} \to \mathscr{P}_{<\Omega}(\mathbf{Seq}_{\Omega})$ and $\mathbf{F}_p : \mathcal{I} \to \mathscr{P}_{<\Omega}(\mathbf{Fun}_{\Omega})$ are arbitrary, and $\mathscr{P}_{<\Omega}$ means all subsets of cardinality $< \Omega$ strictly.

- Note that, unlike product almost-disjoint forcing notions developed in [4,5], $^*\mathbf{P}_{\Omega}$ is **not** a finite-support product;
- If $p \in {}^*\mathbf{P}_{\Omega}$ then we define $|p| = \{\langle n,i \rangle : p(n,i) \neq \langle \varnothing, \varnothing \rangle\}$ and
$$\mathbf{F}_p^{\vee}(n,i) = F_{p(n,i)}^{\vee} = \{f \restriction \xi : f \in \mathbf{F}_p(n,i) \wedge 1 \leq \xi < \Omega\};$$
- If $p,q \in {}^*\mathbf{P}_{\Omega}$ then define $p \wedge q \in {}^*\mathbf{P}_{\Omega}$ by $(p \wedge q)(n,i) = p(n,i) \wedge q(n,i)$, in the sense of Definition 3 in Section 3.1, for all $n,i < \omega$.

Lemma 13. *Conditions $p,q \in {}^*\mathbf{P}_{\Omega}$ are compatible in $^*\mathbf{P}_{\Omega}$ iff $(p \wedge q) \leqslant p$ and $(p \wedge q) \leqslant q$.*

Let an Ω**-system** be any map $U \in \mathbf{L}$, $U : \mathcal{I} \to \mathscr{P}(\mathbf{Fun}_{\Omega})$ such that each set $U(n,i)$ is empty or dense in $\mathbf{Fun}_{\Omega}$. In this case, let $|U| = \{\langle n,i \rangle : U(n,i) \neq \varnothing\}$.

- If U is an Ω-system then $\mathbf{P}[U] = \{p \in {}^*\mathbf{P}_{\Omega} : \forall \langle n,i \rangle \in |p| \, (\mathbf{F}_p(n,i) \subseteq U(n,i))\}$ is the $\mathbf{L}$-product of the sets $P[U(n,i)]$, $n,i < \omega$.

Lemma 14 (in $\mathbf{L}$). *Let U be an Ω-system. Then it is true in $\mathbf{L}$ that $\mathtt{card}\, \mathbf{P}[U] = \Omega^{\oplus}$, and the forcing notion $\mathbf{P}[U]$ is $\Omega^{\ominus}$-closed, hence $\Omega^{\ominus}$-distributive, and satisfies $\Omega^{\oplus}$-CC, and the product $\mathbb{C} \times \mathbf{P}[U]$ satisfies $\Omega^{\oplus}$-CC as well. Moreover $\mathbf{P}[U]$ satisfies $\Omega^{\oplus}$-CC in any generic extension of $\mathbf{L}$ in which $\Omega^{\oplus}$ remains a cardinal.*

Proof. The closed/distributive claims follow from Lemma 11. To prove the antichain claim we observe that if $p,q \in {}^*\mathbf{P}_{\Omega}$ satisfy $\mathbf{S}_p = \mathbf{S}_q$ then p,q are compatible. However the set $\Delta_{\mathbf{S}} = \{\mathbf{S}_p : p \in {}^*\mathbf{P}_{\Omega}\}$ has cardinality $\leq \Omega < \Omega^{\oplus}$ in $\mathbf{L}$ as it consists of all functions $\mathbf{S}_p : \mathcal{I} \to \mathscr{P}_{<\Omega}(\mathbf{Seq}_{\Omega})$. To extend the result to the product $\mathbb{C} \times \mathbf{P}[U]$, note that $\mathtt{card}\, \mathbb{C} = \omega_1^{\mathbf{L}} \leq \Omega$. $\square$

Definition 5. *Suppose that $z \subseteq \mathcal{I}$. If $p \in {}^*\mathbf{P}_{\Omega}$ then define $p' = p \restriction z$ to be the usual restriction, so that $\mathtt{dom}\,(p \restriction z) = z$ and $p'(n,i) = p(n,i)$ for all $\langle n,i \rangle \in z$. A special case: If $n,i < \omega$ then let $p \restriction_{\neq \langle n,i \rangle} = p \restriction z$, where $z = (\mathcal{I} \setminus \{\langle n,i \rangle\})$. If U is an Ω-system then define $U \restriction z$ to be the ordinary restriction as well. Furthermore, if $m < \omega$ then define:*

$$p \restriction^{<m} = p \restriction z \quad \text{and} \quad U \restriction^{<m} = U \restriction z, \quad \text{where} \quad z = \{k : k < m\} \times \omega,$$
$$p \restriction^{\geq m} = p \restriction z \quad \text{and} \quad U \restriction^{\geq m} = U \restriction z, \quad \text{where} \quad z = \{k : k \geq m\} \times \omega,$$
$$p \restriction^{m} = p \restriction z \quad \text{and} \quad U \restriction^{m} = U \restriction z, \quad \text{where} \quad z = \{m\} \times \omega.$$

Finally, if $Q \subseteq {}^*\mathbf{P}_{\Omega}$ then let $Q \restriction z = \{p \restriction z : p \in Q\}$; $Q \restriction z \subseteq {}^*\mathbf{P}_{\Omega} \restriction z$. This will be applied, e.g., in case $Q = \mathbf{P}[U]$, where $U \in \mathbf{L}$ is a Ω-system, and then we get $\mathbf{P}[U] \restriction z = \{p \restriction z : p \in \mathbf{P}[U]\}$, $\mathbf{P}[U] \restriction_{\neq \langle n,i \rangle}$, $\mathbf{P}[U] \restriction^{\geq m}$, etc.

Remark 3. *Suppose that $z \in \mathbf{L}$ in Definition 5. If $p \in {}^*\mathbf{P}_\Omega$, then $p \restriction z$ can be identified with a condition $q \in {}^*\mathbf{P}_\Omega$ such that $q \restriction z = p \restriction z$ and $q(n,i) = \langle \varnothing ; \varnothing \rangle$ for all $\langle n,i \rangle \in \mathcal{I} \setminus z$. For instance, this applies w.r.t. $p \restriction_{\neq \langle n,i \rangle}$, $p \restriction^{\geq m}$, $p \restriction^{<m}$, $p \restriction^{m}$.*

With such an identification, we have ${}^*\mathbf{P}_\Omega \restriction z \subseteq {}^*\mathbf{P}_\Omega$, and $Q \restriction z \subseteq {}^*\mathbf{P}_\Omega$ for $Q \subseteq {}^*\mathbf{P}_\Omega$ (in case $z \in \mathbf{L}$).

However, if $z \notin \mathbf{L}$ then such an identification fails. This is a consequence of our deviation from the finite-support product approach taken in [4,5], which would not work in the setting of this paper.

The same applies for the restrictions $U \restriction z$ of Ω-systems U.

3.3. Structure of Product almost Disjoint Generic Extensions

Arguing under the assumptions and notation of Definition 2, we let U be an Ω-system in $\mathbf{L}$. Consider $\mathbf{P}[U]$ as a forcing notion. We will study $\mathbf{P}[U]$-generic extensions $\mathbf{L}[G]$ of the ground universe $\mathbf{L}$. Define some elements of these extensions. Suppose that $G \subseteq \mathbf{P}[U]$ is a generic set. Let,

$$\mathbf{S}_G(n,i) = S_{G(n,i)} = \bigcup_{p \in G} \mathbf{S}_p(n,i) \quad \text{for any } n,i < \omega,$$

where $G(n,i) = \{p(n,i) : p \in G\} \subseteq P[U(n,i)]$; thus $\mathbf{S}_G(n,i) \subseteq \mathbf{Seq}_\Omega$ and $G \subseteq \mathbf{P}[U]$ splits into the family of sets $G(n,i)$, $n,i < \omega$. This defines a sequence $\vec{S}_G = \{\mathbf{S}_G(n,i)\}_{n,i<\omega}$ of subsets of $\mathbf{Seq}_\Omega$.

If $z \subseteq \mathcal{I}$ then let $G \restriction z = \{p \restriction z : p \in G\}$. If $z \in \mathbf{L}$ then $G \restriction z$ can be identified with $\{p \in G : |p| \subseteq z\}$.

Put $G \restriction_{\neq \langle n,i \rangle} = \{p \in G : \langle n,i \rangle \notin |p|\} = G \restriction (\mathcal{I} \setminus \{\langle n,i \rangle\})$.

Lemma 15. *Let U be an Ω-system in $\mathbf{L}$, and $G \subseteq \mathbf{P}[U]$ be a set $\mathbf{P}[U]$-generic over $\mathbf{L}$. Then:*

(i) $\mathbf{L}[G] = \mathbf{L}[\vec{S}_G]$;

(ii) *If $n,i < \omega$ then the set $G(n,i) = \{p(n,i) : p \in G\} \in \mathbf{L}[G]$ is $P[U(n,i)]$-generic over $\mathbf{L}$, hence if $f \in \mathbf{Fun}_\Omega$ then $f \in U(n,i) \iff \mathbf{S}_G(n,i)$ does not cover f ;*

(iii) *If $X \in \mathbf{L}[G]$, $X \subseteq \Omega$ is bounded, then $X \in \mathbf{L}$;*

(iv) *All $\mathbf{L}$-cardinals are preserved in $\mathbf{L}[G]$, and* GCH *holds in $\mathbf{L}[G]$.*

Proof. To prove (i) apply Lemma 12(ii).

The genericity in (ii) holds by the product forcing theorem, then use Lemma 12(i).

Claim (iii) follows from the $\Omega^\ominus$-closure claim of Lemma 14.

(iv) We conclude from (iii) that all $\mathbf{L}$-cardinals $\leq \Omega$ remain cardinals in $\mathbf{L}[G]$, and GCH holds for all $\mathbf{L}$-cardinals $< \Omega$ strictly. It follows from the $\Omega^\oplus$-CC claim of Lemma 14 that all $\mathbf{L}$-cardinals $\geq \Omega^\oplus$ remain cardinals in $\mathbf{L}[G]$, and since $\mathtt{card}\, \mathbf{P}[U] \leq \Omega^\oplus$ in $\mathbf{L}$, GCH holds for all of them in $\mathbf{L}[G]$. And finally we still have $\exp(\Omega) = \Omega^\oplus$ in $\mathbf{L}[G]$ since by (i) the model $\mathbf{L}[G]$ is an extension of $\mathbf{L}$ by adjoining a subset of Ω obtained by a suitable wrapping of $\vec{S}_G$. □

The next lemma is useful in dealing with combined $(\mathbb{C} \times \mathbf{P}[U])$-generic extensions $\mathbf{L}[\zeta, G]$ of $\mathbf{L}$, where, by the product forcing theorem, $\zeta \in \Xi^\omega$ is $\mathbb{C}$-generic over $\mathbf{L}$ and G is $\mathbf{P}[U]$-generic over $\mathbf{L}[\zeta]$, or equivalently, G is $\mathbf{P}[U]$-generic over $\mathbf{L}$ and ζ is $\mathbb{C}$-generic over $\mathbf{L}[G]$.

Lemma 16. *Let U be an Ω-system in $\mathbf{L}$, and a pair $\langle \zeta, G \rangle$ is $(\mathbb{C} \times \mathbf{P}[U])$-generic over $\mathbf{L}$. Then:*

(i) *All $\mathbf{L}[\zeta]$-cardinals are preserved in $\mathbf{L}[\zeta, G]$, so that $\omega_\xi^{\mathbf{L}[\zeta,G]} = \omega_\xi^{\mathbf{L}[\zeta]} = \omega_{\xi+1}^{\mathbf{L}}$ for all $\xi \geq 1$;*

(ii) GCH *holds in $\mathbf{L}[\zeta, G]$;*

(iii) *If $\Omega \geq \omega_2^{\mathbf{L}}$ and $X \in \mathbf{L}[\zeta, G]$, $X \subseteq \Omega$ is bounded, then $X \in \mathbf{L}[\zeta]$;*

(iv) *If $1 \leq k < \omega$ and $\omega_k^{\mathbf{L}} < \Omega$, then $(\mathbf{H}\Omega)^{\mathbf{L}[\zeta,G]} = (\mathbf{H}\Omega)^{\mathbf{L}[\zeta]}$ and $\mathscr{P}^k(\omega) \cap \mathbf{L}[\zeta, G] = \mathscr{P}^k(\omega) \cap \mathbf{L}[\zeta]$.*

Note that Claims (iii), (iv) are not applicable in case $\Omega = \omega_1^{\mathbf{L}}$.

Proof. To prove (i), (ii) recall that all $\mathbf{L}$-cardinals remain cardinals in $\mathbf{L}[G]$, and GCH holds in $\mathbf{L}[G]$, by Lemma 15(iv). It remains to note that ζ is $\mathbb{C}$-generic over $\mathbf{L}[G]$ and make use of Lemma 8. To prove (iii) apply Lemma 4 with $\vartheta = \Omega^{\ominus}$, $P = \mathbf{P}[U]$, $Q = \mathbb{C}$. Note that $\mathtt{card}\,\mathbb{C} = \omega_1^{\mathbf{L}} \leq \Omega^{\ominus}$ in case $\Omega \geq \omega_2^{\mathbf{L}}$.

Finally Claim (iv) is a routine corollary of (i)–(iii). □

4. The Forcing Notion and the Model

In this Section, we prove Theorem 2 on the base of another result, Theorem 8, see Remark 4 on page 23. The proof of Theorem 8 will follow in the remainder of the paper. The structure of the extension will be presented in Section 4.6, after the definition of the forcing notion involved in Section 4.5. Recall that the $\mathbf{L}$-cardinals:

$$\Omega^{\ominus} = \omega_{\mu}^{\mathbf{L}} \; < \; \Omega = \omega_{\mu+1}^{\mathbf{L}} \; < \; \Omega^{\oplus} = \omega_{\mu+2}^{\mathbf{L}}$$

were introduced by Definition 2 on page 13. They remain to be fixed until Section 4.5, where their value will be specified in terms of the number $\mathbb{M} \geq 1$ we are dealing with in Theorem 2.

4.1. Systems, Definability Aspects

We argue in $\mathbf{L}$ under the assumptions and notation of Definition 2 on page 13.

In continuation of our notation related to Ω-systems in Section 3.2, define the following.

- An Ω-system U is *small*, if each $U(n,i)$ has cardinality $\leq \Omega$ in $\mathbf{L}$;
- An Ω-system U is *disjoint* if $U(n,i) \cap U(k,j) = \varnothing$ whenever $\langle n,i \rangle \neq \langle k,j \rangle$;
- If U, V are Ω-systems and $U(n,i) \subseteq V(n,i)$ for all n, i, then V *extends* U, in symbol $U \preccurlyeq V$;
- If $\{U_\xi\}_{\xi<\lambda}$ is a sequence of Ω-systems then the limit Ω-system $U = \bigvee_{\xi<\lambda} U_\xi$ is defined by $U(n,i) = \bigcup_{\xi<\lambda} U_\xi(n,i)$, for all n, i.

Let $\mathbf{DS}_{\Omega}$ (disjoint systems) be the set of all disjoint Ω-systems, and let $\mathbf{sDS}_{\Omega}$ (*small* disjoint systems) be the set of all small disjoint Ω-systems $U \in \mathbf{DS}_{\Omega}$.

Define $\mathbf{sDS}_{\Omega}\restriction^{\geq m} = \{U\restriction^{\geq m} : U \in \mathbf{sDS}_{\Omega}\}$, and similarly $\mathbf{sDS}_{\Omega}\restriction^{<m}$ etc. by Definition 5.

The sets $\mathbf{DS}_{\Omega}$, $\mathbf{sDS}_{\Omega}$, $\mathbf{sDS}_{\Omega}\restriction^{\geq m}$, $\mathbf{DS}_{\Omega}\restriction^{<m}$ etc., and the order relation $\preccurlyeq$, **belong to** $\mathbf{L}$, of course. Recall that, by (5),

$$\mathbb{H} = (\mathbf{H}\Omega^{\oplus})^{\mathbf{L}} = \{x \in \mathbf{L} : \mathtt{card}\,(\mathrm{TC}\,(x)) \leq \Omega \text{ in } \mathbf{L}\}.$$

Lemma 17 (in $\mathbf{L}$). *The following sets belong to $\Delta_1^{\mathbb{H}}(\{\Omega\})$ and to $\Delta_3^{\mathbb{H}}$: $\{\Omega\}$, $\{\mathbf{Seq}_{\Omega}\}$, $\mathbf{Fun}_{\Omega}$, $^*\mathbf{P}_{\Omega}$, $\mathbf{sDS}_{\Omega}$, $\mathbf{sDS}_{\Omega}\restriction^{\geq m}$, $\mathbf{sDS}_{\Omega}\restriction^{<m}$, the set $\{\langle U, p\rangle : U \in \mathbf{sDS}_{\Omega} \wedge p \in \mathbf{P}[U]\}$, the relation $\preccurlyeq$.*

Proof. All these sets have rather straightforward $\Delta_1^{\mathbb{H}}(\{\Omega\})$ definitions, with $\Omega \in \mathbb{H}$ as the only parameter. To eliminate Ω, it suffices to prove that $\{\Omega\} \in \Delta_3^{\mathbb{H}}$. Note first of all that "ϑ is a cardinal (initial ordinal)" is a Π_1 formula:

$$\vartheta \in \mathbf{Ord} \wedge \forall \alpha < \vartheta \, \forall f \, (f : \alpha \to \vartheta \implies \mathtt{ran}\, f \neq \vartheta).$$

On the other hand, Ω is the largest cardinal in $\mathbb{H}$, hence it holds in $\mathbb{H}$ that:

$$\vartheta = \Omega \iff \forall \varkappa \, (\varkappa \text{ is a cardinal} \implies \varkappa \leq \Omega).$$

We conclude that $\{\Omega\} \in \Pi_2^{\mathbb{H}} \subseteq \Delta_3^{\mathbb{H}}$. Finally, the conversion $\Delta_1^{\mathbb{H}}(\{\Omega\}) \to \Delta_3^{\mathbb{H}}$ is routine. □

4.2. Complete Sequences

We prove a major theorem (Theorem 6) in this Subsection. It deals with $\preccurlyeq$-increasing transfinite sequences in $\mathbf{sDS}_{\Omega}$, satisfying some genericity/definability requirements. This is similar to some constructions in [4] and especially in [5] (Theorem 3). Yet there is a principal difference. Here the notion of extension $\preccurlyeq$ is just the componentwise set theoretic extension, unlike [4,5], and originally [23], where the extension method was designed so that increments had to be finitewise Cohen-style generic over associated transitive models of a certain fragment of **ZFC**. Here the only restriction is that extensions have to obey the disjointness condition as defined in Section 4.1. In other words, if $U \preccurlyeq V$ are Ω-systems in $\mathbf{sDS}_{\Omega}$, then, beside $U(n,i) \subseteq V(n,i)$, the increments $\Delta(n,i) = V(n,i) \smallsetminus U(n,i)$ have to be pairwise disjoint and each $\Delta(n,i)$ to be disjoint with the union $\bigcup_{\langle k,j\rangle \neq \langle n,i\rangle} U(k,j)$.

Such a simplification is made possible here largely because the definability classes of the form $\mathbf{D}_{1m}$ depend only on the highest quantifier order and do not depend on the number and type of the quantifiers involved in the definition of the set considered—unlike e.g., [5], where we dealt with the definability classes Δ^1_n, which obviously depend on the number of the quantifiers involved.

We begin with an auxiliary lemma.

Recall that, by (5), $\mathbb{H} = (\mathbf{H}\Omega^{\oplus})^{\mathbf{L}} = \{x \in \mathbf{L} : \mathtt{card}\,(\mathrm{TC}\,(x)) \leq \Omega \text{ in } \mathbf{L}\} = \mathbf{L}_{\Omega^{\oplus}}$.

Lemma 18 (in $\mathbf{L}$). *Under the assumptions and notation of Definition 2, for any $\alpha < \Omega^{\oplus}$ there exist $m_\alpha < \omega$, $t_\alpha \in \mathbb{H}$, and $U^\alpha \in \mathbf{sDS}_{\Omega}$ such that the sequences $\{m_\alpha\}_{\alpha<\Omega^{\oplus}}$, $\{t_\alpha\}_{\alpha<\Omega^{\oplus}}$, $\{U^\alpha\}_{\alpha<\Omega^{\oplus}}$ belong to $\Delta^{\mathbb{H}}_3$ and, if $m < \omega$, $t \in \mathbb{H}$, and $\{U_\alpha\}_{\alpha<\Omega^{\oplus}}$ is a $\preccurlyeq$-increasing continuous sequence of Ω-systems in $\mathbf{sDS}_{\Omega}$, then any closed unbounded set $C \subseteq \Omega^{\oplus}$ contains an ordinal $\alpha \in C$ such that $m = m_\alpha$, $t = t_\alpha$, $U_\alpha = U^\alpha$.*

Proof. We argue in $\mathbf{L}$, that is, under the assumption of $\mathbf{V} = \mathbf{L}$, the axiom of constructibility. It is known that the diamond principle $\diamondsuit_{\varkappa}$ holds in $\mathbf{L}$ for any regular cardinal $\varkappa$, in particular, for $\varkappa = \Omega^{\oplus}$, see, e.g., Theorem 13.21 and page 442 in [32]. The principle $\diamondsuit_{\varkappa}$ asserts that there is a sequence $\{S_\alpha\}_{\alpha<\Omega^{\oplus}} \in \mathbf{L}$ of sets $S_\alpha \subseteq \alpha$, of definability class $\Delta^{\mathbb{H}}_1$, and such that:

($*$) If $X \subseteq \Omega^{\oplus}$ and $C \subseteq \Omega^{\oplus}$ is a closed unbounded set then there is $\alpha \in C$ such that $X \cap \alpha = S_\alpha$.

Let $h : \Omega^{\oplus} \xrightarrow{\text{onto}} \mathbb{H}$ be any $\Delta^{\mathbb{H}}_1$ bijection. Put $Y_\alpha = \{h(\xi) : \xi \in S_\alpha\}$. Clearly $\{Y_\alpha\}_{\alpha<\Omega^{\oplus}}$ is still a $\Delta^{\mathbb{H}}_1$ sequence. Moreover the following is true:

($\dagger$) If $\{B_\alpha\}_{\alpha<\Omega^{\oplus}}$ is a sequence of sets in $\mathbb{H}$ and $C \subseteq \Omega^{\oplus}$ is a closed unbounded set then there is $\alpha \in C$ with $\{B_\xi\}_{\xi<\alpha} = Y_\alpha$.

Using the sets Y_α, we accomplish the proof of the lemma as follows. Assume that $\alpha < \Omega^{\oplus}$. If Y_α is a sequence of the form $\{y_\xi\}_{\xi<\alpha}$, such that each y_ξ is a triple $\langle m, t, U^\alpha_\xi\rangle$, where both $m \in \omega$ and $t \in \mathbb{H}$ do not depend on ξ whereas $U^\alpha_\xi \in \mathbf{sDS}_{\Omega}$ for each ξ and $\{U^\alpha_\xi\}_{\xi<\alpha}$ is a $\preccurlyeq$-increasing and continuous sequence, then put $m_\alpha = m$, $t_\alpha = t$, and $U^\alpha = \bigvee_{\xi<\alpha} U^\alpha_\xi$. Otherwise put $m_\alpha = t_\alpha = 0$ and let U^α be the null Ω-system, that is, $U^\alpha(n,i) = \varnothing$ for all n, i. It follows from ($\dagger$) (plus a routine analysis of definability based on Lemma 17) that this construction leads to the result required. □

Theorem 6 (in $\mathbf{L}$). *Under the assumptions and notation of Definition 2, there is a $\preccurlyeq$-increasing sequence $\{\mathbb{U}_\alpha\}_{\alpha<\Omega^{\oplus}}$ of Ω-systems in $\mathbf{sDS}_{\Omega}$, such that:*

(i) *The sequence is continuous, so that $\mathbb{U}_\lambda = \bigvee_{\alpha<\lambda} \mathbb{U}_\alpha$ for all limit ordinals $\lambda < \Omega^{\oplus}$;*

(ii) *If $n < \omega$ then the "slice" $\{\mathbb{U}_\alpha \restriction^n\}_{\alpha<\Omega^{\oplus}}$ is $\Delta^{\mathbb{H}}_{n+4}$;*

(iii) *If $m < \omega$ then the "tail" $\{\mathbb{U}_\alpha \restriction^{\geq m}\}_{\alpha<\Omega^{\oplus}}$ is $(m+3)$-complete, in the sense that for any $\Sigma_{m+3}(\mathbb{H})$ set $D \subseteq \mathbf{sDS}_{\Omega} \restriction^{\geq m}$ there is $\xi < \Omega^{\oplus}$ such that the Ω-system $\mathbb{U}_\xi \restriction^{\geq m}$ m-**solves** D, i.e.,*

- *either $\mathbb{U}_\xi \restriction^{\geq m} \in D$;*
- *or there is no Ω-system $U \in D$ with $\mathbb{U}_\xi \restriction^{\geq m} \preccurlyeq U$;*

(iv) *There is a recursive sequence of parameter free $\in$-formulas $\chi_n(\alpha, x)$ such that if $\alpha < \Omega^{\oplus}$ and $x \in \mathbb{H}$ then $\mathbb{H} \models \chi_n(\alpha, x)$ iff $x = \mathbb{U}_\alpha \restriction^n$.*

Here the "slice" $U\restriction^n$ of a system U is essentially equal to the "column" $\{U(n,i)\}_{i<\omega}$ of the whole "matrix" $U = \{U(n,i)\}_{n,i<\omega}$, while the "tail" $U\restriction^{\geq m}$ can be viewed in the union of all columns to the right of m inclusively, see Definition 5.

Proof. We argue in $\mathbf{L}$. One of the difficulties here is that we have to account for different levels of genericity and completeness for different slices of the construction. To cope with this issue, we make use of Lemma 18. Let us fix the sequences of terms m_α, t_α, U^α such as in the lemma.

Let $<_{\mathbf{L}}$ be Gödel's wellordering of $\mathbf{L}$, as in Section 2.2.

For any $m < \omega$, let $\Theta_m \subseteq \mathbb{H} \times \mathbb{H}$ be a fixed universal $\Sigma^{\mathbb{H}}_{m+3}$ set, that is, Θ_m itself is $\Sigma^{\mathbb{H}}_{m+3}$, and if $X \subseteq \mathbb{H}$ is $\Sigma_{m+3}(\mathbb{H})$ (parameters in $\mathbb{H}$ allowed), then there is $t \in \mathbb{H}$ such that $X = \{x : \langle t, x\rangle \in \Theta_m\}$. If $m < \omega$ and $\alpha < \Omega^{\oplus}$, then let $U^{m\alpha}$ be the $<_{\mathbf{L}}$-least Ω-system in $\mathbf{sDS}_\Omega$ satisfying $U^m \preccurlyeq U^{m\alpha}$ and:

(a) $U^{m\alpha}\restriction^{<m} = U^\alpha\restriction^{<m}$, and

(b) The Ω-system $U^{m\alpha}\restriction^{\geq m}$ m-solves the set $D_\alpha = \{V \in \mathbf{sDS}_\Omega : \langle t_\alpha, V\rangle \in \Theta_m\}$.

Making use of 5° of Section 2.2, we conclude that the sequence $\{U^{m\alpha}\}_{\alpha<\Omega^\oplus}$ is $\Delta^{\mathbb{H}}_{m+4}$.

Now we define a sequence of Ω-systems $\mathbb{U}_\xi$, as required by Theorem 6, by induction.

Put $\mathbb{U}_0(n,i) = \varnothing$ for all n, i.

If $\lambda < \Omega^\oplus$ is the limit then by (i) define $\mathbb{U}_\lambda = \bigvee_{\alpha<\lambda} \mathbb{U}_\alpha$.

Suppose that a Ω-system $\mathbb{U}_\alpha$ is defined, and the goal is to define the next one $\mathbb{U}_{\alpha+1}$. Fix n, i and define the components $\mathbb{U}_{\alpha+1}(n,i)$. Note that this definition will depend on the components $\mathbb{U}_\alpha(n,i)$ (with the same n, i) only, but not on the Ω-system $\mathbb{U}_\alpha$ as a whole.

If it is true that:

$$m_\alpha \leq n \quad \text{and} \quad \mathbb{U}_\alpha(n,i) = U^\alpha(n,i) \tag{7}$$

(where U^α is the Ω-system given by Lemma 18), then put $m = m_\alpha$ and $\mathbb{U}_{\alpha+1}(n,i) = U^{m\alpha}(n,i)$. Otherwise, i.e., if (7) fails, just keep it with $\mathbb{U}_{\alpha+1}(n,i) = \mathbb{U}_\alpha(n,i)$.

We assert that this inductive construction of Ω-systems $\mathbb{U}_\alpha$ leads to Theorem 6.

Requirement (i) of the theorem is satisfied by construction.

The definability requirement (ii) of the theorem is subject to routine verification on the base of Lemma 17, which we leave to the reader.

To prove (iii), fix a number m and a $\Sigma_{m+3}(\mathbb{H})$ set $D \subseteq \mathbf{sDS}_\Omega\restriction^{\geq m}$. We have to find an index $\xi < \Omega^\oplus$ such that the Ω-system $\mathbb{U}_\xi\restriction^{\geq m}$ m-solves D. There is an element $t \in \mathbb{H}$ satisfying:

$$D = \{V \in \mathbf{sDS}_\Omega\restriction^{\geq m} : \langle t, V\rangle \in \Theta_m\},$$

where Θ_m is the universal set as above. Pick, by Lemma 18, an ordinal $\alpha < \Omega^\oplus$ satisfying $m = m_\alpha$, $t = t_\alpha$, $\mathbb{U}_\alpha = U^\alpha$. Then (7) holds for all $n \geq m$, and hence by definition we have $\mathbb{U}_{\alpha+1}\restriction^{\geq m} = U^{m\alpha}\restriction^{\geq m}$. Therefore the Ω-system $\mathbb{U}_{\alpha+1}\restriction^{\geq m}$ m-solves the set D by (b), as required.

(iv) Coming back to the choice of universal sets Θ_m in (b), it can be w.l.o.g. assumed that there is a recursive sequence of parameter free $\in$-formulas $\vartheta_n(t,x)$ such that each ϑ_n is a Σ_{n+3} formula and $\Theta_m = \{\langle t, x\rangle \in \mathbb{H} : \mathbb{H} \models \vartheta_n(t,x)\}$. This routinely leads to $\in$-formulas $\chi_n(\alpha, x)$ required. It can be observed that in fact each χ_n is a Σ_{n+4} formula (not important and will not be used).

This completes the proof of Theorem 6. □

4.3. Preservation of the Completeness

The next lemma says that the completeness property (iii) of Theorem 6, of the sequence $\{\mathbb{U}_\xi\}_{\xi<\Omega^\oplus}$, still holds, to some extent, in rather mild generic extensions of $\mathbf{L}$.

Lemma 19. *Under the assumptions and notation of Definition 2, suppose that $\{\mathbb{U}_\alpha\}_{\alpha<\Omega^\oplus} \in \mathbf{L}$ is a $\preccurlyeq$-increasing sequence of Ω-systems in $\mathbf{sDS}_\Omega$ satisfying (i)–(iv) of Theorem 6.*

Let $Q \in \mathbf{L}$ be a forcing notion with $\mathtt{card}\, Q \leq \Omega$ in $\mathbf{L}$, e.g., $Q = \mathbb{C}$. Let $F \subseteq Q$ be a set Q-generic over $\mathbf{L}$.

Assume that $m < \omega$, $\delta < \Omega^{\oplus}$, and a set $D \in \mathbf{L}[F]$, $D \subseteq \mathbf{sDS}_{\Omega}\restriction^{\geq m}$, belongs to $\Sigma_{m+3}(\mathbb{H}[F])$, and is open in $\mathbf{sDS}_{\Omega}\restriction^{\geq m}$ so that any extension of a Ω-system $U \in D$ in $\mathbf{sDS}_{\Omega}\restriction^{\geq m}$ belongs to D itself.

*Then there is an ordinal α, $\delta \leq \alpha < \Omega^{\oplus}$, such that $\mathbb{U}_{\alpha}\restriction^{\geq m}$ **m-solves** D, as in Theorem 6(iii).*

We recall that $\mathbb{H} = (\mathbf{H}\Omega^{\oplus})^{\mathbf{L}}$ and $\mathbb{H}[F] = (\mathbf{H}\Omega^{\oplus})^{\mathbf{L}[F]}$ by (5), (6).

Proof. As obviously $\mathbf{sDS}_{\Omega}\restriction^{\geq m} \subseteq \mathbb{H}$, we conclude by Theorem 5(ii) that there is a $\Sigma_{m+3}(\mathbb{H})$ name $t \in \mathbf{L}$, $t \subseteq Q \times \mathbb{H}$, such that $D = t[F]$.

We argue in $\mathbf{L}$. If $q \in Q$, $U \in \mathbf{sDS}_{\Omega}\restriction^{\geq m}$, and there is such a condition $h \in Q$ that $h \leqslant q$ (meaning h is stronger) and $\langle h, U\rangle \in t$, then write $A(q, U)$. If $b \in Q$ then we define:

$$D(b) = \{U \in \mathbf{sDS}_{\Omega}\restriction^{\geq m} : \exists q \in Q(q \leqslant b \wedge A(q, U))\}.$$

Each of the sets $D(b) \subseteq \mathbb{H}$ belongs to $\Sigma_{m+3}(\mathbb{H})$ by virtue of Lemma 17 and the choice of t. Therefore, by the choice of the sequence of Ω-systems, for every $b \in Q$ there is an ordinal $\alpha(b)$, $\delta < \alpha(b) < \Omega^{\oplus}$, such that the Ω-system $\mathbb{U}_{\alpha(b)}\restriction^{\geq m}$ m-solves the set $D(b)$.

Note that $\delta = \sup_{b \in Q} \alpha(b) < \Omega^{\oplus}$ by the cardinality argument.

We claim that the Ω-system $\mathbb{U}_{\delta}\restriction^{\geq m}$ m-solves D. It suffices to prove that if a Ω-system $U \in D$ extends $\mathbb{U}_{\delta}\restriction^{\geq m}$, then the Ω-system $\mathbb{U}_{\delta}\restriction^{\geq m}$ itself belongs to D. Moreover, as D is open, it suffices to find $b \in Q$, satisfying $\mathbb{U}_{\alpha(b)}\restriction^{\geq m} \in D$.

We argue in $\mathbf{L}$. Consider the set $B = \{b \in Q : \mathbb{U}_{\alpha(b)}\restriction^{\geq m} \in D(b)\}$. If $b \in B$ then pick a particular $q = q(b) \in Q$ such that $q \leqslant b$ and $A(q, \mathbb{U}_{\alpha(b)}\restriction^{\geq m})$ holds. If $b \in Q \smallsetminus B$ then put $q(b) = b$. The set $Q' = \{q(b) : b \in Q\}$ is dense in Q. It follows that there is $b \in Q' \cap F$. On the other hand, as $U \in D$, there is a condition $h \in Q$ with $\langle h, U\rangle \in t$.

Then there exists some $q \in F$ satisfying $q \leqslant h$ and $q \leqslant h(b) \leqslant b$. This implies $U \in D(b)$. It follows, by the choice of $\alpha(b)$, that $\mathbb{U}_{\alpha(b)}\restriction^{\geq m} \in D(b)$, too. However then $b \in B$, and hence we have $A(q(b), \mathbb{U}_{\alpha(b)}\restriction^{\geq m})$. By definition there is a condition $h' \in Q$ with $q(b) \leqslant h'$, such that $\langle h', \mathbb{U}_{\alpha(b)}\restriction^{\geq m}\rangle \in t$. However $h' \in F$ (since $f(b) \in F$). We conclude that $\mathbb{U}_{\delta}\restriction^{\geq m} \in D$, as required. □

4.4. Key Definability Engine

We argue under the assumptions and notation of Definition 2 on page 13. In particular, a successor $\mathbf{L}$-cardinal $\Omega > \omega$ is fixed. We make the following arrangements.

Definition 6 (in $\mathbf{L}$). *We fix a $\preccurlyeq$-increasing sequence of Ω-systems $\{\mathbb{U}^{\Omega}_{\xi}\}_{\xi<\Omega^{\oplus}}$ satisfying conditions (i)–(iv) of Theorem 6 for the particular $\mathbf{L}$-cardinal Ω introduced by Definition 2.*

We define the limit Ω-system $\mathbb{U}^{\Omega} = \bigvee_{\xi<\Omega^{\oplus}} \mathbb{U}^{\Omega}_{\xi}$, the basic forcing notion $\mathbb{P}^{\Omega} = \mathbf{P}[\mathbb{U}^{\Omega}]$, and the subforcings $\mathbb{P}^{\Omega}_{\gamma} = \mathbf{P}[\mathbb{U}^{\Omega}_{\gamma}]$, $\gamma < \Omega^{\oplus}$.

Define restrictions $\mathbb{P}^{\Omega}\restriction z$, $G\restriction z$ $(z \subseteq \mathcal{I}, G \subseteq \mathbb{P}^{\Omega})$, $\mathbb{P}^{\Omega}\restriction_{\neq\langle n,i\rangle}$ etc. as in Section 3.2.

Thus by construction $\mathbb{P}^{\Omega} \in \mathbf{L}$ is the $\mathbf{L}$-product of sets $\mathbb{P}^{\Omega}(n,i) = P[\mathbb{U}^{\Omega}(n,i)]$, $n, i \in \omega$. Lemma 14 implies some cardinal characterictics of $\mathbb{P}^{\Omega}$, namely:

(I) $\mathtt{card}\,\mathbb{P}^{\Omega} = \Omega^{\oplus}$ in $\mathbf{L}$,

(II) $\mathbb{P}^{\Omega}$ satisfies $\Omega^{\oplus}$-CC in $\mathbf{L}$,

(III) $\mathbb{P}^{\Omega}$ is $\Omega^{\ominus}$-closed and $\Omega^{\ominus}$-distributive in $\mathbf{L}$.

Corollary 2. *$\mathbb{P}^{\Omega}$ does not adjoin new reals to $\mathbf{L}$.*

Proof. The result follows from (III) because $\Omega^{\ominus} \geq \omega$ by Definition 2. □

As for definability, the set $\mathbb{U}^{\Omega}$ is not parameter free definable in $\mathbb{H} = (\mathbf{H}\Omega^{\oplus})^{\mathbf{L}}$, yet its slices are:

Lemma 20 (in **L**). *Let $n < \omega$. Then the set $\mathbb{U}^{\Omega}\restriction^{n} = \{\langle i, f\rangle : f \in \mathbb{U}^{\Omega}(n,i)\}$ belongs to $\Sigma^{\mathbb{H}}_{n+4}$. In addition there is a recursive sequence of parameter free $\in$ formulas $u_n(i,f)$ such that, for any $n < \omega$, if $i < \omega$ and $f \in \mathbf{Fun}_{\Omega}$ then $f \in \mathbb{U}^{\Omega}(n,i)$ iff $\mathbb{H} \models u_n(i,f)$.*

Proof. To prove the first claim, apply (ii) of Theorem 6. To prove the additional claim define:

$$u_n(i,f) \;:=\; \exists\,\alpha\,\exists\,x\,\big(\chi_n(\alpha,x) \wedge f \in x(n,i)\big),$$

where χ_n are formulas given by (iv) of Theorem 6. □

We further let formulas Γ^{Ω}_{ni} $(n,i \in \omega)$ be defined as follows:

$$\Gamma^{\Omega}_{ni}(S) :=_{\text{def}} S \subseteq \mathbf{Seq}_{\Omega} \wedge \forall f \in \mathbf{Fun}_{\Omega}\ \big(f \in \mathbb{U}^{\Omega}(n,i) \iff S \text{ does not cover } f\big).$$

The next theorem shows that any real in **L** and even in some generic extensions of **L** can be made parameter free definable in appropriate subextensions of $\mathbb{P}^{\Omega}$-generic extensions, basically by means of the formulas $\Gamma^{\Omega}_{ni}(S)$. We prove this result in a rather general form, which includes the case of a forcing notion $Q = \mathbb{C}$, actually used in this paper, as just a particular case. The proof of the particular case $Q = \mathbb{C}$ would not be any simpler though.

Theorem 7. *Assume that $Q \in \mathbf{L}$ is a forcing notion, $\mathtt{card}\,Q \le \Omega$ in $\mathbf{L}$, a pair $\langle W, G\rangle$ is $(Q \times \mathbb{P}^{\Omega})$-generic over $\mathbf{L}$, $Y \in \mathbf{L}[W]$, and $z \in \mathbf{L}[Y]$, $z \subseteq \mathcal{I} = \omega \times \omega$. Then,*

- (i) *$\Omega^{\oplus}$ is a cardinal in $\mathbf{L}[Y,G]$;*
- (ii) *If $\langle n,i\rangle \in z$ then $S_G(n,i) \in \mathbf{L}[G\restriction z]$ and $\Gamma^{\Omega}_{ni}(S_G(n,i))$ holds, but*
- (iii) *If $\langle n,i\rangle \in \mathcal{I} \smallsetminus z$ then $S_G(n,i) \notin \mathbf{L}[Y, G\restriction z]$; and moreover there is no set $S \subseteq \mathbf{Seq}_{\Omega}$ in $\mathbf{L}[Y,G\restriction z]$ such that $\Gamma^{\Omega}_{ni}(S)$.*
- (iv) *It follows that $z = \{\langle n,i\rangle : \exists S \subseteq \mathbf{Seq}_{\Omega}\, \Gamma^{\Omega}_{ni}(S)\}$ in $\mathbf{L}[Y,G\restriction z]$;*
- (v) *If $n < \omega$ then the n-th slice $(z)_n = \{i : \langle n,i\rangle \in z\}$ belongs to $\Sigma^{\mathbb{T}}_{n+6}$, where $\mathbb{T} = \mathbb{H}[Y,G\restriction z] = (\mathbf{H}\Omega^{\oplus})^{\mathbf{L}[Y,G\restriction z]}$;*
- (vi) *If $1 \le \ell < \omega$, $\Omega^{\oplus} = \omega_{\ell}^{\mathbf{L}[Y,G\restriction z]}$, and GCH holds in $\mathbf{L}[Y,G\restriction z]$ for all cardinals ω_k, $k < \ell - 1$, then it holds in $\mathbf{L}[Y,G\restriction z]$ that $(z)_n \in \mathbf{D}_{1\ell}$ for all $n < \omega$;*
- (vii) *Under the assumptions of (vi), it holds in $\mathbf{L}[Y,G\restriction z]$ that the set z as a whole belongs to $\mathbf{D}_{1,\ell+1}$.*

Proof. (i) $\Omega^{\oplus}$ remains a cardinal in $\mathbf{L}[G]$ by Lemma 15(iv), hence Q still satisfies $\mathtt{card}\,Q < \Omega^{\oplus}$ in $\mathbf{L}[G]$. As W is Q-generic over $\mathbf{L}[G]$, $\Omega^{\oplus}$ remains a cardinal in $\mathbf{L}[W,G]$ and in $\mathbf{L}[Y,G] \subseteq \mathbf{L}[W,G]$.

(ii) If $\langle n,i\rangle \in z$ then by construction:

$$G(n,i) := \{p(n,i) : p \in G\} = \{p'(n,i) : p' \in G\restriction z\} \in \mathbf{L}[G\restriction z],$$

and hence $S_G(n,i) \in \mathbf{L}[G\restriction z]$ as well. Now $\Gamma^{\Omega}_{ni}(S_G(n,i))$ follows from Lemma 15(ii).

(iii) We w.l.o.g. assume that $z = \mathcal{I} \smallsetminus \{\langle n,i\rangle\}$ and $Y = W$. Then $\mathbb{P}^{\Omega}\restriction z = \mathbb{P}^{\Omega}\restriction_{\neq\langle n,i\rangle}$ can be identified with $\{p \in \mathbb{P}^{\Omega} : p(n,i) = \langle\varnothing,\varnothing\rangle\}$, see Remark 3. Suppose towards the contrary that $S \in \mathbf{L}[W, G\restriction_{\neq\langle n,i\rangle}] = \mathbf{L}[W][G\restriction_{\neq\langle n,i\rangle}]$ satisfies $\Gamma^{\Omega}_{ni}(S)$. There is a name $\tau \in \mathbf{L}[W]$, $\tau \subseteq \mathbb{P}^{\Omega}\restriction_{\neq\langle n,i\rangle} \times \mathbf{Seq}_{\Omega}$, such that:

$$S = \tau[G\restriction_{\neq\langle n,i\rangle}] := \{s \in \mathbf{Seq}_{\Omega} : \exists p \in G\restriction_{\neq\langle n,i\rangle}\, (\langle p,s\rangle \in \tau)\}.$$

The forcing $\mathbb{P}^{\Omega}$ remains $\Omega^{\oplus}$-CC in $\mathbf{L}[W]$ by Lemma 14. This allows us to w.l.o.g. assume that $\mathtt{card}\,\tau < \Omega^{\oplus}$ in $\mathbf{L}[W]$, and then $\tau \in \mathbb{H}[W] = (\mathbf{H}\Omega^{\oplus})^{\mathbf{L}[W]}$.

There is a condition $p_0 \in G$ which $(\mathbb{P}^{\Omega} \restriction_{\neq\langle n,i\rangle})$-forces $\Gamma_{ni}(\tau[\underline{G} \restriction_{\neq\langle n,i\rangle}])$ over $\mathbf{L}[W]$. If $s \in \mathbf{Seq}_{\Omega}$ then put $A_s = \{p : \langle p,s\rangle \in \tau\}$; $A_s \subseteq \mathbb{P}^{\Omega} \restriction_{\neq\langle n,i\rangle}$.

We argue in $\mathbf{L}$. As $\operatorname{card} \tau < \Omega^{\oplus}$, there is an ordinal $\gamma < \Omega^{\oplus}$ such that $\tau \subseteq (\mathbb{P}^{\Omega}_{\gamma} \restriction_{\neq\langle n,i\rangle}) \times \mathbf{Seq}_{\Omega}$ and $p_0 \in \mathbb{P}^{\Omega}_{\gamma} \restriction_{\neq\langle n,i\rangle}$. Consider the set D of all Ω-systems $U \in \mathbf{sDS}_{\Omega}$ extending $\mathbb{U}^{\Omega}_{\gamma}$ and such that there exists a condition $p' \in \mathbf{P}[U] \restriction_{\neq\langle n,i\rangle}$, $p' \leqslant p_0$, an element $f \in U^{\complement}(n,i) = \bigcup_{\langle k,j\rangle \neq \langle n,i\rangle} U(k,j)$, and an ordinal $\mu < \Omega$, such that p' contradicts to every $p \in \bigcup_{\mu \le \alpha < \Omega} A_{f \restriction \alpha}$. Then D is $\Delta^{\mathbb{H}}_3$ by Lemma 17 (and Theorem 5(i), to transfer the definability properties from $\mathbb{H}$ to $\mathbb{H}[W]$), with $\tau \in \mathbb{H}[W]$ as a parameter. Therefore, by Lemma 19, there is an ordinal $\eta < \Omega^{\oplus}$ such that the pair $\mathbb{U}^{\Omega}_{\eta}$ 0-solves D as in Theorem 6(iii). We have two cases.

Case 1: $\mathbb{U}^{\Omega}_{\eta} \in D$. Let this be witnessed by p', f, μ as indicated. Then $f \in (\mathbb{U}^{\Omega}_{\eta})^{\complement}(n,i)$, therefore $f \notin \mathbb{U}^{\Omega}(n,i)$. By definition $\mathbb{U}^{\Omega}_{\gamma} \preccurlyeq \mathbb{U}^{\Omega}_{\eta}$, hence $\gamma \le \eta$. Furthermore, if $s = f \restriction \xi$, $\mu \le \xi < \omega_1$, then the condition p' $(\mathbb{P}^{\Omega} \restriction_{\neq\langle n,i\rangle})$-forces $s \notin \tau[\underline{G} \restriction_{\neq\langle n,i\rangle}]$ over $\mathbf{L}[W]$. We conclude that p' forces $\tau[\underline{G}]/f < \mu < \Omega$ over $\mathbf{L}[W]$. Note that p_0 forces $\tau[\underline{G} \restriction_{\neq\langle n,i\rangle}]/f = \Omega$ because $f \notin \mathbb{U}(n,i)$. However $p' \leqslant p_0$. This is a contradiction.

Case 2: There is no Ω-system $U \in D$ extending $\mathbb{U}^{\Omega}_{\eta}$. We can assume that $\gamma \le \eta$, since if $\eta < \gamma$ then the Ω-system $\mathbb{U}^{\Omega}_{\gamma}$ has the same property. Easily there exists δ, $\eta < \delta < \omega_1$, such that $\mathbb{U}^{\Omega}_{\delta}(n,i) \smallsetminus \mathbb{U}^{\Omega}_{\eta}(n,i) \neq \varnothing$. (To prove this claim note that the set D' of all Ω-systems $U \in \mathbf{sDS}_{\Omega}$ satisfying $U(n,i) \smallsetminus \mathbb{U}^{\Omega}_{\eta}(n,i) \neq \varnothing$ is dense in $\mathbf{sDS}_{\Omega}$ therefore, any U that 0-solves D' belongs to D'.)

Take any $f \in \mathbb{U}^{\Omega}_{\delta}(n,i) \smallsetminus \mathbb{U}^{\Omega}_{\eta}(n,i)$. Then $f \in \mathbb{U}^{\Omega}(n,i)$, and hence p_0 forces $\tau[\underline{G}]/f < \Omega$ over $\mathbf{L}[W]$ by the choice of p_0. It follows that there exists a condition $p' \in \mathbb{P}^{\Omega} \restriction_{\neq\langle n,i\rangle}$, $p' \leqslant p_0$, and an ordinal $\mu < \omega_1$, such that for any $\alpha \ge \mu$, p' forces $s \notin \tau[\underline{G} \restriction_{\neq\langle n,i\rangle}]$ over $\mathbf{L}[W]$, where $s = f \restriction \alpha$. Thus p' contradicts to each condition $p \in \bigcup_{\mu \le \alpha < \Omega} A_{f \restriction \alpha}$. We may w.l.o.g. assume that $p' \in \mathbb{P}^{\Omega}_{\delta} \restriction_{\neq\langle n,i\rangle}$ (otherwise increase δ appropriately). Under these assumptions, define a Ω-system U so that:

$$U(n,i) = \mathbb{U}^{\Omega}_{\delta}(n,i) \smallsetminus \{f\}, \quad U(n,i+1) = \mathbb{U}^{\Omega}_{\delta}(n,i+1) \cup \{f\},$$

and $U(k,j) = \mathbb{U}^{\Omega}_{\delta}(k,j)$ for all pairs of indices $\langle k,j\rangle$ other than $\langle n,i\rangle$ and $\langle n,i+1\rangle$. Obviously U extends $\mathbb{U}^{\Omega}_{\eta}$, and $p' \in \mathbf{P}[U]$. Therefore $U \in D$. But this contradicts the Case 2 hypothesis.

Claim (iv) is an immediate corollary of (ii) and (iii).

To prove (v), note that (*) $(z)_n = \{i : \exists S \subseteq \mathbf{Seq}_{\Omega}\, \Gamma^{\Omega}_{ni}(S)\}$ by (iv). However with n fixed the relation $f \in \mathbb{U}^{\Omega}(n,i)$ with i, f as arguments is $\Sigma^{\mathbb{H}}_{n+4}$ by Lemma 20, hence $\Sigma^{\mathbb{T}}_{n+4}$ by Theorem 5(i). Now $(z)_n \in \Sigma^{\mathbb{T}}_{n+6}$ follows by (*).

To prove (vi), make use of (v) and Theorem 3.

Let us finally prove (vii). Detalizing the proof of (v) and (vi) on the base of formulas $u_n(f,i)$ of Lemma 20, we obtain a recursive sequence of parameter free $\in$-formulas $\varphi_n(i)$ such that if $n, i < \omega$ then $i \in (z)_n$ iff $\mathbb{T} \models \varphi_n(i)$. The proof of Theorem 3 is obviously effective enough to obtain another recursive sequence of parameter free type-theoretic formulas $\psi_n(i)$ of order $\le \ell$ such that it holds in $\mathbf{L}[Y, G \restriction z]$ that: $i \in (z)_n$ iff $\psi_n(i)$, that is, $z = \{\langle n,i\rangle : \psi_n(i)\}$.

However it is known that the truth of formulas of order $\le \ell$ can be uniformly expressed by a suitable formula of order $\ell + 1$, see e.g., [18]. In other words, there is a parameter free type theoretic formula $\Psi(n,i)$ of order $\le \ell + 1$ such that it holds in $\mathbf{L}[Y, G \restriction z]$ that: $i \in (z)_n$ iff $\Psi(n,i)$, that is, $z = \{\langle n,i\rangle : \Psi(n,i)\}$. We conclude that z is definable in $\mathbf{L}[Y, G \restriction z]$ by a type-theoretic formula of order $\le \ell + 1$. In other words, $z \in \mathbf{D}_{1,\ell+1}$ in $\mathbf{L}[Y, G \restriction z]$, as required. □

4.5. We Specify Ω

We come back to Theorem 2. Now it is time to specify the value of the $\mathbf{L}$-cardinal Ω, so far left rather arbitrary by Definition 2 on page 13.

Definition 7 (in $\mathbf{L}$). *Recall that* $1 \le \mathbb{M} < \omega$ *is a number considered in Theorem 2.*

We let $\Omega = \omega_{\mathbb{M}}^{\mathbf{L}}$, *and accordingly define* $\Omega^{\ominus} = \omega_{\mathbb{M}-1}^{\mathbf{L}}$, $\Omega^{\oplus} = \omega_{\mathbb{M}+1}^{\mathbf{L}}$,

$$\mathbb{H} = (\mathbf{H}\Omega^{\oplus})^{\mathbf{L}} = (\mathbf{H}\omega_{\mathbb{M}+1}^{\mathbf{L}})^{\mathbf{L}} = \{x \in \mathbf{L} : \mathtt{card}\,(\mathrm{TC}\,(x)) < \omega_{\mathbb{M}+1}^{\mathbf{L}} \text{ in } \mathbf{L}\}$$

by Definition 2. Applying Definition 6 with $\Omega = \omega_{\mathbb{M}}^{\mathbf{L}}$, *we accordingly fix:*

- *A* $\preccurlyeq$*-increasing sequence of* Ω*-systems* $\{\mathbb{U}_{\xi}^{\Omega}\}_{\xi<\Omega^{\oplus}}$ *satisfying (i), (ii), (iii), (iv) of Theorem 6 for the chosen* **L**-*cardinal* $\Omega = \omega_{\mathbb{M}}^{\mathbf{L}}$,
- *The limit* Ω*-system* $\mathbb{U}^{\Omega} = \bigvee_{\xi<\Omega^{\oplus}} \mathbb{U}_{\xi}^{\Omega}$,
- *The basic forcing notion* $\mathbb{P}^{\Omega} = \mathbf{P}[\mathbb{U}^{\Omega}]$, *and the subforcings* $\mathbb{P}_{\gamma}^{\Omega} = \mathbf{P}[\mathbb{U}_{\gamma}^{\Omega}]$, $\gamma < \Omega^{\oplus}$,

and define restrictions $\mathbb{P}^{\Omega} \restriction z\ (z \subseteq \mathcal{I})$, $\mathbb{P}^{\Omega} \restriction^{\geq n}$, $\mathbb{P}^{\Omega} \restriction^{<n}$, $\mathbb{P}^{\Omega} \restriction_{\neq \langle n,i \rangle}$ *etc. as in Section 3.2.*

4.6. The Model

To prove Theorem 2 we make use of a certain submodel of a $(\mathbb{C} \times \mathbb{P}^{\Omega})$-generic extension of **L**. First of all, if $g : \omega \to \mathscr{P}(\omega)$ is any function then we put:

$$\boldsymbol{w}[g] = \{\langle k, j \rangle : k < \omega \wedge j \in g(k)\}. \tag{8}$$

Now consider a pair $\langle \zeta, G \rangle$, $(\mathbb{C} \times \mathbb{P}^{\Omega})$-generic over **L**. Thus $\zeta : \omega \xrightarrow{\text{onto}} \Xi$ is a generic collapse function, while the set $G \subseteq \mathbb{P}^{\Omega}$ is $\mathbb{P}^{\Omega}$-generic over $\mathbf{L}[\zeta]$. The set:

$$\boldsymbol{w}[\zeta] = \{\langle k, j \rangle : k < \omega \wedge j \in \zeta(k)\} \subseteq \mathcal{I} = \omega \times \omega \tag{9}$$

obviously belongs to the model $\mathbf{L}[\zeta] = \mathbf{L}[\boldsymbol{w}[\zeta]]$, but not to **L**. Therefore the restrictions $\mathbb{P}^{\Omega} \restriction \boldsymbol{w}[\zeta]$, $G \restriction \boldsymbol{w}[\zeta]$ in the next theorem have to be understood in the sense of Definition 5 on page 15, ignoring Remark 3 since, definitely $\boldsymbol{w}[\zeta] \notin \mathbf{L}$. Thus $\mathbb{P}^{\Omega} \restriction \boldsymbol{w}[\zeta]$ is a forcing notion in $\mathbf{L}[\zeta]$, not in **L**.

The following theorem describes the structure of such generic models.

Theorem 8. *Under the assumptions of Definition 7, let a pair* $\langle \zeta, G \rangle$ *be* $(\mathbb{C} \times \mathbb{P}^{\Omega})$*-generic over* **L**. *Then:*

(i) $G \restriction \boldsymbol{w}[\zeta]$ *is a set* $(\mathbb{P}^{\Omega} \restriction \boldsymbol{w}[\zeta])$*-generic over* $\mathbf{L}[\zeta]$,

(ii) $\omega_{\gamma}^{\mathbf{L}[\zeta, G \restriction \boldsymbol{w}[\zeta]]} = \omega_{1+\gamma}^{\mathbf{L}}$ *for all ordinals* $\gamma \geq 1$, *in particular,* $\Omega^{\oplus} = \omega_{\mathbb{M}}^{\mathbf{L}[\zeta, G \restriction \boldsymbol{w}[\zeta]]}$;

and it is true in the model $\mathbf{L}[\zeta, G \restriction \boldsymbol{w}[\zeta]]$ *that*

(iii) *If* $\mathbb{M} \geq 2$ *then* $\Omega = \omega_{\mathbb{M}-1}$ *and* $\Omega^{\oplus} = \Omega^{+} = \omega_{\mathbb{M}}$, *whereas if* $\mathbb{M} = 1$ *then* $\omega < \Omega = \Omega^{\oplus} = \omega_1$;

(iv) GCH *holds;*

(v) *Every constructible real belongs to* $\mathbf{D}_{1\mathbb{M}}$,

(vi) *If* $1 \leq m < \omega$ *and* $m \neq \mathbb{M}$ *then* $\mathbf{D}_{1m} \not\subseteq \mathbf{D}_{2m}$, *and*

(vii) *every real in* $\mathbf{D}_{1\mathbb{M}}$ *is constructible.*

Remark 4. *Theorem 8 implies Theorem 2 via the model* $\mathbf{L}[\zeta, G \restriction \boldsymbol{w}[\zeta]]$, *of course. As for Theorem 8 itself, its proof follows below in this paper. Claims (i)–(vi) will be established right now, and Claim (vii) is accomplished in Section 6.6, based on the substantial work in Sections 5 and 6.*

Proof (Claims (i)–(vi) of Theorem 8). To prove that $G \restriction \boldsymbol{w}[\zeta]$ is $(\mathbb{P}^{\Omega} \restriction \boldsymbol{w}[\zeta])$-generic over $\mathbf{L}[\zeta]$, note that $G \subseteq \mathbb{P}^{\Omega}$ is $\mathbb{P}^{\Omega}$-generic over $\mathbf{L}[\zeta]$ by the product forcing theorem w.r.t. the product $\mathbb{C} \times \mathbb{P}^{\Omega}$. However $\mathbb{P}^{\Omega}$ can be naturally identified with the product $(\mathbb{P}^{\Omega} \restriction \boldsymbol{w}[\zeta]) \times (\mathbb{P}^{\Omega} \restriction z)$ in $\mathbf{L}[\zeta]$, where $z = \mathcal{I} \smallsetminus \boldsymbol{w}[\zeta]$. This implies the result by another application of the product forcing theorem.

To establish (ii), (iii), and (iv), it suffices to apply Lemma 16, as $\mathbf{L}[\zeta] \subseteq \mathbf{L}[\zeta, G \restriction \boldsymbol{w}[\zeta]] \subseteq \mathbf{L}[\zeta, G]$.

To prove Claim (v), let $x \in \mathbf{L}$, $x \subseteq \omega$. By the genericity of ζ, there is a number $n_0 < \omega$ such that $x = \zeta(n_0)$. Then, for any i, we have $\langle n_0, i\rangle \in \boldsymbol{w}[\zeta]$ iff $i \in x$. By Theorem 7(vi) (with $Q = \mathbb{C}$, $z = \boldsymbol{w}[\zeta]$, $Y = \zeta$, $\ell = \mathbb{M}$), it is true in $\mathbf{L}[\zeta, G \restriction \boldsymbol{w}[\zeta]]$ that x belongs to $\mathbf{D}_{1\mathbb{M}}$, as required.

To prove Claim (vi), assume that $1 \le m < \omega$ and $m \ne \mathbb{M}$; we have to show that $\mathbf{D}_{1m} \notin \mathbf{D}_{2m}$ in $\mathbf{L}[\zeta, G \restriction \boldsymbol{w}[\zeta]]$. We have two cases.

Case 1: $m > \mathbb{M}$. Consider the set $z = \boldsymbol{w}[\zeta]$ defined by (9) in Section 4.6. By definition $z \subseteq \omega \times \omega$, $z \in \mathbf{L}[\zeta]$. It follows from Theorem 7(vii) (with $Q = \mathbb{C}$, $z = \boldsymbol{w}[\zeta]$, $Y = \zeta$, $\ell = \mathbb{M}$), that $z \in \mathbf{D}_{1,\mathbb{M}+1}$, hence $z \in \mathbf{D}_{1,\mathbb{M}+1}$ as $\mathbb{M}+1 \le m$. Now suppose to the contrary that $\mathbf{D}_{1m} \in \mathbf{D}_{2m}$ in $\mathbf{L}[\zeta, G \restriction z]$. As $\omega_1^{\mathbf{L}[z]} = \omega_1^{\mathbf{L}[\zeta, G \restriction z]} = \omega_2^{\mathbf{L}}$, there exist real $x \in \mathbf{L}[z]$, $x \subseteq \omega$, which do not belong to $\mathbf{D}_{1m}$; let x_0 be the least of them in the sense of the Gödel well ordering of $\mathbf{L}[z]$. Then x_0 itself belongs to $\mathbf{D}_{1m}$ by 5° of Section 2.2, since so does z by the above, which is a contradiction.

Case 2: $1 \le m < \mathbb{M}$. It suffices to apply Lemma 2 on page 8 because $m < \mathbb{M}$ and $\mathbf{D}_{1\mathbb{M}} = \mathscr{P}(\omega) \cap \mathbf{L}$ holds in $\mathbf{L}[\zeta, G \restriction \boldsymbol{w}[\zeta]]$ by Claims (v) and (vii). We may note that this short argument refers to Claim (vii) that will be conclusively established only in Section 6.6.

An independent proof is as follows. If $1 \le m < \mathbb{M}$, then $\mathbb{M} \ge 2$, and hence Theorem 8(iii) implies:

$$\mathscr{P}^m(\omega) \cap \mathbf{L}[\zeta] = \mathscr{P}^m(\omega) \cap \mathbf{L}[\zeta, G \restriction w] = \mathscr{P}^m(\omega) \cap \mathbf{L}[\zeta, G].$$

We conclude that the sets $\mathbf{D}_{1m}$ and $\mathbf{D}_{2m}$ are the same in these models, and hence it suffices to prove that $\mathbf{D}_{1m} \notin \mathbf{D}_{2m}$ in the $\mathbb{C}$-generic extension $\mathbf{L}[\zeta]$. Now we apply the fact that collapse forcing notions similar to $\mathbb{C}$ are homogeneous enough for any parameter free formula either be forced by every condition, or be negated by every condition. In our case, it follows that $(\mathbf{D}_{1m})^{\mathbf{L}[\zeta]} \in \mathbf{L}$ and $(\mathbf{D}_{1m})^{\mathbf{L}[\zeta]}$ is countable in $\mathbf{L}$. Therefore if, to the contrary, $\mathbf{D}_{1m} \in \mathbf{D}_{2m}$ in $\mathbf{L}[\zeta]$, then taking the Gödel-least $x \in (\mathscr{P}(\omega) \cap \mathbf{L}) \setminus \mathbf{D}_{1m}$ in $\mathbf{L}[\zeta]$, we routinely get $x \in \mathbf{D}_{1m}$ in $\mathbf{L}[\zeta]$ via 5° of Section 2.2, with a contradiction.

This completes the proof of Claims (i)–(vi) of Theorem 8. □

5. Forcing Approximation

We argue under the assumptions and notation of Definition 7 on page 22.

Beginning here a lengthy proof of Claim (vii) of Theorem 8, our plan will be to establish the following, somewhat unexpected result. Recall that, by Theorem 8(ii), it is true in $\mathbf{L}[\zeta, G \restriction \boldsymbol{w}[\zeta]]$ that $\Omega = \omega_{\mathbb{M}-1}$ and $\Omega^{\oplus} = \Omega^{+} = \omega_{\mathbb{M}}$ in case $\mathbb{M} \ge 2$, whereas $\omega < \Omega = \Omega^{\oplus} = \omega_1$ in case $\mathbb{M} = 1$.

Theorem 9. *Assume that a pair* $\langle \zeta, G\rangle$ *is* $(\mathbb{C} \times \mathbb{P}^{\Omega})$*-generic over* $\mathbf{L}$, *and* $a \in \mathbf{L}[\zeta, G \restriction \boldsymbol{w}[\zeta]]$, $a \subseteq \omega$, *and it is true in* $\mathbf{L}[\zeta, G \restriction \boldsymbol{w}[\zeta]]$ *that:*

either $\mathbb{M} \ge 2$ *and* a *is* $\in$*-definable in* $\langle \mathscr{P}(\Omega); \in, \mathbb{p}\rangle$ (see Section 2.4);
or $\mathbb{M} = 1$ *and* a *is* $\in$*-definable in* $\langle \mathscr{P}(\omega); \in\rangle$.

Then $a \in \mathbf{L}[G]$.

Remark 5. *Theorem 9 implies Claim (vii) of Theorem 8.*

Indeed, **arguing in** $\mathbf{L}[\zeta, G \restriction \boldsymbol{w}[\zeta]]$, *suppose that* $a \subseteq \omega$, $a \in \mathbf{D}_{1\mathbb{M}}$. *If* $\mathbb{M} = 1$ *then we immediately have the "or" case of Theorem 9. Thus suppose that* $\mathbb{M} \ge 2$. *Theorem 3 is applicable by Theorem 8(iv), therefore* x *is* $\in$*-definable in* $\mathbf{H}\omega_{\mathbb{M}}$, *that is, in* $\mathbf{H}\Omega^{\oplus}$ *by Theorem 8(iii). Then Theorem 4 is applicable as well, and hence we have the "either" case of Theorem 9. We conclude that* $a \in \mathbf{L}[G]$ *by Theorem 9. However, by Lemma 14, the forcing notion* $\mathbb{P}$ *is* $\Omega^{\ominus}$*-closed in* $\mathbf{L}$, *and this property is sufficient for* $\mathbb{P}$*-generic sets not to add new subsets of* ω, *so* $a \in \mathbf{L}$, *as required by (vii) of Theorem 8.*

Thus Theorem 9 completes the proof of Theorem 8 as a whole because other claims of Theorem 8 have been already established, see Section 4.6.

To prove Theorem 9, we are going to define a forcing-like relation **forc** similar to approximate forcing relations considered in [4,5], and earlier in [3] and some other papers on the base of forcing notions not of an almost-disjoint type. Then we exploit certain symmetries of objects related to **forc**.

Definition 8. *Extending Definition 7 on page 22, let us fix a pair* $\langle \zeta, \mathbf{G} \rangle$, $(\mathbb{C} \times \mathbb{P}^{\Omega})$*-generic over* $\mathbf{L}$ *for the remainder of the text. We consider generic extensions:*

$$\mathbf{L}[\zeta] \subseteq \mathbf{L}[\zeta, \mathbf{G} \restriction w[\zeta]] \subseteq \mathbf{L}[\zeta, \mathbf{G}].$$

We shall assume that $\mathbb{M} \geq 2$ *(the "either" case of Theorem 9). The "or" case* $\mathbb{M} = 1$ *is pretty similar:* Ω *is changed to* ω *during the course of the proof.*

5.1. Language

We argue under the assumptions and notation of Definitions 7 and 8.

- Assume that $z \in \mathbf{L}[\zeta]$, $z \subseteq \mathcal{I} = \omega \times \omega$. Then let $\mathbf{Nam}^z_\zeta \in \mathbf{L}[\zeta]$ be the set of all sets $\tau \in \mathbf{L}[\zeta]$, $\tau \subseteq ({}^*\mathbf{P}_\Omega \restriction z) \times \Omega$, with $\operatorname{card} \tau < \Omega^{\oplus}$ in $\mathbf{L}[\zeta]$.

Note that ${}^*\mathbf{P}_\Omega$, a bigger forcing notion, is used instead of $\mathbb{P}^{\Omega}$ in this definition. One of the advantages is that ${}^*\mathbf{P}_\Omega$ is $\in$-definable in $\mathbb{H}$ by Lemma 17.

If $\tau \in \mathbf{Nam}^z_\zeta$ and $G \subseteq {}^*\mathbf{P}_\Omega \restriction z$ then put $\tau[G] = \{\alpha < \Omega : \exists p \in G\,(\langle p, \alpha \rangle \in \tau)\}$.

Lemma 21. $\mathscr{P}(\Omega) \cap \mathbf{L}[\zeta, G \restriction w[\zeta]] = \{\tau[\mathbf{G} \restriction w[\zeta]] : \tau \in \mathbf{Nam}^{w[\zeta]}_\zeta\}$.

Proof. Let $X \in \mathbf{L}[\zeta, G \restriction w[\zeta]]$, $X \subseteq \Omega$. The set $\mathbf{G} \restriction w[\zeta]$ is $(\mathbb{P}^{\Omega} \restriction w[\zeta])$-generic over $\mathbf{L}[\zeta]$ by the product forcing theory. Therefore, by a well-known property of generic extensions (see, e.g., [32]), there is a name $t \in \mathbf{L}[\zeta]$, $t \subseteq (\mathbb{P}^{\Omega} \restriction w[\zeta]) \times \Omega$, such that $X = t[\mathbf{G} \restriction w[\zeta]]$. To reduce t to a name τ with the same property, satisfying $\operatorname{card} \tau < \Omega^{\oplus}$, apply Lemma 14. □

Now, arguing in $\mathbf{L}[\zeta]$, we introduce a language that will help us to study analytic definability in the generic extensions considered. We argue under the assumptions and notation of Definition 8.

Let $\mathcal{L}$ be the 2nd order language, with variables $\alpha, \beta, \dots$, assumed to vary over ordinals $< \Omega$, and $X, Y, \dots$, varying over the subsets of Ω. Atomic formulas of the following types are allowed:

$$\alpha < \beta, \quad \alpha = \beta, \quad \alpha \in X, \quad \mathbb{p}(\alpha, \beta) = \gamma.$$

(See Section 2.4 on $\mathbb{p}$.) Only the connectives $\wedge$ and $\neg$ and quantifiers $\exists \alpha$ and $\exists X$ are allowed, the other connectives and $\forall$ are treated as shortcuts, and, to reduce the number of cases, the equality $X = Y$ will be treated as a shortcut for $\forall \alpha(\alpha \in X \iff \alpha \in Y)$.

The *complexity* $\#(\varphi)$ of an $\mathcal{L}$-formula φ is defined by induction so that:

- $\#(\varphi) = 0$ for all atomic formulas,
- $\#(\varphi \wedge \psi) = \max\{\#(\varphi), \#(\psi)\}$,
- $\#(\exists \alpha\, \varphi(\alpha)) = \#(\varphi(\alpha))$ and $\#(\exists X\, \varphi(X)) = \#(\varphi(X))$,
- Finally, $\#(\neg \varphi) = \#(\varphi) + 1$.

Note that the complexity of quantifier-free formulas can be as high as one wants.

If $z \in \mathbf{L}[\zeta]$, $z \subseteq \omega \times \omega$, then let $\mathcal{L}(z)$ be the extension of $\mathcal{L}$ by:

- Ordinals $\alpha < \Omega$ to substitute variables over Ω,
- Names in $\mathbf{Nam}^z_\zeta$ to substitute variables $X, Y, \dots$ over $\mathscr{P}(\Omega)$.

If $G \subseteq {}^*\mathbf{P}_M \restriction z$, then the *valuation* $\varphi[G]$ of such a formula φ is defined by substitution of $\tau[G]$ for any name $\tau \in \mathbf{Nam}^z_\zeta$ that occurs in φ, and relativizing each quantifier $\exists\,\alpha$ or $\exists\,X$ to resp. Ω, $\mathscr{P}(\Omega)$. Thus $\varphi[G]$ is a formula of $\mathcal{L}$ with parameters in $\mathscr{P}(\Omega) \cap \mathbf{L}[\zeta, G]$ and quantifiers relativized as above, that is, to Ω and to $\mathscr{P}(\Omega)$, and $\varphi[G]$ can contain $\mathbb{p}$ interpreted as $\mathbb{p} \restriction (\Omega \times \Omega)$. (See Section 2.4 on $\mathbb{p}$.)

5.2. Forcing Approximation

We still argue under the assumptions and notation of Definitions 7 and 8.

Our next goal is to define, in $\mathbf{L}[\zeta]$, a forcing-style relation p $\mathbf{forc}^z_U$ φ. In case $z = w[\zeta]$ and $U = \mathbb{U}^\Omega$, the relation $\mathbf{forc}^z_U$ will be compatible with the truth in the model $\mathbf{L}[\zeta, \mathbf{G} \restriction w[\zeta]] = \mathbf{L}[\zeta][\mathbf{G} \restriction w[\zeta]]$, viewed as a $(\mathbb{P}^\Omega \restriction w[\zeta])$-generic extension of $\mathbf{L}[\zeta]$. But, perhaps unlike the true forcing relation associated with $\mathbb{P}^\Omega \restriction w[\zeta]$, the relation $\mathbf{forc}^z_U$ will be invariant under certain transformations.

The definition goes on in $\mathbf{L}[\zeta]$ by induction on the complexity of φ.

(F1) When writing p $\mathbf{forc}^z_U$ φ, it will always be assumed that $U \in \mathbf{sDS}_\Omega$, $z \in \mathbf{L}[\zeta]$, $z \subseteq \omega \times \omega$, $p \in \mathbf{P}[U] \restriction z$, φ is a closed formula in $\mathcal{L}_z$.

(F2) If $U \in \mathbf{sDS}_\Omega$, $z \in \mathbf{L}[\zeta]$, $z \subseteq \omega \times \omega$, $p \in \mathbf{P}[U] \restriction z$, and $\alpha, \beta, \gamma < \Omega$, then: p $\mathbf{forc}^z_U$ $\alpha + \beta = \gamma$ iff in fact $x^0 + y^0 = z^0$, and the same for the formulas $\alpha + \beta = \gamma$ and $\mathbb{p}(\alpha, \beta) = \gamma$.

(F3) If U, p, z are as above, $\alpha < \Omega$, $Y \in \mathbf{Nam}^z_\zeta$, then: p $\mathbf{forc}^z_U$ $\alpha \in Y$ iff there exists a condition $q \in \mathbf{P}[U] \restriction z$ such that $\langle q, \alpha \rangle \in Y$ and $p \leqslant q$.

(F4) If U, p, z are as above, then: p $\mathbf{forc}^z_U$ $(\varphi \wedge \psi)$ iff p $\mathbf{forc}^z_U$ φ and p $\mathbf{forc}^z_U$ ψ.

(F5) If U, p, z are as above, then p $\mathbf{forc}^z_U$ $\exists\,\alpha\, \varphi(\alpha)$ iff there is $\alpha < \Omega$ such that p $\mathbf{forc}^z_U$ $\varphi(\alpha)$.

(F6) If U, p, z are as above, then p $\mathbf{forc}^z_U$ $\exists\,Y\, \varphi(Y)$ iff there exists a name $\tau \in \mathbf{Nam}^z_\zeta$ such that p $\mathbf{forc}^z_U$ $\varphi(Y)$.

We precede the last item with another definition. If $n < \omega$ then let $\mathbf{sDS}[n]$ be the set of all Ω-systems $U \in \mathbf{sDS}_\Omega$ such that $U \restriction^{<n} = \mathbb{U}^\Omega_\xi \restriction^{<n}$ for some $\xi < \Omega^\oplus$. Thus $\mathbf{sDS}[0] = \mathbf{sDS}_\Omega$.

(F8) If U, p, z are as in (F1), φ is a closed $\mathcal{L}_z$ formula, $n = \#(\varphi)$, then p $\mathbf{forc}^z_U$ $\neg\,\varphi$ iff there is no Ω-system $U' \in \mathbf{sDS}[n]$ extending U, and no $q \in \mathbf{P}[U'] \restriction z$, $q \leqslant p$, such that q $\mathbf{forc}^z_{U'}$ φ.

Lemma 22 (in $\mathbf{L}[\zeta]$)**.** *Let U, p, z, φ satisfy (F1) above. Then:*

(i) *If p $\mathbf{forc}^z_U$ φ, $U' \in \mathbf{sDS}_\Omega$ extends U, and $q \in \mathbf{P}[U'] \restriction z$, $q \leqslant p$, then q $\mathbf{forc}^z_{U'}$ φ;*

(ii) *If $U \in \mathbf{sDS}[n]$, $\#(\varphi) \le n$, and p $\mathbf{forc}^z_U$ φ, then p $\mathbf{forc}^z_U$ $\neg\,\varphi$ fails.*

Proof. The proof of (i) by straightforward induction is elementary. As for (ii), make use of (F8). □

Now let us evaluate the complexity of the relation $\mathbf{forc}$. Given a parameter free $\mathcal{L}$-formula $\varphi(\alpha, \beta, \ldots, X, Y, \ldots)$ with any set of free variables allowed in $\mathcal{L}$, we define, in $\mathbf{L}[\zeta]$, the set:

$$\begin{aligned}\mathbf{Forc}(\varphi) \;=\; \{\langle z, U, p, \alpha, \beta, \ldots, \tau_X, \tau_Y, \ldots\rangle : & \; U \in \mathbf{sDS}_\Omega \wedge z \subseteq \omega \times \omega \\ & \wedge p \in \mathbf{P}[U] \restriction z \wedge \alpha, \beta, \cdots < \Omega \wedge \tau_X, \tau_Y, \cdots \in \mathbf{Nam}^z_\zeta \\ & \wedge p \;\mathbf{forc}^z_U\; \varphi(\alpha, \beta, \ldots, \tau_X, \tau_Y, \ldots) \;\}.\end{aligned}$$

Lemma 23 (in $\mathbf{L}[\zeta]$)**.** *If φ is a parameter free $\mathcal{L}$-formula and $n = \#(\varphi)$, then $\mathbf{Forc}(\varphi)$ is $\Sigma^{\mathbb{H}[\zeta]}_{n+3}$.*

Proof. The set $\mathbf{sDS}_\Omega$ is $\Delta^{\mathbb{H}}_3$ by Lemma 17, and hence $\Delta^{\mathbb{H}[\zeta]}_3$ as well by Theorem 5(i) in Section 2.6. The relations $p \in \mathbf{P}[U] \restriction z$, $\alpha < \Omega$, $\tau \in \mathbf{Nam}^z_\zeta$, with arguments resp. p, U, z; α; τ, z, are routinely checked to be $\Delta^{\mathbb{H}[\zeta]}_3$, too. (Note that bounded quantifiers preserve $\Delta^{\mathbb{H}[\zeta]}_3$.) After this remark, prove the lemma by induction on the structure of φ.

The case of atomic formulas of type (F2) is immediately clear. (The pairing function $\mathbb{p} \restriction (\Omega \times \Omega)$ in (F2) is $\Delta_1^{\mathbb{H}}$ by Lemma 3.) The result for atomic formulas of type (F3) amounts to the formula $\exists q \in {}^*\mathbf{P}_\Omega \restriction z\,(\langle q, \alpha\rangle \in Y \wedge p \leqslant q)$, which is $\Sigma_3^{\mathbb{H}[\zeta]}$ by the above. The step (F4) amounts to the intersection of two sets is quite obvious. And so are steps (F5) and (F6) (a $\exists$-quantification on the top of a given $\Sigma^{\mathbb{H}[\zeta]}_{\#(\varphi)+3}$).

To carry out the step (F8), note that $\mathbf{sDS}[n]$ is $\Sigma^{\mathbb{H}}_{n+3}$ by Lemma 20, therefore $\Sigma^{\mathbb{H}[\zeta]}_{n+3}$ by Theorem 5(i) in Section 2.6. This if $\mathbf{Forc}(\varphi)$ is $\Sigma^{\mathbb{H}[\zeta]}_{n+3}$ then $\mathbf{Forc}(\neg\,\varphi)$ is $\Pi^{\mathbb{H}[\zeta]}_{n+3}$, hence $\Sigma^{\mathbb{H}[\zeta]}_{n+4}$, as required. □

5.3. Consequences for the Complete Forcing Notions

We continue to argue under the assumptions and notation of Definitions 7 on page 22 and 8 on page 25. Coming back to the sequence of Ω-systems $\mathbb{U}^\Omega_\xi \in \mathbf{sDS}_\Omega$ given by Definition 7, we note that every Ω-system $\mathbb{U}^\Omega_\xi$ belongs to $\bigcap_m \mathbf{sDS}[m]$.

Let $\mathbf{forc}^z_\xi$ be $\mathbf{forc}^z_{\mathbb{U}^\Omega_\xi}$, and let $p\ \mathbf{forc}^z_\infty\ \varphi$ mean: $\exists \xi < \Omega^\oplus\ (p\ \mathbf{forc}^z_\xi\ \varphi)$. Note that $p\ \mathbf{forc}^z_\xi\ \varphi$ implies $p \in \mathbb{P}^\Omega_\xi \restriction z$, whereas $p\ \mathbf{forc}^z_\infty\ \varphi$ implies $p \in \mathbb{P}^\Omega \restriction z$. Lemma 22 takes the following form:

Lemma 24 (in $\mathbf{L}[\zeta]$). *Assume that $z \subseteq \omega \times \omega$, φ is a closed $\mathcal{L}_z$ formula, $p \in \mathbb{P}^\Omega \restriction z$. Then:*

(i) *If $p\ \mathbf{forc}^z_\xi\ \varphi$ and $\xi \le \eta < \Omega^\oplus$, $q \in \mathbb{P}^\Omega_\eta \restriction z$, $q \leqslant p$, then $q\ \mathbf{forc}^z_\eta\ \varphi$, and accordingly, if $p\ \mathbf{forc}^z_\infty\ \varphi$ and $q \in \mathbb{P}^\Omega \restriction z$, $q \leqslant p$, then $q\ \mathbf{forc}^z_\infty\ \varphi$;*

(ii) *$p\ \mathbf{forc}^z_\infty\ \varphi$ and $p\ \mathbf{forc}^z_\infty\ \neg\,\varphi$ contradict to each other.*

The following result will be very important.

Lemma 25 (in $\mathbf{L}[\zeta]$). *If $z \subseteq \omega \times \omega$, φ is a closed $\mathcal{L}_z$ formula, $p \in \mathbb{P}^\Omega \restriction z$, then there is a condition $q \in \mathbb{P}^\Omega \restriction z$, $q \leqslant p$, such that either $q\ \mathbf{forc}^z_\infty\ \varphi$, or $q\ \mathbf{forc}^z_\infty\ \neg\,\varphi$.*

Proof. Let $n = \#(\varphi)$. There is an ordinal $\eta < \Omega^\oplus$ such that $p \in \mathbb{P}^\Omega_\eta \restriction z$. Consider the set D of all Ω-systems $U' \in \mathbf{sDS}_\Omega \restriction^{\ge n}$ such that there is a Ω-system $U \in \mathbf{sDS}[n]$ that extends $\mathbb{U}^\Omega_\eta$ and satisfies $U \restriction^{\ge n} = U'$, and there is also a condition $q \in \mathbf{P}[U] \restriction z$, $q \leqslant p$, satisfying $q\ \mathbf{forc}^z_U\ \varphi$. The set D belongs to $\Sigma_{n+3}(\mathbb{H}[\zeta])$ (with $\mathbb{U}^\Omega_\eta$, $\mathbb{V}^\Omega_\eta$, p as definability parameters) by Lemma 23. Therefore by Lemma 19 there is an ordinal ξ, $\eta \le \xi < \Omega^\oplus$, such that the Ω-system $\mathbb{U}^\Omega_\xi \restriction^{\ge n}$ n-solves D. We have two cases.

Case 1: $\mathbb{U}^\Omega_\xi \restriction^{\ge m} \in D$. Then there exist: a Ω-system $U \in \mathbf{sDS}[n]$ extending $\mathbb{U}^\Omega_\eta$ and satisfying $U \restriction^{\ge n} = \mathbb{U}^\Omega_\xi \restriction^{\ge n}$, and a condition $q \in \mathbf{P}[U] \restriction z$, $q \leqslant p$, with $q\ \mathbf{forc}^z_U\ \varphi$. By definition there is an ordinal $\vartheta < \Omega^\oplus$ such that $U \restriction^{<n} = \mathbb{U}^\Omega_\vartheta \restriction^{<n}$. Now let $\mu = \max\{\xi, \vartheta\}$. Then $U \preccurlyeq \mathbb{U}^\Omega_\mu$, hence $q\ \mathbf{forc}^z_\mu\ \varphi$ and $q\ \mathbf{forc}^z_\infty\ \varphi$.

Case 2: There is no Ω-system $U \in D$ that extends $\mathbb{U}^\Omega_\xi \restriction^{\ge n}$. Prove that $p\ \mathbf{forc}^z_\xi\ \neg\,\varphi$. Suppose towards the contrary that this fails. Then, by (F8) in Section 5.2, there exists a Ω-system $U \in \mathbf{sDS}[n]$ extending $\mathbb{U}^\Omega_\xi$, and a condition $q \in \mathbf{P}[U]$, $q \leqslant p$, such that $q\ \mathbf{forc}^z_U\ \varphi$. Define $U' = U \restriction^{\ge n}$. Then by definition the Ω-system U' belongs to $\mathbf{sDS}_\Omega \restriction^{\ge n}$, and moreover the Ω-system U witnesses that $U' \in D$. But this contradicts the Case 2 assumption. □

5.4. Truth Lemma

According to the next theorem ("the truth lemma"), the truth in the generic extensions considered is connected in the usual way with the relation $\mathbf{forc}_\infty$. We continue to argue under the assumptions and notation of Definitions 7 on page 22 and 8 on page 25.

Theorem 10. *Assume that $\mathbf{z} = \boldsymbol{w}[\zeta]$ and φ is a $\mathcal{L}_\mathbf{z}$-formula. Then $\varphi[\mathbf{G} \restriction \mathbf{z}]$ is true in $\mathbf{L}[\zeta, \mathbf{G} \restriction \mathbf{z}]$ iff there is a condition $p \in \mathbf{G} \restriction \mathbf{z}$ such that $p\ \mathbf{forc}^\mathbf{z}_\infty\ \varphi$.*

Proof. We proceed by induction. Suppose that φ is an atomic formula of type (F3) of Section 5.2. (The case of formulas as in (F2) is pretty elementary.) To prove the implication $\Longleftarrow$, assume that $p \in \mathbf{G}\restriction z$ and p $\mathbf{forc}_\infty^z$ $\alpha \in \tau$, where $\alpha < \Omega$ and $\tau \in \mathbf{Nam}_\zeta^z$. Then by definition ((F3) in Section 5.2) there exists a condition $q \in \mathbb{P}^\Omega\restriction z$ satisfying $p \leqslant q$ and $\langle q, \alpha\rangle \in \tau$. There are conditions $p', q' \in \mathbb{P}^\Omega$ such that $p = p'\restriction z$ and $q = q'\restriction z$, but not necessarily $p' \leqslant q'$. We only know that $p'(n,i) \leqslant q'(n,i)$ for all $\langle n,i\rangle \in z$. Therefore $z \subseteq Z = \{\langle n,i\rangle : p'(n,i) \leqslant q'(n,i)\}$. The set Z belongs to $\mathbf{L}$ since so do p', q' as elements of $\mathbb{P}^\Omega \in \mathbf{L}$ (whereas about z we only assert that $z \in \mathbf{L}[\zeta]$). Therefore a condition $q'' \in \mathbb{P}^\Omega$ can be defined by:

$$q''(n,i) = \begin{cases} q(n,i), & \text{in case} \quad \langle n,i\rangle \in Z, \\ p(n,i), & \text{in case} \quad \langle n,i\rangle \notin Z, \end{cases}$$

and we still have $q''\restriction z = q'\restriction z$ and $p' \leqslant q''$. It follows that $q'' \in \mathbf{G}$ by genericity, hence $q''\restriction z = q'\restriction z \in \mathbf{G}\restriction z$. But then $\alpha \in \tau[\mathbf{G}\restriction z]$, as required.

To prove the converse, assume that $\alpha \in \tau[\mathbf{G}\restriction z]$. There exists a condition $p \in \mathbf{G}\restriction z$ such that $\langle q, \alpha\rangle \in \tau$, and we have p $\mathbf{forc}_\infty^z$ $\alpha \in \tau$, as required.

Rather simple inductive steps (F4), (F5) of Section 5.2 are left for the reader.

Let us carry out step (F6). Let $\varphi := \exists X\, \psi(X)$. Suppose that $p \in \mathbf{G}\restriction z$ and p $\mathbf{forc}_\infty^z$ φ. By definition there exists a name $\tau \in \mathbf{Nam}_\zeta^z$ such that p $\mathbf{forc}_\infty^z$ $\psi(\tau)$. The formula $\psi(\tau)[\mathbf{G}\restriction z]$ is then true in $\mathbf{L}[\zeta, \mathbf{G}\restriction z]$ by the inductive hypothesis. But $\psi(\tau)[\mathbf{G}\restriction z]$ coincides with $\psi[\mathbf{G}\restriction z](Y)$, where $Y = \tau[\mathbf{G}\restriction z] \in \mathbf{L}[\zeta, \mathbf{G}\restriction z]$, $Y \subseteq \Omega$. We conclude that $\exists X\, \psi(X)[\mathbf{G}\restriction z]$ is true in $\mathbf{L}[\zeta, \mathbf{G}\restriction z]$, as required.

To prove the converse, let $\varphi[\mathbf{G}\restriction z]$, that is, $\exists X\, \psi(X)[\mathbf{G}\restriction z]$, be true in $\mathbf{L}[\zeta, \mathbf{G}\restriction z]$. As X is relativized to $\mathscr{P}(\Omega)$, there is a set $X \in \mathscr{P}(\Omega)$ in $\mathbf{L}[\zeta, \mathbf{G}\restriction z]$ satisfying $\varphi(X)[\mathbf{G}\restriction z]$ in $\mathbf{L}[\zeta, \mathbf{G}\restriction z]$. By Lemma 21, there is a name $\tau \in \mathbf{Nam}_\zeta^z$ with $X = \tau[\mathbf{G}\restriction z]$, so $\psi(\tau)[\mathbf{G}\restriction z]$ holds in $\mathbf{L}[\zeta, \mathbf{G}\restriction z]$. The inductive hypothesis implies that some $p \in \mathbf{G}\restriction z$ satisfies p $\mathbf{forc}_\infty^z$ $\psi(\tau)$, hence p $\mathbf{forc}_\infty^z$ φ, as required.

Finally, let us carry out step (F8), which is somewhat less trivial. Prove the lemma for a $\mathcal{L}_z$ formula $\neg\varphi$, assuming that the result holds for φ. If $\neg\varphi[\mathbf{G}\restriction z]$ is false in $\mathbf{L}[\zeta, \mathbf{G}]$ then $\varphi[\mathbf{G}\restriction z]$ is true. Thus by the inductive hypothesis, there is a condition $p \in \mathbf{G}\restriction z$ such that p $\mathbf{forc}_\infty^z$ φ. Then q $\mathbf{forc}_\infty^z$ $\neg\varphi$ for any $q \in \mathbf{G}\restriction z$ is impossible by Lemma 24 above.

Conversely suppose that p $\mathbf{forc}_\infty^z$ $\neg\varphi$ holds for no $p \in \mathbf{G}\restriction z$. Then by Lemma 25 there exists $q \in \mathbf{G}\restriction z$ such that q $\mathbf{forc}_\infty^z$ φ. It follows that $\varphi[\mathbf{G}\restriction z]$ is true by the inductive hypothesis, therefore $\varphi[\mathbf{G}\restriction z]$ is false. □

6. Invariance

The goal of this section is to prove Theorem 9 on page 24, and thereby accomplish the proof of Theorem 8, and the proof of Theorem 2 (the main theorem) itself. The proof makes use of the relation **forc** introduced in Section 5, and exploits certain symmetries in **forc**, investigated in Section 6.5.

6.1. *Hidden Invariance*

Theorem 9 belongs to a wide group of results on the structure of generic models which assert that such-and-such elements of a given generic extension belong to a smaller and/or better shaped extension. One of possible methods to prove such results is to exploit the homogeneity of the forcing notion considered, or in different words, its invariance w. r. t. a sufficiently large system of order-preserving transformations. In particular, for a straightforward proof of Theorem 11 below, which is our key technical step in the proof of Theorem 9, the invariance of the forcing notion $\mathbb{P}^\Omega$ under permutations of indices in $\mathcal{I} = \omega \times \omega$ (to permute areas z and $\widehat{z}$) would be naturally required, whereas $\mathbb{P}^\Omega$ is definitely not invariant w. r. t. permutations.

On the other hand, the auxiliary forcing relation **forc** is invariant w.r.t. permutations. Theorem 10 in Section 5.4 conveniently binds the relation **forc** with the truth in $\mathbb{P}^{\Omega}$-generic extensions by means of a forcing-style association. This principal association was based on the $\mathbb{M}$-completeness property (Definition 7 on page 22 and Theorem 6). Basically it occurs that some transformations, that is, permutations, are *hidden* in construction of $\mathbb{P}^{\Omega}$, so that they do not act explicitly, but their influence is preserved and can be recovered via the relation **forc**.

This method of *hidden invariance*, that is, invariance properties (of an auxiliary forcing-type relation like **forc**) hidden in $\mathbb{P}^{\Omega}$ by a suitable generic-style construction of $\mathbb{P}^{\Omega}$, was introduced in Harrington's notes [22] in in the context of the almost disjoint forcing (in a somewhat different terminology from what is used here). It was introduced independently by one of the authors in [37] in the context of the Sacks forcing and its Jensen's modification in [38]; see e.g., [3,28,39] for further research in this direction based on product and iterated versions of the Sacks and Jensen forcing earlier studied in detail in [40–47].

6.2. The Invariance Theorem

We still argue under the assumptions and notation of Definitions 7 on page 22 and 8 on page 25.

Let Π be the group of all *finite permutations* of ω, that is, all bijections $\pi : \omega \overset{\text{onto}}{\longrightarrow} \omega$ since the set $|\pi| = \{k : \pi(k) \neq k\}$ is finite. If $m < \omega$ then the subgroup Π_m consists of all $\pi \in \Pi$ satisfying $\pi(k) = k$ for all $k < m$. If $\pi \in \Pi$, and $z \subseteq \omega \times \omega$ then put $\pi z = \{\langle \pi(n), i\rangle : \langle n, i\rangle \in z\}$.

If in addition $g : \omega \to \Xi$ then define $\pi g : \omega \to \Xi$ by $\pi g(\pi(n)) = g(n)$, all n.

Similarly if $e \in \mathbb{C}$ and $|\pi| \subseteq \mathtt{lh}\, e$, then define $e' = \pi e \in \mathbb{C}$ such that $\mathtt{lh}\, e' = \mathtt{lh}\, e$ and $e'(\pi(n)) = e(n)$ for all $n < \mathtt{lh}\, e$. The following is the invariance theorem.

Theorem 11 (in $\mathbf{L}[\zeta]$). *Assume that $\mathbf{z} = \boldsymbol{w}[\zeta]$, $\pi \in \Pi_m$, $\mathbf{z}' = \pi \mathbf{z}$, φ is a closed parameter free formula of $\mathcal{L}_{\mathbf{z}}$, $\#(\varphi) \le m$, and $p_0 \in \mathbb{P}^{\Omega}$. Then $p_0 \restriction \mathbf{z}\ \mathbf{forc}^{\mathbf{z}}_{\infty} \neg \varphi$ iff $p_0 \restriction \mathbf{z}'\ \mathbf{forc}^{\mathbf{z}'}_{\infty} \neg \varphi$.*

A lengthy proof of Theorem 11 follows below in this Section.

6.3. Proof of Theorem 9 from the Invariance Theorem

Under the assumptions of Theorem 9, consider an arbitrary set $a \in \mathbf{L}[\zeta, \mathbf{G} \restriction \boldsymbol{w}[\zeta]]$, $a \subseteq \omega$, and assume that $\mathbb{M} \ge 2$ (see Definition 8) and it is true in $\mathbf{L}[\zeta, \mathbf{G} \restriction \boldsymbol{w}[\zeta]]$ that a is parameter free definable in $\langle \mathscr{P}(\Omega); \in, \mathbb{p}\rangle$, i.e., $a = \{j < \omega : \neg \varphi(j)\}$, where $\varphi(\cdot)$ is a parameter free $\mathcal{L}_{\mathbf{z}}$-formula. Let $m = \#(\varphi)$ and $\boldsymbol{w} = \boldsymbol{w}[\zeta]$. The goal is to prove that $a \in \mathbf{L}[\mathbf{G}]$. This is based on the next lemma.

Lemma 26. *The set $T = \{\langle p, j\rangle : p \in \mathbb{P}^{\Omega} \wedge p \restriction \boldsymbol{w}\ \mathbf{forc}^{\boldsymbol{w}}_{\infty} \neg \varphi(j)\}$ belongs to $\mathbf{L}$.*

Proof. Note that, by Lemma 23, the set:

$$K = \{\langle z, p, j\rangle : p \in \mathbb{P}^{\Omega} \wedge z \in \mathbf{L}[\zeta] \wedge z \subseteq \omega \times \omega \wedge j < \omega \wedge p \restriction z\ \mathbf{forc}^{z}_{\infty} \neg \varphi(j)\}$$

is definable in $\mathbf{L}[\zeta]$ by a formula with sets in $\mathbf{L}$ as parameters, say $K = \{\langle z, p, j\rangle : \vartheta(z, p, j, S)\}$ in $\mathbf{L}[\zeta]$, where $S \in \mathbf{L}$ is a sole parameter. Recall that $\zeta \in \Xi^{\omega}$ is $\mathbb{C}$-generic over $\mathbf{L}$, and $\boldsymbol{w} = \boldsymbol{w}[\zeta] = \{\langle n, j\rangle : j \in \zeta(n)\}$. Let $\breve{\zeta}$ be a canonical $\mathbb{C}$-name for ζ, and $\Vdash$ be the $\mathbb{C}$-forcing relation over $\mathbf{L}$. We claim that:

$$\vartheta(\boldsymbol{w}, p, j, S) \text{ holds in } \mathbf{L}[\zeta] \quad \text{iff} \quad \zeta \restriction m \Vdash \vartheta(\boldsymbol{w}[\breve{\zeta}], p, j, S)\,; \tag{10}$$

$\zeta \restriction m$ belongs to $\mathbb{C}$, of course. The direction $\Longleftarrow$ is obvious.

To establish $\Longrightarrow$, assume that the right-hand side fails. Then there is a condition $e_0 \in \mathbb{C}$ such that $\zeta \restriction m \subseteq e_0$ and $e_0 \Vdash \neg \vartheta(\boldsymbol{w}[\breve{\zeta}], p, j, S)$. We note that the set:

$$D = \{e \in \mathbb{C} : \zeta \restriction m \subseteq e \wedge \exists \pi \in \Pi_m\, (|\pi| \subseteq \mathtt{dom}\, e \wedge e_0 \subseteq \pi e)\}$$

is dense in $\mathbb{C}$ over $\zeta \upharpoonright m$. Therefore, by the genericity of ζ, there exists a number $k > m$ such that $e = \zeta \upharpoonright k \in D$. Accordingly, there is a permutation $\pi \in \Pi_m$ satisfying $|\pi| \subseteq k$ and $e_0 \subseteq \pi e$.

We put $\zeta' = \pi\zeta$; this is still a $\mathbb{C}$-generic element of Ξ^ω, with $\mathbf{L}[\zeta'] = \mathbf{L}[\zeta]$ since $\pi \in \mathbf{L}$, and we have $e_0 \subseteq \pi e \subset \zeta'$. It follows, by the choice of e_0, that $\vartheta(w[\zeta'], p, j, S)$ *fails* in $\mathbf{L}[\zeta'] = \mathbf{L}[\zeta]$, and hence $\langle w[\zeta'], p, j\rangle \notin K$ by the choice of ϑ. However $w[\zeta'] = \pi \cdot w[\zeta] = \pi w$, thus we have $\langle \pi w, p, j\rangle \notin K$.

We conclude that $p \upharpoonright \pi w \ \mathbf{forc}_\infty^{\pi w} \neg \varphi(j)$ *fails* by the definition of K. Therefore $p \upharpoonright w \ \mathbf{forc}_\infty^{w} \neg \varphi(j)$ *fails* as well by Theorem 11, so we have $\langle w, p, j\rangle \notin K$, and hence $\vartheta(w, p, j, S)$ fails in $\mathbf{L}[\zeta'] = \mathbf{L}[\zeta]$, as required. This completes the proof of (10). Now, coming back to the lemma, we deduce the equality $T = \{\langle p, j\rangle \in \mathbf{L} : \zeta \upharpoonright m \Vdash \vartheta(w[\check{\zeta}], p, j, S)\}$ from (10). This implies $T \in \mathbf{L}$. □

It remains to notice that, by Theorem 10,

$$\begin{aligned} j \in a \iff \mathbf{L}[\zeta, \mathbf{G} \upharpoonright w[\zeta]] \models \neg \varphi(j) &\iff \exists p \in \mathbf{G} \upharpoonright w (p \ \mathbf{forc}_\infty^{w} \neg \varphi(j)) \\ &\iff \exists p \in \mathbf{G} (p \upharpoonright w \ \mathbf{forc}_\infty^{w} \neg \varphi(j)). \end{aligned}$$

Therefore $j \in a \iff \exists p \in \mathbf{G} (\langle p, j\rangle \in T)$. But $T \in \mathbf{L}$ by Lemma 26. We conclude that $a \in \mathbf{L}[\mathbf{G}]$, as required.

This completes the proof of Theorem 9 from Theorem 11.

6.4. The Invariance Theorem: Setup

We still argue under the assumptions and notation of Definitions 7 on page 22 and 8 on page 25.

Here we begin **the proof of Theorem** 11. It will be completed in Section 6.6.

We fix m, $\pi \in \Pi_m$, p_0, $\mathbf{z} = w[\zeta]$, $\widehat{\mathbf{z}} = \pi \mathbf{z}$, and φ with $\#(\varphi) \le m$, as in Theorem 11. Suppose towards the contrary that $p_0 \upharpoonright \widehat{\mathbf{z}} \ \mathbf{forc}_\infty^{\widehat{\mathbf{z}}} \neg \varphi$, but $p_0 \upharpoonright \mathbf{z} \ \mathbf{forc}_\infty^{\mathbf{z}} \neg \varphi$ *fails*. By definition there is an ordinal $\mu < \Omega^\oplus$ such that $p_0 \upharpoonright \widehat{\mathbf{z}} \ \mathbf{forc}_\mu^{\widehat{\mathbf{z}}} \neg \varphi$, but $p_0 \upharpoonright \mathbf{z} \ \mathbf{forc}_\mu^{\mathbf{z}} \neg \varphi$ *fails*. Then we have:

(A) a Ω-system $U^1 \in \mathbf{sDS}[m]$ with $\mathbb{U}_\mu^\Omega \preccurlyeq U^1$, and a condition $p_1 \in \mathbf{P}[U^1]$, $p_1 \leqslant p_0$, such that $p_1 \upharpoonright \mathbf{z} \ \mathbf{forc}_{U^1}^{\mathbf{z}} \varphi$, but $p_1 \upharpoonright \widehat{\mathbf{z}} \ \mathbf{forc}_{U^1}^{\widehat{\mathbf{z}}} \neg \varphi$ still holds by Lemma 22.

We now recall that any condition $p \in {}^*\mathbf{P}_\Omega$ is a map $p \in \mathbf{L}$, defined on $\mathcal{I} = \omega \times \omega$, and each value $p(n,i) = \langle \mathbf{S}_p(n,i); \mathbf{F}_p(n,i)\rangle$ is a pair of a set $\mathbf{S}_p(n,i) \subseteq \mathbf{Seq}_\Omega$ and $\mathbf{F}_p(n,i) \subseteq \mathbf{Fun}_\Omega$, with $\mathtt{card}\,(\mathbf{S}_p(n,i) \cup \mathbf{F}_p(n,i)) < \Omega$ strictly, in $\mathbf{L}$. We define the *support*:

$$\|p\| = \bigcup_{n,i<\omega} \|p\|_{ni}, \text{ where } \|p\|_{ni} = \{s(0) : s \in \mathbf{S}_p(n,i)\} \cup \{f(0) : f \in \mathbf{F}_p(n,i)\};$$

then $\|p\| \in \mathbf{L}$, $\|p\| \subseteq \Omega$, and $\mathtt{card}\,\|p\| < \Omega$ strictly, so that $\|p\|$ is a bounded subset of Ω. In particular, $\|p_1\|$ is a bounded subset of Ω in $\mathbf{L}$. Therefore there is:

(B) A bijection $b \in \mathbf{L}$, $b : \Omega \xrightarrow{\text{onto}} \Omega$, such that $\|p_1\| \cap (b''\|p_1\|) = \varnothing$ and $b = b^{-1}$.

Furthermore, as $U^1 \in \mathbf{sDS}[m]$, the Ω-system U^1 is Ω-size, and hence the set $J = \bigcup_{n,i<\omega} U^1(n,i) \in \mathbf{L}$ satisfies $\mathtt{card}\,J \le \Omega$ in $\mathbf{L}$. It follows that there exists:

(C) A sequence $\{F_\alpha\}_{\alpha<\Omega} \in \mathbf{L}$ of bijections $F_\alpha : \Omega \xrightarrow{\text{onto}} \Omega$, such that $F_0 = b$ (see above), $F_\alpha = {F_\alpha}^{-1}$, and if $f, g \in J$ then there is an ordinal $\alpha < \Omega$ such that $f(\alpha) \ne F_\alpha(g(\alpha))$.

6.5. Transformation

In continuation of **the proof of Theorem** 11, we now define an automorphism acting on several different domains in $\mathbf{L}$. It will be based on π and F_α of Section 6.4 and its action will be denoted by $\widehat{\ }$. Along the way we will formulate properties (D)–(H) of the automorphism, a routine check of which is left to the reader.

We argue under the assumptions and notation of Definitions 7 on page 22 and 8 on page 25.

If $\alpha \leq \Omega$ and $f : \alpha \to \Omega$ then $\widehat{f} : \alpha \to \Omega$ is defined by $\widehat{f}(\gamma) = F_\gamma(f(\gamma))$ for all $\gamma < \alpha$. In particular, $\widehat{f}(0) = F_0(f(0)) = b(f(0))$. This defines $\widehat{s} \in \mathbf{Seq}_\Omega$ and $\widehat{f} \in \mathbf{Fun}_\Omega$ for all $s \in \mathbf{Seq}_\Omega$ and $f \in \mathbf{Fun}_\Omega$.

(D) $f \longmapsto \widehat{f}$ is a bijection $\mathbf{Seq}_\Omega \xrightarrow{\text{onto}} \mathbf{Seq}_\Omega$ and $\mathbf{Fun}_\Omega \xrightarrow{\text{onto}} \mathbf{Fun}_\Omega$, and if $f, g \in J = \bigcup_{n,i<\omega} U^1(n,i)$ then $\widehat{f} \neq g$ by (C).

If $u \subseteq \mathbf{Fun}_\Omega$ then let $\widehat{u} = \{\widehat{f} : f \in u\}$. If $S \subseteq \mathbf{Seq}_\Omega$ then let $\widehat{S} = \{\widehat{s} : s \in S\}$.
If U is a Ω-system then define a Ω-system $\widehat{U}$, such that:

$$\begin{array}{rcll} \widehat{U}(n,i) & = & U(n,i), & \text{in case} \quad n < m\,; \\ \widehat{U}(\pi(n),i) & = & \widehat{U(n,i)} = \{\widehat{f} : f \in U(n,i)\}, & \text{in case} \quad n \geq m\,. \end{array}$$

If $p \in {}^*\mathbf{P}_\Omega$ then let $\widehat{p} \in {}^*\mathbf{P}_\Omega$ be defined so that:

$$\begin{array}{rcll} \widehat{p}(n,i) & = & p(n,i), & \text{in case} \quad n < m\,; \\ \widehat{p}(\pi(n),i) & = & \langle \widehat{\mathbf{S}_p(n,i)}; \widehat{\mathbf{F}_p(n,i)} \rangle, & \text{in case} \quad n \geq m\,; \end{array}$$

where $\widehat{\mathbf{S}_p(n,i)} = \{\widehat{s} : s \in \mathbf{S}_p(n,i)\}$ and $\widehat{\mathbf{F}_p(n,i)} = \{\widehat{s} : s \in \mathbf{F}_p(n,i)\}$ by the above. These are consistent definitions because $\pi \in \Pi_m$.

(E) $\widehat{U} \restriction^{<m} = U \restriction^{<m}$ for any Ω-system U. The map $U \longmapsto \widehat{U}$ is a bijection of $\mathbf{sDS}_\Omega$ onto itself and $\mathbf{sDS}[k]$ onto itself for any $k \leq m$.

(F) $\widehat{p} \restriction^{<m} = p \restriction^{<m}$ for any $p \in {}^*\mathbf{P}_\Omega$. The map $p \longmapsto \widehat{p}$ is a $\leqslant$-preserving bijection of $\mathbf{P}[U]$ onto $\mathbf{P}[\widehat{U}]$.

If in addition $z \subseteq \omega \times \omega$ (not necessarily $z \in \mathbf{L}$), then if conditions $p, q \in {}^*\mathbf{P}_\Omega$ satisfy $p \restriction z = q \restriction z$, then easily $\widehat{p} \restriction \widehat{z} = \widehat{q} \restriction \widehat{z}$, where $\widehat{z} = \pi \cdot z = \{\langle \pi(n), i \rangle : \langle n, i \rangle \in z\}$. This allows us to define $\widehat{r} = \widehat{p} \restriction \widehat{z}$ for every $r \in {}^*\mathbf{P}_\Omega \restriction z$, where $p \in {}^*\mathbf{P}_\Omega$ is any condition satisfying $r = p \restriction z$.

(G) If $z \subseteq \omega \times \omega$ then $p \longmapsto \widehat{p}$ is a $\leqslant$-preserving bijection of $\mathbf{P}[U] \restriction z$ onto $\mathbf{P}[\widehat{U}] \restriction \widehat{z}$.

If $z \subseteq \omega \times \omega$ and $\tau \in \mathbf{Nam}^z_\zeta$ (see Section 5.1) then we define $\widehat{\tau} = \{\langle \widehat{p}, \alpha \rangle : \langle p, \alpha \rangle \in \tau\}$, and accordingly if φ is a $\mathcal{L}_z$-formula then $\widehat{\varphi}$ is obtained by substituting $\widehat{\tau}$ for each name τ in φ.

(H) If $z \subseteq \omega \times \omega$, $z \in \mathbf{L}[\zeta]$, then the mapping $\tau \longmapsto \widehat{\tau}$ is a bijection of $\mathbf{Nam}^z_\zeta$ onto $\mathbf{Nam}^{\widehat{z}}_\zeta$ and a bijection of $\mathcal{L}_z$-formulas onto $\mathcal{L}_{\widehat{z}}$-formulas.

Remark 6. *The action of* $\widehat{\ }$ *is idempotent, so that e.g.,* $\widehat{\widehat{f}} = f$ *for any* $f \in \mathbf{Fun}_\Omega$ *etc. This is because we require that* $b^{-1} = b$ *and* $F_\alpha^{-1} = F_\alpha$ *for all* $\alpha < \Omega$.

The action of $\widehat{\ }$ *is constructible on* $\mathbf{Seq}_\Omega$, $\mathbf{Fun}_\Omega$, *Ω-systems,* ${}^*\mathbf{P}_\Omega$, *since both* π *and the sequence of maps* F_α *belong to* $\mathbf{L}$ *by (B), (C).*

If $z \in \mathbf{L}[\zeta]$ *then the action of* $\widehat{\ }$ *on* ${}^*\mathbf{P}_\Omega \restriction z$ *and names in* $\mathbf{Nam}^z_\zeta$ *belongs to* $\mathbf{L}[\zeta]$, *since the extra parameter* $z \in \mathbf{L}[\zeta]$ *does not necessarily belong to* $\mathbf{L}$.

It is not unusual that transformations of a forcing notion considered lead to this or another invariance. The next lemma is exactly of this type.

Lemma 27 (in $\mathbf{L}[\zeta]$). *Assume that* $U \in \mathbf{sDS}_\Omega$, $z = w[\zeta]$, $\pi \in \Pi_m$, $\widehat{z} = \pi z$, $p \in \mathbf{P}[U] \restriction z$, *and* φ *is a closed formula of* $\mathcal{L}_z$, $\#(\varphi) \leq m+1$. *Then* $p \ \mathbf{forc}^z_U \ \Phi$ *iff* $\widehat{p} \ \mathbf{forc}^{\widehat{z}}_{\widehat{U}} \ \widehat{\Phi}$.

Proof. We argue by induction on the structure of Φ. Routine cases of atomic formulas (F2) and steps (F4) and (F5) of Section 5.2 by means of (D)–(H) are left to the reader. Thus we concentrate on atomic formulas of type (F3) and steps (F6) and (F8) in Section 5.2. In all cases we take care of only one direction of the equivalence of the lemma, as the other direction is entirely similar via Remark 6 just above.

Formulas of type (F3). Let Φ be $\alpha \in \tau$, where $\alpha < \Omega$ and $\tau \in \mathbf{Nam}^{z}_{\xi}$. Assume that $p\ \mathbf{forc}^{z}_{U}\ \alpha \in \tau$. Then by definition there is a condition $q \in \mathbf{P}[U]\restriction z$ such that $p \leqslant q$ and $\langle q, \alpha\rangle \in \tau$. Then $\widehat{q}$ and $\widehat{p}$ belong to $\mathbf{P}[\widehat{U}]\restriction \widehat{z}$, $\widehat{p} \leqslant \widehat{q}$, and $\langle \widehat{q}, \alpha\rangle \in \widehat{\tau}$, so we have $\widehat{p}\ \mathbf{forc}^{\widehat{z}}_{\widehat{U}}\ \alpha \in \widehat{\tau}$, as required.

Step (F6). Let $\Phi := \exists X\, \Psi(X)$. Suppose that $p\ \mathbf{forc}^{z}_{U}\ \Phi$. By definition there exists a name $\tau \in \mathbf{Nam}^{z}_{\xi}$ such that $p\ \mathbf{forc}^{z}_{\infty}\ \Psi(\tau)$, Then we have $\widehat{p}\ \mathbf{forc}^{\widehat{z}}_{\widehat{U}}\ \widehat{\Psi(\tau)}$ by the inductive hypothesis. But $\widehat{\Psi(\tau)}$ coincides with $\widehat{\Psi}(\widehat{\tau})$, where $\widehat{\tau} \in \mathbf{Nam}^{\widehat{z}}_{\xi}$ by (H) above. We conclude that $\widehat{p}\ \mathbf{forc}^{\widehat{z}}_{\widehat{U}}\ \exists X\, \widehat{\Psi(X)}$, that is, $\widehat{p}\ \mathbf{forc}^{\widehat{z}}_{\widehat{U}}\ \widehat{\Phi}$, as required.

Step (F8). Prove the lemma for a $\mathcal{L}_z$ formula $\Phi := \neg\Psi$, assuming that the result holds for Ψ. Note that $\#(\Phi) \leqslant m+1$, hence $\#(\Psi) \leqslant m$. Suppose that $p\ \mathbf{forc}^{z}_{U}\ \neg\Psi$ *fails*. By definition there is a Ω-system $U' \in \mathbf{sDS}[m]$ extending U', and a condition $q \in \mathbf{P}[U']\restriction z$, $q \leqslant p$, such that $q\ \mathbf{forc}^{z}_{U'}\ \Psi$. Then $\widehat{q}\ \mathbf{forc}^{\widehat{z}}_{\widehat{U'}}\ \widehat{\Psi}$ by the inductive hypothesis. Yet $\widehat{U'}$ belongs to $\mathbf{sDS}_{\Omega}$, extends $\widehat{U}$, and satisfies $\widehat{U'}\restriction^{<m} = U'\restriction^{<m}$ by (E), hence belonging even to $\mathbf{sDS}[m]$ by the choice of U', and in addition $\widehat{q} \in \mathbf{P}[\widehat{U'}]\restriction \widehat{z}$ and $\widehat{q} \leqslant \widehat{p}$ by (F). We conclude, by definition, that $\widehat{p}\ \mathbf{forc}^{\widehat{z}}_{\widehat{U}}\ \neg\widehat{\Psi}$ fails too, as required. □

6.6. Finalization

We continue to argue under the assumptions and notation of Definitions 6 on page 20 and 8 on page 25. The goal of this Section is to accomplish the proof of Theorem 11 in Section 6.2 that was started in Section 6.4. We return to objects introduced in (A), (B), (C) of Section 6.2.

Let $q_1 = p_1\restriction z$, so that $q_1 \in \mathbf{P}[U^1]\restriction z$ and $q_1\ \mathbf{forc}^{z}_{U^1}\ \varphi$ by (A). We have:

$$\widehat{U^1} \in \mathbf{sDS}[m] \ \wedge\ \widehat{p_1} \in \mathbf{P}[\widehat{U^1}] \ \wedge\ \widehat{q_1} = \widehat{p_1}\restriction \widehat{z} \in \mathbf{P}[\widehat{U^1}]\restriction \widehat{z} \ \wedge\ \widehat{q_1}\ \mathbf{forc}^{\widehat{z}}_{\widehat{U^1}}\ \varphi \tag{11}$$

by Lemma 27. (Here φ, as a parameter free formula, coincides with $\widehat{\varphi}$.) Let a Ω-system U be defined by $U(n,i) = U^1(n,i) \cup \widehat{U^1}(n,i)$.

Lemma 28. *The Ω-system U belongs to $\mathbf{sDS}[m]$ and extends both U^1 and $\widehat{U^1}$.*
Conditions p_1 and $\widehat{p_1}$ belong to $\mathbf{P}[U]$ and are compatible in $\mathbf{P}[U]$.

Proof (Lemma). It follows by (D) (last claim) that U is a disjoint Ω-system. It follows by (E) that $U\restriction^{<m} = U^1\restriction^{<m} = \widehat{U^1}\restriction^{<m}$. Therefore U belongs to $\mathbf{sDS}[m]$ because so does U^1.

To prove compatibility, it suffices to check that if $n, i < \omega$ then either $p_1(n,i) = \widehat{p_1}(n,i)$ or $\|p_1\|_{ni} \cap \|\widehat{p_1}\|_{ni} = \varnothing$. If $n < m$ then we have the 'either' case because by definition $p_1\restriction^{<m} = \widehat{p_1}\restriction^{<m}$. Suppose that $n \geq m$. Let $k = \pi^{-1}(n)$; thus still $k \geq m$ (as $\pi \in \Pi_m$), $n = \pi(k)$, and $\widehat{p_1}(n,i) = \langle \widehat{\mathbf{S}_p(k,i)}; \widehat{\mathbf{F}_p(k,i)}\rangle$. It follows that $\|\widehat{p_1}\|_{ni}$ is the F_0-image, hence the b-image of the set $\|p_1\|_{ki}$. However $\|p_1\|_{ki} \cup \|p_1\|_{ni} \subseteq \|p_1\|$. We conclude that $\|p_1\|_{ni} \cap \|\widehat{p_1}\|_{ni} = \varnothing$ by Claim (B) of Section 6.4, as required. □

To finalize **the proof of Theorem** 11, let, by Lemma 28, $r \in \mathbf{P}[U]\restriction \widehat{z}$ satisfy both $r \leqslant p_1\restriction \widehat{z}$ and $r \leqslant \widehat{p_1}\restriction \widehat{z} = \widehat{q_1}$. However $\widehat{q_1}\ \mathbf{forc}^{\widehat{z}}_{\widehat{U^1}}\ \varphi$ by (11). We conclude that $r\ \mathbf{forc}^{\widehat{z}}_{U}\ \varphi$ by Lemma 28 and Lemma 22. On the other hand, $p_1\restriction \widehat{z}\ \mathbf{forc}^{\widehat{z}}_{U^1}\ \neg\varphi$ by (A) of Section 6.4, therefore we have $r\ \mathbf{forc}^{\widehat{z}}_{U}\ \neg\varphi$. It remains to remind that $\#(\varphi) \leq m$ and $U \in \mathbf{sDS}[m]$ by Lemma 28—and we still get a contradiction by Lemma 22(ii). The contradiction completes the proof of Theorem 11.

Finalization.

Theorem 11 just proved implies Theorem 9, see Section 6.3.

Theorem 9 ends the proof of Theorem 8 of Section 4.6, see Remark 5 on page 24.

This completes the proof of Theorem 2, the main result of this paper, see Remark 4 on page 23.

7. Conclusions and Discussion

In this study, the method of almost-disjoint forcing was employed to the problem of getting a model of **ZFC** in which the set $\mathbf{D}_{1m}$ of all reals, definable by a parameter free type-theoretic formula with the highest quantifier order not exceeding a given natural number $\mathbb{M} \geq 1$, belongs to $\mathbf{D}_{2\mathbb{M}}$, that is, it is itself definable by a formula of the same quantifier order. Moreover, we have $\mathbf{D}_{1\mathbb{M}} = \mathbf{L} \cap \mathbb{R}$ in the model, that is, the set $\mathbf{D}_{1\mathbb{M}}$ is equal to the set of all Gödel-constructible reals.

The problem of getting a model for $\mathbf{D}_{1\mathbb{M}} \in \mathbf{D}_{2\mathbb{M}}$ was posed in Alfred Tarski's article [18]. Its particular case $\mathbb{M} = 1$ (analytical definability), that is, the problem of getting models for $\mathbf{D}_{11} \in \mathbf{D}_{21}$, or stronger, $\mathbf{D}_{11} = \mathbf{L} \cap \mathbb{R}$, has been known since the early years of forcing, see e.g., problem 87 in Harvey Friedman's survey [21], and problems 3110, 3111, and 3112 in another early survey [20] by A. Mathias. As mentioned in [20,21], the particular case $\mathbb{M} = 1$ was solved by Leo Harrington, and a sketch of the proof, related to a model for $\Delta^1_3 = \mathbf{L} \cap \mathbb{R}$, can be found in Harrington's handwritten notes [22]. Our paper presents a full proof of the comprehensive result (Theorem 2) that finally solves the Tarski problem.

From this study, it is concluded that the hidden invariance technique (as outlined in Section 6.1) allows one to solve the problem by providing a generic extension of $\mathbf{L}$ in which the constructible reals are precisely the $\mathbf{D}_{1\mathbb{M}}$ reals, for a chosen value $\mathbb{M} \geq 1$. The hidden invariance technique has also been applied in recent papers [3–5,28] for the problem of getting a set theoretic structure of this or another kind at a preselected projective level. We finish with a short list of related problems.

1. If $x \subseteq \omega$ then let $\mathbf{D}_{pm}(x)$ be the set of all objects of order p, definable by a formula with x as the only parameter, whose all quantified variables are over orders $\leq m$. (Compare to Definition 1 on page 2.) One may be interested in getting a model for:

$$\forall\, x \subseteq \omega\, (\mathbf{D}_{1m}(x) \in \mathbf{D}_{2m}(x), \ \text{ or stronger, } \ \mathbf{D}_{1m}(x) = \mathscr{P}(\omega) \cap \mathbf{L}). \tag{12}$$

This is somewhat similar to Problem 87′ in [21]: Find a model of:

$$\mathbf{ZFC} + \text{“ for any reals } x, y, \text{ we have: } \ x \in \mathbf{L}[y] \implies x \text{ is } \Delta^1_3 \text{ in } y\text{”}. \tag{13}$$

Problem (13) was known in the early years of forcing, see, e.g., problem 3111 in [20] or (3) in [23] (Section 6.1). Problem (13) was positively solved by René David [48,49], where the question is attributed to Harrington. The proof makes use of a tool known as David's trick, see S. D. Friedman [27] (Chapters 6, 8).

So far it is unknown whether the result of David [48] generalizes to higher projective classes Δ^1_n, $n \geq 4$, or Δ^1_ω, whether it can be strengthened towards $\Longleftrightarrow$ instead of $\Longrightarrow$, and whether it can lead to an even partial solution of (12). This is a very interesting and perhaps difficult question.

2. Coming back to Harvey Friedman's Δ^1_n problem of getting a model for the sentence:

$$\text{the set } \ d_n = \mathscr{P}(\omega) \cap \Delta^1_n \ \text{ is equal to } \ \mathscr{P}(\omega) \cap \mathbf{L}, \tag{14}$$

(Section 1.2), it is clear that, unlike $\mathbf{D}_{1m} \in \mathbf{D}_{2m}$, if (14) holds for some $n \geq 3$ then it definitely fails for any $n' \neq n$. But we can try to weaken (14) to just:

$$d_n \in \Pi^1_n, \tag{15}$$

and then ask whether there is a generic extension of $\mathbf{L}$ satisfying $\forall\, n\, (d_n \in \Pi^1_n)$. It holds by rather routine estimations that $d_1 \in \Pi^1_1 \smallsetminus \Sigma^1_1$, $d_2 \in \Sigma^1_2 \smallsetminus \Pi^1_2$, and if all reals are constructible then $d_n \in \Sigma^1_n \smallsetminus \Pi^1_n$ for all $n \geq 3$ as well, so Π^1_n looks rather suitable in (15).

3. Recall that Theorem 2 implies the consistency of $\mathbf{D}_{1m} \in \mathbf{D}_{2m}$ for each particular $m \geq 1$.

But what about the consistency of the sentence "$\mathbf{D}_{1m} \in \mathbf{D}_{2m}$ holds for all $m \geq 1$"? Perhaps a method developed in [50] can be useful to solve this problem.

4. It would be interesting to define a generic extension of $\mathbf{L}$ in which, for instance, $\mathbf{D}_{1m} \in \mathbf{D}_{2m}$ holds for all even $m \geq 1$ but fails for all odd $m \geq 1$, or vice versa.

Lemma 2 on page 8 presents a possible difficulty: If we have $\mathbf{D}_{1n} \in \mathbf{D}_{2n}$ for some $n \geq 1$ by means of the equality $\mathbf{D}_{1n} = \mathscr{P}(\omega) \cap \mathbf{L}$, then $\mathbf{D}_{1m} \in \mathbf{D}_{2m}$ definitely *fails* for all $m < n$.

5. Another question considered by Tarski in [18] is related to the sets $\mathbf{D}_k = \bigcup_m \mathbf{D}_{km}$ (all elements of order k, definable by a formula of any order). Tarski proves that $\mathbf{D}_k \notin \mathbf{D}_{k+1}$ for all $k \geq 2$, and leaves open the question whether $\mathbf{D}_1 \in \mathbf{D}_2$. Similarly to the problem $\mathbf{D}_{1m} \in \mathbf{D}_{2m}$ in Section 1.1, the *negative* answer $\mathbf{D}_1 \notin \mathbf{D}_2$ follows from the axiom of constructibility $\mathbf{V} = \mathbf{L}$, and hence is consistent with **ZFC**.

Prove the consistency of the sentences $\mathbf{D}_1 \in \mathbf{D}_2$ and $\mathbf{D}_1 = \mathscr{P}(\omega) \cap \mathbf{L}$.

Supplementary Materials: The following are available online at http://www.mdpi.com/2227-7390/8/12/2214/s1 .

Author Contributions: Conceptualization, V.K. and V.L.; methodology, V.K. and V.L.; validation, V.K. and V.L.; formal analysis, V.K. and V.L.; investigation, V.K. and V.L.; writing original draft preparation, V.K.; writing review and editing, V.K.; project administration, V.L.; funding acquisition, V.L. All authors have read and agreed to the published version of the manuscript.

Funding: This research was funded by the Russian Foundation for Basic Research RFBR grant number 18-29-13037.

Acknowledgments: We thank the anonymous reviewers for their thorough review and highly appreciate the comments and suggestions, which significantly contributed to improving the quality of the publication.

Conflicts of Interest: The authors declare no conflict of interest. The funders had no role in the design of the study; in the collection, analyses, or interpretation of data; in the writing of the manuscript, or in the decision to publish the results.

References

1. Kanovei, V.; Lyubetsky, V. A definable E_0 class containing no definable elements. *Arch. Math. Logic* **2015**, *54*, 711–723. [CrossRef]
2. Golshani, M.; Kanovei, V.; Lyubetsky, V. A Groszek—Laver pair of undistinguishable E_0 classes. *Math. Logic Q.* **2017**, *63*, 19–31. [CrossRef]
3. Kanovei, V.; Lyubetsky, V. Definable E_0 classes at arbitrary projective levels. *Ann. Pure Appl. Logic* **2018**, *169*, 851–871. [CrossRef]
4. Kanovei, V.; Lyubetsky, V. Models of set theory in which nonconstructible reals first appear at a given projective level. *Mathematics* **2020**, *8*, 910. [CrossRef]
5. Kanovei, V.; Lyubetsky, V. On the Δ^1_n problem of Harvey Friedman. *Mathematics* **2020**, *8*, 1477. [CrossRef]
6. Hadamard, J.; Baire, R.; Lebesgue, H.; Borel, E. Cinq lettres sur la théorie des ensembles. *Bull. Soc. Math. Fr.* **1905**, *33*, 261–273. [CrossRef]
7. Karagila, A. The Bristol model: An abyss called a Cohen reals. *J. Math. Log.* **2018**, *18*, 1850008. [CrossRef]
8. Antos, C.; Friedman, S.D. Hyperclass forcing in Morse-Kelley class theory. *J. Symb. Log.* **2017**, *82*, 549–575. [CrossRef]
9. Antos, C.; Friedman, S.D.; Honzik, R.; Ternullo, C. (Eds.) *The Hyperuniverse Project and Maximality*; Birkhäuser: Cham, Switzerland, 2018; xi + 270p.
10. Cummings, J.; Friedman, S.D.; Magidor, M.; Rinot, A.; Sinapova, D. Ordinal definable subsets of singular cardinals. *Isr. J. Math.* **2018**, *226*, 781–804. [CrossRef]
11. Chan, W. Ordinal definability and combinatorics of equivalence relations. *J. Math. Log.* **2019**, *19*, 1950009. [CrossRef]
12. Fischer, V.; Schrittesser, D. A Sacks indestructible co-analytic maximal eventually different family. *Fundam. Math.* **2020**. [CrossRef]
13. Enayat, A.; Kanovei, V. An unpublished theorem of Solovay on OD partitions of reals into two non-OD parts, revisited. *J. Math. Log.* **2020**, 1–22. [CrossRef]
14. Tarski, A. Der Wahrheitsbegriff in den formalisierten Sprachen. *Studia Philos.* **1935**, *1*, 261–401.

15. Murawski, R. Undefinability of truth. The problem of priority: Tarski vs Gödel. *Hist. Philos. Log.* **1998**, *19*, 153–160. [CrossRef]
16. Addison, J.W. Tarski's theory of definability: Common themes in descriptive set theory, recursive function theory, classical pure logic, and finite-universe logic. *Ann. Pure Appl. Logic* **2004**, *126*, 77–92. [CrossRef]
17. Tarski, A. Sur les ensembles définissables de nombres réels. I. *Fundam. Math.* **1931**, *17*, 210–239. [CrossRef]
18. Tarski, A. A problem concerning the notion of definability. *J. Symb. Log.* **1948**, *13*, 107–111. [CrossRef]
19. Gödel, K. *The Consistency of the Continuum Hypothesis*; Annals of Mathematics Studies, No. 3; Princeton University Press: Princeton, NJ, USA, 1940; p. 66. [CrossRef]
20. Mathias, A.R.D. Surrealist landscape with figures (a survey of recent results in set theory). *Period. Math. Hung.* **1979**, *10*, 109–175.
21. Friedman, H. One hundred and two problems in mathematical logic. *J. Symb. Log.* **1975**, *40*, 113–129. [CrossRef]
22. Harrington, L. The Constructible Reals Can Be Anything. Preprint dated May 1974 with several addenda dated up to October 1975: (A) Models Where Separation Principles Fail, May 74; (B) Separation without Reduction, April 75; (C) The Constructible Reals Can Be (Almost) Anything, Part II, May 75. Available online: http://logic-library.berkeley.edu/catalog/detail/2135 (accessed on 9 December 2020).
23. Jensen, R.B.; Solovay, R.M. Some applications of almost disjoint sets. In *Math. Logic Found. Set Theory, Proc. Int. Colloqu., Jerusalem 1968*; Studies in Logic and the Foundations of Mathematics; Bar-Hillel, Y., Ed.; North-Holland: Amsterdam, The Netherlands; London, UK, 1970; Volume 59, pp. 84–104.
24. Hinman, P.G. *Recursion-Theoretic Hierarchies*; Perspectives in Mathematical Logic; Springer: Berlin/Heidelberg, Germany, 1978; x + 480p.
25. Harrington, L. Long projective wellorderings. *Ann. Math. Logic* **1977**, *12*, 1–24. [CrossRef]
26. Friedman, S.D. Constructibility and class forcing. In *Handbook of Set Theory. In 3 Volumes*; Springer: Dordrecht, The Netherlands, 2010; pp. 557–604.
27. Friedman, S.D. *Fine Structure and Class Forcing*; De Gruyter Series in Logic and Its Applications; de Gruyter: Berlin, Germany, 2000; Volume 3, x + 221p. [CrossRef]
28. Kanovei, V.; Lyubetsky, V. Definable minimal collapse functions at arbitrary projective levels. *J. Symb. Log.* **2019**, *84*, 266–289. [CrossRef]
29. Kanovei, V.; Lyubetsky, V. Non-uniformizable sets with countable cross-sections on a given level of the projective hierarchy. *Fundam. Math.* **2019**, *245*, 175–215.
30. Kanovei, V.; Lyubetsky, V. Models of set theory in which separation theorem fails. *Izvestiya: Math.* **2021**, *85*, to appear.
31. Barwise, J. (Ed.) *Handbook of Mathematical Logic*; Studies in Logic and the Foundations of Mathematics; North-Holland: Amsterdam, The Netherlands, 1977; Volumed 90, p. 1165.
32. Jech, T. *Set Theory*; The Third Millennium Revised and Expanded Ed.; Springer: Berlin/Heidelberg, Germany, 2003. xiii + 769p. [CrossRef]
33. Addison, J.W. Some consequences of the axiom of constructibility. *Fundam. Math.* **1959**, *46*, 337–357. [CrossRef]
34. Jensen, R.B.; Johnsbraten, H. A new construction of a non-constructible Δ^1_3 subset of ω. *Fundam. Math.* **1974**, *81*, 279–290.
35. Kunen, K. *Set Theory*; Studies in Logic; College Publications: London, UK, 2011; Volume 34, viii + 401p.
36. Shoenfield, J.R. Unramified forcing. In *Axiomatic Set Theory. Proc. Sympos. Pure Math.*; Scott, D.S., Ed.; AMS: Providence, RI, USA, 1971; Volume 13, Part 1, pp. 357–381. [CrossRef]
37. Kanovei, V. On the nonemptiness of classes in axiomatic set theory. *Math. USSR, Izv.* **1978**, *12*, 507–535.
38. Jensen, R. Definable sets of minimal degree. In *Math. Logic Found. Set Theory, Proc. Int. Colloqu., Jerusalem 1968*; Studies in Logic and the Foundations of Mathematics; Bar-Hillel, Y., Ed.; North-Holland: Amsterdam, The Netherlands; London, UK, 1970; Volume 59, pp. 122–128. [CrossRef]
39. Friedman, S.D.; Gitman, V.; Kanovei, V. A model of second-order arithmetic satisfying AC but not DC. *J. Math. Log.* **2019**, *19*, 1850013. [CrossRef]
40. Kanovei, V.; Lyubetsky, V. Counterexamples to countable-section Π^1_2 uniformization and Π^1_3 separation. *Ann. Pure Appl. Logic* **2016**, *167*, 262–283. [CrossRef]
41. Abraham, U. A minimal model for $\neg CH$: iteration of Jensen's reals. *Trans. Am. Math. Soc.* **1984**, *281*, 657–674. [CrossRef]
42. Abraham, U. Minimal model of "$\aleph_1^L$ is countable" and definable reals. *Adv. Math.* **1985**, *55*, 75–89. [CrossRef]

43. Kanovei, V.; Lyubetsky, V. Non-uniformizable sets of second projective level with countable cross-sections in the form of Vitali classes. *Izv. Math.* **2018**, *82*, 61–90. [CrossRef]
44. Groszek, M.; Jech, T. Generalized iteration of forcing. *Trans. Amer. Math. Soc.* **1991**, *324*, 1–26. [CrossRef]
45. Groszek, M.; Laver, R. Finite groups of OD-conjugates. *Period. Math. Hung.* **1987**, *18*, 87–97. [CrossRef]
46. Kanovei, V. On non-wellfounded iterations of the perfect set forcing. *J. Symb. Log.* **1999**, *64*, 551–574. [CrossRef]
47. Versaci, M.; di Barba, P.; Morabito, F.C. Curvature-Dependent Electrostatic Field as a Principle for Modelling Membrane-Based MEMS Devices. A Review. *Membranes* **2020**, *10*. [CrossRef]
48. David, R. Δ^1_3 reals. *Ann. Math. Logic* **1982**, *23*, 121–125. [CrossRef]
49. David, R. A very absolute Π^1_2 real singleton. *Ann. Math. Logic* **1982**, *23*, 101–120. [CrossRef]
50. Kanovei, V.; Lyubetsky, V. The full basis theorem does not imply analytic wellordering. *Ann. Pure Appl. Logic* **2020**, Online. [CrossRef]

mathematics

MDPI

Article

Linear Time Additively Exact Algorithm for Transformation of Chain-Cycle Graphs for Arbitrary Costs of Deletions and Insertions

Konstantin Gorbunov and Vassily Lyubetsky *

Institute for Information Transmission Problems of the Russian Academy of Sciences, Bolshoi Karetnyi, 19, Moscow 127994, Russia; gorbunov@iitp.ru
* Correspondence: lyubetsk@iitp.ru; Tel.: +7-910-464-6917

Received: 18 October 2020; Accepted: 5 November 2020; Published: 10 November 2020

Abstract: We propose a novel linear time algorithm which, given any directed weighted graphs a and b with vertex degrees 1 or 2, constructs a sequence of operations transforming a into b. The total cost of operations in this sequence is minimal among all possible ones or differs from the minimum by an additive constant that depends only on operation costs but not on the graphs themselves; this difference is small as compared to the operation costs and is explicitly computed. We assume that the double cut and join operations have identical costs, and costs of the deletion and insertion operations are arbitrary strictly positive rational numbers.

Keywords: discrete optimization; exact algorithm; additively exact algorithm; graph transformation; graph of degree 2; chain-cycle graph; operation cost; minimization of total cost

---

Dedicated to the 70-th anniversary of A. L. Semenov.

1. Introduction and Basic Definitions

We consider the problem of constructing an algorithm for efficient solution of the below problem. We are given directed graphs a and b in which each vertex has degree either 1 or 2 and each edge is assigned with its *unique* name, a natural number (in this sense, a graph is referred to as a weighted graph with unique names). We consider a vertex in a graph as two *joined* (identified) endpoints of the adjacent edges. The following operations over such graphs are well known: cut any vertex (*Cut*) or join two currently free (i.e., of degree 1) ends (*OM*); cut a vertex and join one of the thus formed free ends with any currently free end (*SM*); cut two vertices and join the four thus formed free ends (*DM*). The latter two operations are compositions of the two former ones, but they are considered as independent operations. These four operations were defined in [1] and are traditionally referred to as *DCJ* (i.e., double cut and join) *operations*; they were depicted in [2] (Figure 1).

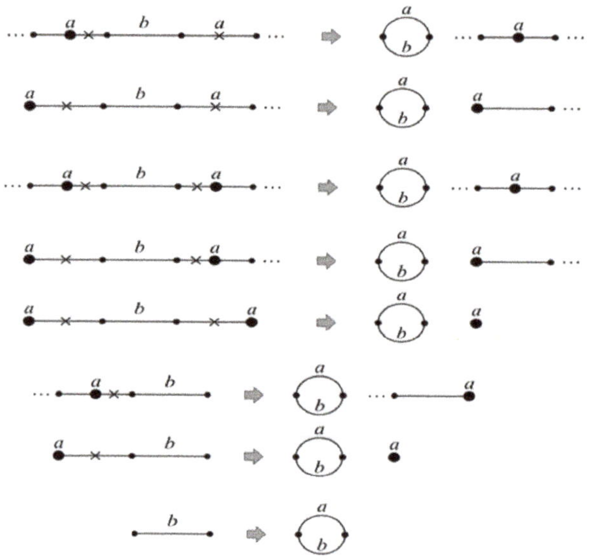

Figure 1. Cutting out a conventional *b*-edge. Singular vertices are marked by large circles. The cases differ by the form of edges adjacent to the edge to be cut out. From top to bottom: singular nonhanging and conventional; hanging and conventional; both singular nonhanging; hanging and singular nonhanging; both hanging; only one singular nonhanging; only one hanging; no adjacent edges.

There are two more operations: *remove* (*Rem*) a connected fragment of edges with names in *a* but not in *b* or, vice versa, *insert* (*Ins*) such a fragment with names not in *a* but in *b*. When removing a fragment, the arising free ends are joined; when inserting a fragment, first, some vertex is cut (if it is not extreme) and, then, after the insertion, two pairs of arising free ends are joined. These operations are analogous to standard deletion of a subword in a word and to insertion of a word as a subword. Each operation is assigned with its cost, a strictly positive rational number.

It is required to find a sequence (composition) of these operations with the minimum total cost which transforms the given graphs a and b, the first into the second. Such a sequence is said to be the shortest. In [2], previously obtained results were stated in detail and references on this problem were presented. Among these references, we point out the two latest papers [3,4], which, in turn, contain further historical references. The previous results concerning this problem and including its applied aspects, were overviewed in [5] (Chapter 10) and [6]. A principal distinction between our Theorem 1 and preceding results is the fact that we do not assume the condition of equal costs of deletion and insertion operations, which essentially simplified the problem. In our paper, we assume no restrictions on costs of these operations. As before, we still keep the condition of equal costs of DCJ operations. Thus, in this paper we prove the following:

Theorem 1. *If DCJ operations have the same cost w and the deletion and insertion operations have arbitrary costs w_d and w_i, then the **Algorithm** described below in Section 2, Section 4, and Section 5 outputs an additively shortest sequence of operations and has linear time complexity.*

We denote the additive exactness constant by k. The proof of the exactness splits into ***three cases***: **(I)** both costs w_d and w_i are not less than w (then, $k \leq 2w$), **(II)** both are not greater than w (then $k = 0$), or **(III)** one of them is less and the other greater than w (an expression for k is given at the beginning of Section 4). From the description of a problem equivalent to this (Section 2, Stage 0), it follows that the two possibilities in the last case are symmetric and, therefore, do not differ from each other. In each of the cases, the description of the Algorithm, given in Section 2, slightly changes. The linear time complexity of each of these versions of the Algorithm directly follows from its description. The arXiv publication [2] is preparatory to the present paper and contains figures illustrating the algorithm operation.

2. Description of the Algorithm for Case II

We list the stages of this algorithm, which are then explained in sequence and in detail below, in this section (see also Table 1).

Table 1. Basic parameters predefined in or computed by the Algorithm for Case II. *OM, single merging; SM, sesquialteral intermerging; DM, double intermerging.*

Notation	Meaning
w	Equal costs of the *Cut*, *OM*, *SM*, and *DM* operations
$w_d = w_a$	Cost of the *Rem* operation
$w_i = w_b$	Cost of the *Ins* operation
x_s	Number of elements corresponding to the 2-interaction s in the sought-for domain M
t	Type of an argument in any 2-interaction; this t is one of the following types:$1a$, $1b$, $2a$, $2b^*$, $2b'$, $3a^*$, $3a'$, $3b$, 2, 3.
c_{ts}	Number of occurrences of the type t in arguments of the 2-interaction s
l_t	Number of chains of the type t in the breakpoint graph G'
$P(s)$	Quality of the 2-interaction s

Stage 0: Transform initial given graphs a and b into the new graph, denoted by $a + b$ below.

Stage 1.: Cut out conventional edges in $a + b$.

Stage 2.0: Solve the integer linear programming problem which outputs the set of disjoint elements (each of them consisting of pairs, triples, or quadruples of chains of the current graph) with the largest aggregate quality.

Stage 2.1: For each element in this set, perform the interaction between its chains; the interaction is uniquely determined by the element and combines the chains of the element into a single chain.

Stage 3: Circularize chains of sizes strictly larger than 0 into cycles, and then break all cycles into many cycles of size 2; then, remove all singular vertices and loops.

Now, we pass to a detailed description of each stage.

Stage 0: An initial pair of graphs a and b is transformed in linear time into a new («breakpoint») graph $a + b$ such that the original problem is equivalent to the problem of reducing $a + b$ to the simplest form, referred to as final. A *final form* is a graph consisting of cycles of length 2 (one edge of each cycle is labeled by a, and the other, by b) and isolated conventional vertices; the definition of «conventional» is given below. The proof of the equivalence of these two problems literally repeats the proof of Corollary 5 in [7]; that proof used the equality of costs of DCJ operations only, which is assumed in Theorem 1. In [7], there was also given a linear time algorithm that transforms a solution of the second problem into a solution of the first (and vice versa).

The definition of the graph $a + b$ is given in [7]; in other variants, it was known from earlier works, for instance, from [8,9]. We recall the definition from [7], which describes vertices and edges of $a + b$. Vertices in $a + b$ are all endpoints of edges that occur in both a and b (they are denoted by n_i, where n is the edge name, $i = 1$ for a tail of an edge, and $i = 2$ for its head) and also vertices uniquely corresponding to every maximal connected segment of edges («block») in either $a \backslash b$ or $b \backslash a$, which we label by a or b, respectively. Vertices of the first type are referred to as *conventional*, and those of the second type are referred to as *singular*. Edges in $a + b$ connect vertices whenever the latter are joined in a or in b or if an extremity of a block is joined with a vertex in a or in b; edges are labeled by a or b, respectively. Edges of the first type are said to be *conventional*, and those of the second type are said to be *singular*. If a singular endpoint in $a + b$ is of degree 1, then both the edge and the endpoint itself are said to be *hanging*.

An operation o' over $a + b$ is defined through an operation o over a by commutativity: $o'(a + b) = o(a) + b$. Thus, over a breakpoint graph, the five operations listed below are allowed, which are shown in Figure 5 in [2]. In fact, the number of operations is six, since the fifth one, removal of a singular vertex, divides into removal of an *a-singular* vertex, which corresponds to deleting a block in a, and removal of a *b-singular* vertex, which is inverse to inserting a block in b.

Double intermerging (**DM**): deletion of two edges with the same label and joining the four thus formed endpoints by two new edges with this label. If the operation involves a loop, then its vertex is regarded as having two endpoints. If two adjacent singular vertices are formed, then the edge connecting them is deleted and they are *merged* into a single vertex.

Sesquialteral intermerging (**SM**): deletion of an edge and adding an edge with the same label connecting one of the thus formed free ends with a free end of an edge with an alternative label, or with a hanging or singular isolated vertex with the same label (with possible *merging* of adjacent singular vertices).

Single merging (**OM**): adding an a-edge between free vertices such that each of them is either a conventional vertex incident to a b-edge or an isolated vertex, a-hanging or a-isolated, with possible *merging* of adjacent singular vertices. A similar definition is given with b instead of a.

Cut (**Cut**): deletion of any edge.

The above-listed operations are referred to as DCJ (or sometimes *standard*) operations. A supplementary operation is *removal* (**Rem**) of a singular vertex. Namely, if it is of degree 2, it is removed and the edges incident to it are joined into an edge with the same label; if it is hanging, it is removed together with the edge incident to it; if it isolated or has a loop, the vertex and the loop are removed.

Each standard operation is assigned with its original *cost* w; removal of an a-singular vertex has original *cost* $w_a = w_d$, and removal of a b-singular vertex has original *cost* $w_b = w_i$.

An inclusion-maximal connected fragment of conventional edges is called a *segment*; depending on its length, it can be either even or odd.

The *size* of a component in $a + b$ is the number of conventional edges plus half the number of singular nonhanging edges in it. For conventional isolated vertices and loops, the size is defined to be 0, and, for singular isolated vertices, the size is defined to be −1.

Our algorithm successively generates graphs G starting from $a + b$; all these G are of the form $c + d$ for their initial graphs c and d; all these G together form a sequence that begins with $a + b$ and ends with a graph of a final form. Now, we pass to the description of the Algorithm, which consists of three consecutive stages.

Stage 1: From all components other than cycles of size 2, *cut out* conventional edges, i.e., apply a DM to a pair of edges adjacent to the edge to be cut out, or similarly apply an SM or OM if one or two of the adjacent edges do not exist. Such a derived operation is called *cutting out* (conventional edges) (Figure 1).

A chain of an odd (even) size is called *odd* (*even*); 0 is an even number, and −1 is odd. The definition of a *type of a chain* plays a crucial role. First, assume that a chain *does not contain conventional edges*. Then, $1a$ is an odd chain with one hanging b-edge; $2a^*$ is an odd chain with two hanging b-edges; $2a'$ is a b-singular isolated vertex; $2a$ denotes type $2a^*$ or $2a'$. Type $3a^*$ is an odd chain without hanging edges, with two extremal a-edges, and having a b-singular vertex; $3a'$ is a chain aa; $3a$ denotes type $3a^*$ or $3a'$. Type $1_a{}^*$ is an even chain with one hanging a-edge that has a b-singular vertex; $1_a{}'$ is a hanging a-edge; 1_a denotes type $1_a{}^*$ or $1_a{}'$. Similar definitions are given with b instead of a. Type 2^* is an even chain with two hanging edges nonincident to each other; $2'$ is two hanging edges incident to a common conventional vertex; 2 denotes type 2* or 2′. Type 3 is an even chain without hanging edges but having singular vertices; 0 is a chain without singular vertices. The chain types were presented in Figure 6 in [2]. The type of a chain *with conventional edges* is defined as the type of a chain obtained by cutting them out; it does not depend on the order of cuttings (see Lemma 6 in [7]).

A *hanging extremity* of a chain not of type 0 is an extremity with an adjacent hanging edge or odd segment. After cutting out this segment, a hanging edge appears. A chain 2*a* (or 2*b*, 3*a*, 3*b*) is a chain of type 2*a*′ (respectively, 2*b*′, 3*a*′, 3*b*′) if and only if it contains *b*-singular vertices only (*a*-singular only, *a*-singular only, *b*-singular only). A chain of type 2 is a chain of type 2′ if and only if it contains a conventional vertex with only *a*-singular vertices on one side and only *b*-singular vertices on the other.

An *interaction* in *G* is a chain of operations successively applied to *G*. In Section 3, we introduce a key notion of *P*(*s*), the ***quality*** of an interaction *s*. We also give there a convenient formula for computing it; see Equation (1b), which is consistent with Equation (1a). A term equality $1a + 1b = 1^*_b$ means that an interaction is applied to two chains of types 1*a* and 1*b* and outputs a chain of type 1^*_b. The same applies for other term equalities that are defined with the help of type designations, the + sign, and parentheses. Each interaction below corresponds to its term equality, which can be regarded as a designation (name) of this interaction. In square brackets, we give the interaction quality.

Stage 2: *2-interactions* are SM applied to two different chains with the following term equalities (a cut chain is always given the first; on the right-hand side, we do not present conventional isolated vertices): $1a + 1b = 1^*_b\ [w_a + w_b]$, $3a^* + 2b^* = 1_a\ [w_b]$, $3a^* + 2b' = 1_a\ [w_a]$, $3a' + 2b^* = 1_a\ [w_a]$, $3a' + 2b' = 1'_a\ [w_a]$, $3b + 2a = 1_b\ [w_b]$, $3 + 2 = 1^*_b\ [w_a + w_b - 1]$, $(1a + 2b^*) + 3 = 1^*_b\ [w_a + 2w_b - 1]$, $(1a + 2b') + 3 = 1^*_b\ [2w_a + w_b - 1]$, $(1b + 2a) + 3 = 1^*_b\ [w_a + 2w_b - 1]$, $(3a^* + 1b) + 2 = 1^*_b\ [w_a + 2w_b - 1]$, $(3a' + 1b) + 2 = 1^*_b\ [2w_a + w_b - 1]$, $(3b + 1a) + 2 = 1^*_b\ [w_a + 2w_b - 1]$, $1a + 2 = 2a^*\ [w_a + w_b - 1]$, $1b + 2 = 2b^*\ [w_a + w_b - 1]$, $3 + 1a = 3a^*\ [w_a + w_b - 1]$, $3 + 1b = 3b^*\ [w_a + w_b - 1]$, $(3b + 1a) + (1a + 2b^*) = 1^*_b\ [w_a + 3w_b - 1]$, $(3b + 1a) + (1a + 2b') = 1^*_b\ [2w_a + 2w_b - 1]$, $(3a^* + 1b) + (1b + 2a) = 1^*_b\ [w_a + 3w_b - 1]$, $(3a' + 1b) + (1b + 2a) = 1^*_b\ [2w_a + 2w_b - 1]$, $1a + (1a + 2b^*) = 2a^*\ [w_a + 2w_b - 1]$, $1a + (1a + 2b') = 2a^*\ [2w_a + w_b - 1]$, $1b + (1b + 2a) = 2b^*\ [w_a + 2w_b - 1]$, $(3b + 1a) + 1a = 3a^*\ [w_a + 2w_b - 1]$, $(3a^* + 1b) + 1b = 3b^*\ [w_a + 2w_b - 1]$, $(3a' + 1b) + 1b = 3b^*\ [2w_a + w_b - 1]$, $1a + 2b^* = 2\ [w_b]$, $1a + 2b' = 2\ [w_a]$, $1b + 2a = 2\ [w_b]$, $3a^* + 1b = 3\ [w_b]$, $3a' + 1b = 3\ [w_a]$, $3b + 1a = 3\ [w_b]$, $3 + ((3 + 2b^*) + 2a) = 1^*_b\ [w_a + 3w_b - 2]$, $3 + ((3 + 2b') + 2a) = 1^*_b\ [2w_a + 2w_b - 2]$, $(3a^* + (3b + 2)) + 2 = 1^*_b\ [w_a + 3w_b - 2]$, $(3a' + (3b + 2)) + 2 = 1^*_b\ [2w_a + 2w_b - 2]$, $(3a^* + 2) + 2 = 2a^*\ [w_a + 2w_b - 2]$, $(3a' + 2) + 2 = 2a^*\ [2w_a + w_b - 2]$, $(3b + 2) + 2 = 2b^*\ [w_a + 2w_b - 2]$, $3 + (3 + 2a) = 3a^*\ [w_a + 2w_b - 2]$, $3 + (3 + 2b^*) = 3b^*\ [w_a + 2w_b - 2]$, $3 + (3 + 2b') = 3b^*\ [2w_a + w_b - 2]$, $(3 + 2b^*) + 2a = 2^*\ [2w_b - 1]$, $(3 + 2b') + 2a = 2^*\ [w_a + w_b - 1]$, $3a^* + (3b + 2) = 3\ [2w_b - 1]$, $3a' + (3b + 2) = 3\ [w_a + w_b - 1]$, and OM with the equalities $1a + 1a = 3a^*\ [w_a + w_b - 1]$, $1b + 1b = 3b^*\ [w_a + w_b - 1]$.

Two additional *2-interactions* are SM with the equalities $3a^* + 3b = 3\ [w_b - w_a]$ and $2a + 2b^* = 2^* + 1'_a\ [w_b - w_a]$ (see Figure 2). Lemma 1 below demonstrates that the *quality of a 2-interaction depends only on types* of chains on the left-hand side of a term equality.

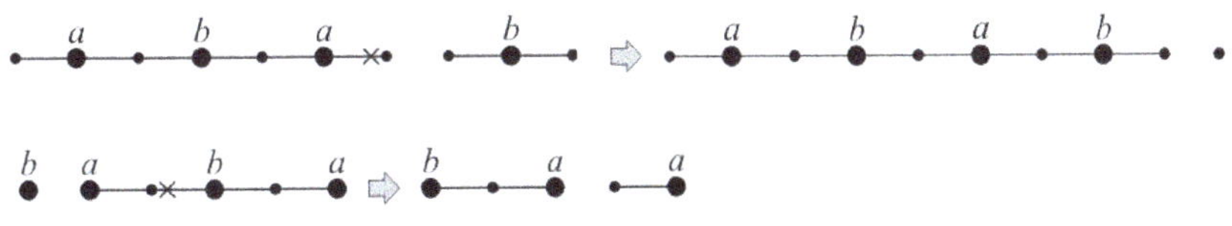

Figure 2. Interactions 3*a** + 3*b*′ = 3 and 2*a*′ + 2*b** = 2 + $1'_a$. In both cases, *sesquialteral intermerging* (SM) is applied.

Unlike the algorithms in [2,7], where interactions are performed in the same order as they are listed, now, the order of interactions is described in the following nontrivial way:

Notice 1. Denote the graph obtained after Stage 1 by *G*′. For each of the above-listed 2-interactions *s*, we call its *element* an unordered pair, triple, or quadruple (depending on the number of arguments in the composition of *s*, which we also denote by *s*) of chains in *G*′ that have the same types as the types of arguments in s. The pairs, triples, and quadruples of such *types* themselves will be referred to as a *polytypes;* a type of a chain in a polytype may occur several times. For instance, for the interaction $f = (1a + 1b = 1^{*}_{b})$, its element is any pair of chains of types $1a$ and $1b$, and its polytype is the pair $\{1a,1b\}$ of these types. To each element, precisely one 2-interaction corresponds, which is specified by its polytype. The *quality of an element* is defined to be the quality of this 2-interaction. By a *domain*, we call any set of elements (usually, from different interactions) where the elements are disjoint. The ***quality of a domain*** M is the aggregate quality of its elements. A domain with the maximum quality is called *maximal*.

Thus, Stage 2 consists of applying, simultaneously and independently of each other, 2-interactions corresponding to some maximal domain M for G' to elements of M.

Let us find a maximal domain M for G'. To this end, we use integer linear programming (ILP) with at most 51 variables and at most 10 nontrivial (i.e., not of the form $x \geq 0$) constraints. Namely, to each 2-interaction s we assign a nonnegative integer-valued variable x_s whose value must be equal to the number of elements of this 2-interaction in the sought-for domain M. This condition on the vector $\{x_s\}$ is expressed by the following linear relations: for every type t of a chain occurring in G' and corresponding to an argument of one of the 2-interactions, we impose the constraint $\sum_{s} c_{ts} \cdot x_s \leq l_t$, where c_{ts} is the number of occurrences t in the arguments of any 2-interaction s (c_{ts} can be 0, 1, or 2), and l_t is the number of chains of type t in G'. *Maximize* the target function $F(\{x_s\}) = \sum_{s} P(s) \cdot x_s$, where $P(s)$ is the quality of a 2-interaction s and the summation is over all 2-interactions. A solution to this ILP problem gives a maximal domain M.

Define *autonomous reduction* as the following sequence of operations: cut out all conventional edges. Circularize chains of sizes strictly larger than 0 into cycles using an OM or SM operation (after SM, there remains one extremal vertex or one extremal edge) (see Figure 3a). When circularizing a chain of type 2*, choose a variant with joining two b-singular vertices; after circularizing a chain of type $3a^*$ or $3b^*$, cut out the arising conventional edge. Then, break all cycles into cycles of size 2 using a DM operation, which cuts out a cycle of size 2 with an a-singular vertex from a cycle (see Figure 3b). Remove all singular vertices and loops. The *autonomous cost* $A(G)$ of a graph G is defined as the total cost of the sequence of operations in the autonomous reduction of G (see Lemma 1 below).

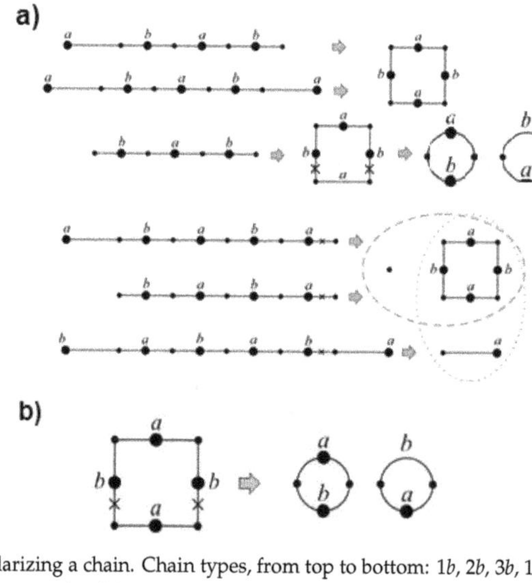

Figure 3. (**a**) Circularizing a chain. Chain types, from top to bottom: $1b$, $2b$, $3b$, $1a$, 3, 2. (**b**) Cutting out a cycle of size 2 from a cycle of size 4.

Stage 3: Perform autonomous reduction of the remaining components to a final form.

End of the Algorithm description.

It follows from the proof presented in Section 3 that, for the graph G'' obtained after Stage 2, there exists no interaction with a strictly positive quality, which is by no means evident a priori.

Below, we *consider the case* $w_a \leq w_b$, since the description of the Algorithm (up to interchanging a and b) and the subsequent proof do not depend on which of the removal costs is smaller, w_a for an a-singular vertex or w_b for a b-singular vertex.

3. Proof of Theorem 1 (Case II)

First, we make an obvious remark concerning the **linear runtime** of the Algorithm. For Stage 0, this follows from the fact that $a + b$ is constructed by one-time examination of all components in a and b. Stage 1 requires one-time examination of all components in $a + b$. The number of interactions executed at Stage 2 is linear, since each of them reduces the number of chains in the corresponding graph, and each interaction is executed in constant time. Similarly, the number of operations executed at Stage 3 is linear, since each of them reduces the number of singular vertices or the number of chains. Linear runtime of solving the ILP problem at Stage 2 follows from [10], where it was shown that the time required for solving an ILP problem with a fixed number of variables and constraints is polynomial in the logarithm of the maximum absolute value of a coefficient of the problem. In our problem, this coefficient is not greater than the problem size.

The rest of this section is devoted to the proof of additive exactness of the Algorithm. If, when executing a standard operation, singular vertices are joined, the operation is said to be *special;* otherwise, it is *nonspecial*; a removal operation is special by definition.

Clearly, we may assume that $w = 1$. For a graph G, we use the following *notation*: d is the total size of all components in it (we call it the *size* of G), f is the number of odd chains, c is the number of cycles (excluding loops), B is the number of singular vertices, S is the sum of integral parts of halved segment lengths plus the number of extremal (on a chain) odd segments minus the number of cyclic segments, D is the number of chains of types $1a$, $1b$, $3a$, $3b$, and 3, and K_b is the number of components containing a b-singular vertex.

Lemma 1. *Let w_a and w_b be the removal costs for singular a- and b-vertices, $w_a \leq w_b$, and let all other operations have cost* 1. *Then, the autonomous cost $A(G)$ of a graph G is*

$$A(G) = (1 - w_a)\cdot(0.5d + 0.5f - c) + w_a\cdot(B + S+D) + (w_b - w_a)\cdot K_b. \quad (1a)$$

Proof. Denote the right-hand side of this equality by $A'(G)$. Let us check the equality for each component of the graph separately and then sum up the obtained equalities. Denote by I_b the indicator function which is 1 if G contains a b-singular vertex and 0 otherwise. □

(1) For a conventional isolated vertex, the equality is trivial, since $A'(G)$ is equal to w_a for an a-loop and w_b for a b-loop. For a singular isolated vertex (odd chain), we have $d = -1, f = 1$, and $B = 1$; for a b-isolated vertex, $K_b = 1$ and the other quantities are zero. $A'(G)$ is equal to w_a for an a-isolated vertex and w_b for a b-isolated vertex, as well as $A(G)$.

(2) For a cycle without singular edges (and hence without singular vertices), we have $d > 0$ and d even, $c = 1$, $S = 0.5d - 1$; the other quantities are zero, and $A'(G) = 0.5d - 1$. Autonomous reduction includes only cutting out conventional edges by a DM operation. Each cutting reduces a large cycle by 2 edges; the last cycle requires no operation. Therefore, $0.5d - 1$ DM operations are executed.

(3) For a cycle with singular edges, we have $d > 0$ and d even, $B > 0$, $S \geq 0$, $c = 1$, $K_b = I_b$, and the other quantities are zero; $A'(G) = (1 - w_a)(0.5d - 1) + w_a(B + S) + (w_b - w_a)I_b$. When cutting out conventional edges, the size d of a graph does not change. Each cutting out a conventional edge or a cycle of size 2 reduces the size of a large cycle by 2; the last cycle requires no operation. Thus, $0.5d - 1$ DM operations are executed. The number of nonspecial operations among them is S, since every

segment of length l for an even l requires $0.5l$ nonspecial cuttings, and, for an odd l, $l - 0.5(l - 1)$ nonspecial cuttings and one special cutting. Therefore, the number of Rem operations is $B - (0.5d - 1 - S)$ (the number of b-removals among them is I_b), and $A(G) = 0.5d - 1 + w_a(B - (0.5d - 1 - S) - I_b) + w_bI_b = A'(G)$.

(4) For an odd chain without singular edges, we have $d > 0$ and d odd, $f = 1$, $S = 0.5(d + 1)$ (such a chain is an extremal odd segment), and the other quantities are zero. Its autonomous reduction requires only DM operations, i.e., cutting out conventional edges, and, at the end, when a conventional edge without adjacent conventional edges remains, one OM. Each DM separates a final *ab* cycle from a current chain, and the chain length (equal to its size) reduces by 2. Therefore, $0.5(d + 1)$ operations are required in total, which equals $A(G)$, since the cost of a standard operation is 1. For $A'(G)$, we obtain the same quantity.

(5) For an odd chain with singular edges, we have $d > 0$ and d odd, $f = 1$, $B > 0$, D is either 0 or 1, $S \geq 0$, $K_b = I_b$, and the other quantities are zero. Cutting out conventional edges requires S nonspecial DM or SM operations. After the cuttings, a chain either has an odd size strictly greater than 0 or turns into a singular isolated vertex. In the first case, the chain is circularized into a cycle by a special OM (if $D = 0$) or by a nonspecial OM (if $D = 1$), after which the obtained cycle is finalized by special operations (this cycle may contain only one conventional edge, namely, the one by which the chain was circularized into a cycle). Every standard operation in autonomous reduction increases the number of cycles in the graph by 1, an OM increases the graph size by 1, and a DM or SM does not change it. Thus, finally, the graph size becomes $d + 1$, and it contains $0.5(d + 1)$ cycles of size 2. In the second case, the graph size remains to be d, and it also contains $0.5(d + 1)$ cycles of size 2 formed when making cuttings (a singular isolated vertex is of size -1 and $D = 0$). Since the initial graph had no cycles, $0.5(d + 1)$ standard operations are made in total, $S + D$ among them being nonspecial. Therefore, the number of Rem operations is $B - (0.5d + 0.5 - S - D)$ (the number of b-removals among them is I_b), and $A(G) = 0.5(d + 1) + w_a(B - (0.5d + 0.5 - S - D) - I_b) + w_bI_b = A'(G)$.

(6) For an even chain without singular edges, we have $d > 0$ and d even, $S = 0.5d$, and the other quantities are zero. Its autonomous reduction requires only DM operations, cuttings of conventional edges. Each of them separates an *ab* cycle from the current chain; therefore, the length of the chain, equal to its size, reduces by 2. Therefore, in total, $A(G) = 0.5d$ operations are executed. For $A'(G)$, we obtain the same quantity.

(7) For an even chain with singular edges, we have $d > 0$ and d even, $B > 0$, D is either 0 or 1, $S \geq 0$, $K_b = I_b$, and the other quantities are zero. Cutting out conventional edges requires S nonspecial DM or SM operations. After a cutting, the chain has an even size either strictly greater than 0 or equal to 0. In the first case, the chain is circularized by a special SM (if $D = 0$) or a nonspecial SM (if $D = 1$), after which the obtained cycle is finalized by special operations. Every standard operation in the autonomous reduction increases the number of cycles in the graph by 1, and a DM or SM does not change the size of the graph. Therefore, finally, the graph size remains to be d, and it contains $0.5d$ cycles of size 2. The same happens in the second case. Since the initial graph had no cycles, in total, $0.5d$ standard operations are applied, $S + D$ among them being nonspecial. Thus, the number of Rem operations is $B - (0.5d - S - D)$ (the number of b-removals among them is I_b), and $A(G) = 0.5d + w_a(B - (0.5\mathrm{d} - \mathrm{S} - \mathrm{D}) - I_b) + w_bI_b = A'(G)$.

Recall that an interaction in G is a sequence s of operations successively applied to G. The ***quality*** *$P(s)$ of an interaction s* is defined as

$$P(G,s) = A(G) - A(s(G)) - c(s), \tag{1b}$$

where $s(G)$ is the graph obtained after applying the composition s to G, and $c(s)$ is the *total cost of operations* in s. The quality shows what total cost can be saved by using the interaction s as against the autonomous reduction of G, i.e., without using s. By examining all 2-interactions (see the example below), we check that $P(G,s)$ is determined by the *polytype* of the 2-interaction, consisting of the types of its arguments or, equivalently, by the corresponding term equality. Accordingly, the quality of an element α (for the substitution in s) is determined by its polytype, which coincides with the polytype of s itself.

In this way, we computed the qualities of 2-interactions that were presented above, after each of them, in square brackets. For instance, let us find the quality of the interaction $1a + 1b = 1_b^*$ (Figure 4), $c(s) = 1$, and the operation is special (generally, all operations of Stage 2 are special). This 2-interaction does not change the values of d, c, and S, reduces B and K_b by 1, and reduces D and f by 2. Hence, $P(s) = w_a + w_b$.

Figure 4. Interaction $1a + 1b = 1_b^*$, an SM operation.

Let E be any reducing sequence for a graph G, which is arbitrarily divided into connected fragments of operations; we call them interactions here and denote by s. Denote by $T(R,E)$ the *total cost of operations* in a reducing subsequence which starts in E with some intermediate graph R and continues to a final form. Denote by $P(R,E)$ the aggregate quality of all interactions in E starting from R, where $P(s)$ is formally defined in Equation (1b).

Lemma 2. *For any graph G, we have* $T(G,E) = A(G) - P(G,E)$.

Proof. It suffices to check $A(R) - T(R,E) = P(R,E)$. This is proven by induction on the number of interactions from the end of the sequence E. If R is of a final form, all the three quantities are zero. Let $A(R') - T(R',E) = P(R',E)$ for a subsequence starting with a graph R' next to R. Let s be an interaction taking R into R', $c(s)$ be the total cost of its operations, and $P(s)$ be its quality, defined formally by (1′). Then $P(R,E) = P(R',E) + P(s)$, $T(R,E) = T(R',E) + c(s)$, and $P(s) = A(R) - A(R') - c(s)$. Hence, $A(R) - T(R,E) = A(R') + c(s) + P(s) - T(R',E) - c(s) = A(R') - T(R',E) + P(s) = P(R',E) + P(s) = P(R,E)$. □

If E is a sequence of operations constructed by the Algorithm, then it contains 2-interactions, and all other operations (i.e., those from Stages 1 and 3) by Equation (1b) have formal quality 0 (for uniformity, these operations can be viewed as formal interactions).

Notice 2. Thus, $P(G)$ is the ***aggregate quality of 2-interactions*** (equivalently, interactions in all stages of the algorithm), equal (!) to the quality of the maximal domain on chains in the initial $a + b$ (equivalently, on chains after all cuttings of Stage 1).

By Lemmas 1 and 2,

$$T = T(G) = (1 - w_a)\cdot(0.5d + 0.5f - c) + w_a\cdot(B + S + D) - P(G) + (w_b - w_a)\cdot K_b \quad (2)$$

is the ***total cost of operations of the algorithm*** on a graph G. Denote the first and second term in this expression by T' and T'' respectively.

Now, we use the following important equivalence:

Exactness of the algorithm is equivalent to the ***triangle inequality***: for any operation o and any graph G, we have

$$c(o) \geq T(G) - T(o(G)),$$

where $o(G)$ is the result of applying o to G.

The proof is performed by induction on $C(G)$, the *minimum total cost* over all reducing sequences. If G is of a final form, then $T(G) \leq C(G)$ (let a reducing sequence be empty). Consider some nonempty shortest reducing sequence for G, and denote the first operation in it by o. By the induction hypothesis, $T(o(G)) \leq C(o(G))$. Then, by this inequality, $T(G) = c(o) + T(o(G)) \leq c(o) + C(o(G)) = C(G)$, i.e., $T(G) = C(G)$. The converse is even simpler, but we use only this implication. Note that this argument works for any algorithm, not only for the one described above.

Thus, to complete the proof of Theorem 1, it remains to check the triangle inequality for each of the above-mentioned operations o.

1. o **is the Rem operation**. When passing from G to $o(G)$, B reduces by 1. Consider several cases.

1.1. An isolated singular vertex is removed (chain of type $2a'$ or $2b'$). Then S, D, and c do not change; d increases by 1; f reduces by 1. P does not increase when removing any chain. Therefore, $T' + T'' - P$ reduces by at most w_a. If a b-vertex is removed, then K_b reduces by 1 and T reduces by at most w_b.

1.2. A singular vertex is removed from a cycle, or a loop is removed. Then, S does not reduce.

1.3. An interior singular vertex (i.e., on both sides of it there are other singular vertices) is removed from a chain. The type of the chain does not change, changes from $3b^*$ to $3b'$, or changes from $2a^*$ to $2a'$. Then, P does not increase, since no element increases its quality when making this change.

1.4. A hanging vertex which is the only singular vertex in a chain is removed from the chain. Then, S and D do not reduce, and K_b reduces by 1 if a b-vertex is removed.

1.5. A hanging vertex that is not unique in a chain is removed from the chain. If, when passing from G to $o(G)$, S does not change (the segment adjacent to the hanging edge is even), then the hanging extremity becomes nonhanging, and the following changes in the type of the chain are possible: $1a$ changes into $3a$, $1b$ changes into $3b$, 2 changes into 1, $2a$ changes into $1a$, $2b$ changes into $1b$, or 1 changes into 3. In the first three cases, D does not change and P does not increase. Indeed, all elements containing type $1a$ do not increase their quality when making the change from $1a$ to $3a$; the same applies to the two other changes. In the last three cases, D increases by 1, P either does not change or increases by at most w_a (consider the inverse change), and all other quantities do not change. Then, $T' + T'' - P$ reduces by at most w_a. If S increases by one (the segment adjacent to the hanging edge is odd), then the hanging extremity remains to be hanging, and the type of the chain does not change. Therefore, D and P do not change.

1.6. A nonhanging exterior singular vertex (on the right or on the left of it there are no other singular vertices) is removed from a chain, and it is the only singular vertex in the chain. Then, the chain has one of the types $3a$, $3b$, $2a$, $2b$, or 1, since, after cutting out conventional edges, we obtain a chain with one singular vertex, and P does not increase. If the chain is of type $3a$ or $3b$, then S increases by 1, since both segments adjacent to the removed vertex are even and the resulting segment is odd, and D reduces by 1. If the chain is of type $2a$, $2b$, or 1, then D and S do not change (in the cases of $2a$ or $2b$, two odd segments are replaced with one odd; in the case of 1, one even and one odd segments are replaced with one even). Therefore, $T' + T'' - P$ reduces by at most w_a.

1.7. A nonhanging exterior singular vertex is removed from a chain, and it is not a unique singular vertex in the chain. If, when passing from G to $o(G)$, S does not change (the extremal segment is odd, and the next one is even), then the hanging extremity becomes nonhanging, and we repeat the arguments from case 1.5. If S increases by 1, then either the type of the chain does not change (then, $T' + T'' - P$ does not change) or (if both segments adjacent to the removed vertex are even) the nonhanging extremity becomes hanging. Then, one of the following changes in the type of the chain is possible: $1a$ changes into $2a$, $1b$ changes into $2b$, 3 changes into 1, $3a$ changes into $1a$, $3b$ changes into $1b$, or 1 changes into 2. In the first three cases, D reduces by 1 and P does not increase. In the last three cases, D does not change and P either does not change or increases by at most w_a. Thus, in all cases, $T' + T'' - P$ reduces by at most w_a.

2. o **is an OM**.

2.1. Extremities of one chain are joined. This is possible for odd chains only. Then, d increases by 1, f reduces by 1, and c either increases by 1 or does not change, such that T' either reduces by $1 - w_a$ or does not change. Consider possible chain types. For a chain of type 0, S reduces by 1, since an odd segment turns into a cyclic one of length greater by 1. Then, T reduces by 1 (now, $c(o) = 1$). For a chain of type $1a$ or $1b$, B and S do not change, D reduces by 1, and P either does not change or reduces by at most $w_a + w_b$. Therefore, T changes by at most 1. For types $2a$ or $2b$, D does not change and P either does not change or reduces by at most w_b. If at both ends of the chain there are hanging edges, then after joining their endpoints, B reduces by 1 and S does not change. If at least on one side there is an odd extremal segment, then B does not change and S reduces by 1. For types $3a$ or $3b$, B and S do not change, D reduces by 1, and P either does not change or reduces by at most w_b. Therefore, T changes by at most 1.

2.2. Extremities of different chains are joined. In Table 2, we present results of joining the endpoints of nonhanging edges labeled with a or of hanging edges labeled with b (the added edge is labeled by b). Such a join is referred to as a *b-join* (similarly for an *a-join*). The result of an a-join coincides with the result of a b-join obtained by interchanging the labels a and b in chains.

Table 2. Results of joining extremities of chains with types specified in a row and column; the type of resulting chain is given at the bottom of a cell.

	0a	**0**	**1a**	**1a′**	**2a*,2a′**	**3a*,3a′**	**1,1\b**	**1′,1′\b**	**2**	**3**
0a	(0) [0] 0a	(0) [0] 0	(0) [0] 1a	(0) [0] 1a′	(0) [0] 2a*,2a′	(0) [0] 3a*,3a′	(0) [0] 1	(0) [0] 1	(0) [0] 2	(0) [0] 3
0	(0) [0] 0	(+1) [0] 0b	(0 … +1) [0 … −1] 3	(0 … +1) [0 … −1] 2	(0 … $+w_{ba}$) [0 … $-w_{ba}$] 1	(0 … $+w_{ba}$) [0 … $-w_{ba}$] 1	(+1 … $1-w_{ba}$) [0 … $+w_{ba}$] 3b*,3b′	(+1 … $1-w_{ba}$) [0 … $+w_{ba}$] 2b*,2b′	(+1 … 0) [0 … +1] 1b	(+1 … 0) [0 … +1] 1b
1a	(0) [0] 1a	(0 … +1) [0 … −1] 3	(−1 … $+w_a$) [$1-w_a-w_b$ … $-w_a-1$] 3a*	($-w_b$ … $+w_a$) [0 … $-w_a-w_b$] 1a	($-w_{ba}$ … 0) [0 … $-w_{ba}$] 1a	($-w_{ba}$ … $+w_a$b) [0 … $-w_a-w_b$] 3a*	($-w_{ba}$ … $1-w_{ba}$) [0 … −1] 3	($-w_{ba}$ … $+w_a$b) [0 … $-w_a-w_b$] 1	(−1 … 0) [$1-w_a-w_b$ … $-w_a-w_b$] 1	($-w_b$ … $+w_a$) [0 … $-w_a-w_b$] 3
1a′	(0) [0] 1a′	(0 … +1) [0 … −1] 2	($-w_b$ … $+w_a$) [0 … $-w_a-w_b$] 1a	(−1 … $+w_a$) [$1-w_a-w_b$ … $-w_a-1$] 2a*	($-w_{ba}$ … $+w_a$b) [0 … $-w_a-w_b$] 2a*	($-w_{ba}$ … 0) [0 … $-w_{ba}$] 1a	($-w_{ba}$ … $+w_a$b) [0 … $-w_a-w_b$] 1	($-w_{ba}$ … $1-w_{ba}$) [0 … −1] 2	($-w_b$ … $+w_a$) [0 … $-w_a-w_b$] 2	(−1 … 0) [$1-w_a-w_b$ … $-w_a-w_b$] 1
2a*,2a′	(0) [0] 2a*2a′	(0 … $+w_{ba}$) [0 … $-w_{ba}$] 1	($-w_{ba}$ … 0) [0 … $-w_{ba}$] 1a	($-w_{ba}$ … $+w_a$b) [0 … $-w_a-w_b$] 2a*	($-w_{ba}$ … 0) [0 … $-w_{ba}$] 2a*,2a′	($-w_{ba}$ … w_a) [w_a … $-w_{ba}$] 1a	($-w_{ba}$ … 0) [0 … $-w_{ba}$] 1	(w_ab … $-w_{ba}$) [$-w_{ba}$ … w_ab] 2	($-w_{ba}$ … 0) [0 … $-w_{ba}$] 2	($-w_{ba}$ … w_ab) [0 … $-w_a-w_b$] 1
3a*,3a′	(0) [0] 3a*3a′	(0 … $+w_{ba}$) [0 … $-w_{ba}$] 1	($-w_{ba}$ … $+w_a$b) [0 … $-w_a-w_b$] 3a*	($-w_{ba}$ … 0) [0 … $-w_{ba}$] 1a	($-w_{ba}$ … w_a) [w_a … $-w_{ba}$] 1a	($-w_{ba}$ … 0) [0 … $-w_{ba}$] 3a*,3a′	(w_ab … $-w_{ba}$) [$-w_{ba}$ … w_ab] 3	($-w_{ba}$ … 0) [0 … $-w_{ba}$] 1	($-w_{ba}$ … w_ab) [0 … $-w_a-w_b$] 1	($-w_{ba}$ … 0) [0 … $-w_{ba}$] 3
1,1\b	(0) [0] 1	(+1 … $1-w_{ba}$) [0 … $+w_{ba}$] 3b*,3b′	($-w_{ba}$ … $1-w_{ba}$) [0 … −1] 3	($-w_{ba}$ … $+w_a$b) [0 … $-w_a-w_b$] 1	($-w_{ba}$ … 0) [0 … $-w_{ba}$] 1	(w_ab … $-w_{ba}$) [$-w_{ba}$ … w_ab] 3	($1-w_{ba}$ … $1-2w_{ba}$) [0 … $+w_{ba}$] 3b*,3b′	($1-w_{ba}+w_a$ … $1-w_{ba}-w_b$) [0 … w_a+w_b] 1b	($1-w_{ba}$ … $-w_{ba}$) [0 … +1] 1b	($1-w_{ba}$ … $1-w_{ba}-w_b$) [0 … w_b] 3b
1′,1′\b	(0) [0] 1	(+1 … $1-w_{ba}$) [0 … $+w_{ba}$] 2b*,2b′	($-w_{ba}$ … $+w_a$b) [0 … $-w_a-w_b$] 1	($-w_{ba}$ … $1-w_{ba}$) [0 … −1] 2	(w_ab … $-w_{ba}$) [$-w_{ba}$ … w_ab] 2	($-w_{ba}$ … 0) [0 … $-w_{ba}$] 1	($1-w_{ba}+w_a$ … $1-w_{ba}-w_b$) [0 … w_a+w_b] 1b	($1-w_{ba}$ … $1-2w_{ba}$) [0 … $+w_{ba}$] 2b*,2b′	($1-w_{ba}$ … $1-w_{ba}-w_b$) [0 … w_b] 2b*	($1-w_{ba}$ … $-w_{ba}$) [0 … +1] 1b
2	(0) [0] 2	(+1 … 0) [0 … +1] 1b	(−1 … 0) [$1-w_a-w_b$ … $-w_a-w_b$] 1	($-w_b$ … $+w_a$) [0 … $-w_a-w_b$] 2	($-w_{ba}$ … 0) [0 … $-w_{ba}$] 2	($-w_{ba}$ … w_ab) [0 … $-w_a-w_b$] 1	($1-w_{ba}$ … $-w_{ba}$) [0 … +1] 1b	($1-w_{ba}$ … $1-w_{ba}-w_b$) [0 … w_b] 2b*	($1-w_b$ … −1) [$-w_a$ … $2-w_a-w_b$] 2b*	($-w_b$ … w_a) [1 … $1-w_a-w_b$] 1b
3	(0) [0] 3	(+1 … 0) [0 … +1] 1b	($-w_b$ … $+w_a$) [0 … $-w_a-w_b$] 3	(−1 … 0) [$1-w_a-w_b$ … $-w_a-w_b$] 1	($-w_{ba}$ … w_ab) [0 … $-w_a-w_b$] 1	($-w_{ba}$ … 0) [0 … $-w_{ba}$] 3	($1-w_{ba}$ … $1-w_{ba}-w_b$) [0 … w_b] 3b	($1-w_{ba}$ … $-w_{ba}$) [0 … +1] 1b	($-w_b$ … w_a) [1 … $1-w_a-w_b$] 1b	($1-w_b$ … −1) [$-w_a$ … $2-w_a-w_b$] 3b

Define *types of extremities*: $0a$ is an extremity of an odd chain of type 0, 0 is an extremity of an even chain of type 0, $1a$ is a hanging extremity of a chain of type $1a$, $1a'$ is a nonhanging extremity of a chain of type $1a$, $2a^*$ is an extremity of a chain of type $2a^*$ (respectively, $2a'$), $3a^*$ is an extremity of a chain of type $3a^*$ (respectively, $3a'$), 1 is a hanging extremity of a chain of type 1 with singular b-vertices, $1'$ is a nonhanging extremity of a chain of type 1 with singular b-vertices (respectively, $1\backslash b$ and $1'\backslash b$ are extremities of chains of type 1 without singular b-vertices), 2 is an extremity of a chain of type 2, and 3 is an extremity of a chain of type 3.

Denote by ΔX the *increment* of X, i.e., the difference of values of X when passing from G to $o(G)$, $\Delta X = X(o(G)) - X(G)$. In Table 2, we present *number segments* which *contain* all possible increments of T (in parentheses) and P (in square brackets) from Equations (1a)–(2) when joining an extremity of a chain with the row name and an extremity of a chain with the column name, i.e., when passing from G to $o(G)$. These segments are merely lower and upper estimates for T and P, which is sufficient for the following step. The increments are given as functions of the costs w_a and w_b. At the bottom of the same cell, we give the type of the chain obtained as a result of this passing, i.e., joining the initial chains. For instance, joining the extremity $2a^*$ or $2a'$ with any extremity has the same increment of all arguments in Equation (1a) except for ΔK_b, and the corresponding ΔP is given in the cells of row $\{2a^*, 2a'\}$ in Table 2. Here, $w_{b,a}$ is equal to w_b for $2a^*$ and w_a for $2a'$. Similarly, this applies to the pairs $3a^*$ and $3a'$, 1 and $1\backslash b$, and $1'$ and $1'\backslash b$. Next, (0) means that the corresponding quantity "does not change", $(-w_a)$ means "reduces by at most w_a", $(+w_a)$ means "increases by at most w_a", $(1 - w_a)$ means "increases by at most $1 - w_a$", etc.; an ellipsis denotes numbers between the ends of the given number segments. Bounds for ΔT are easily computed through bounds for ΔP and increments of other arguments in Equation (2), with the latter being easy to find.

Thus, the problem consists precisely of estimating ΔP. For that, we use a functional $P^-(M,M_1)$, where M is a set consisting of two chains in G whose types are the one from the row name and the one from the column name, while M_1 consists of a single chain in G with the name given in the corresponding cell. The functional itself is the *minimum* increment of P when replacing M with M_1 over all graphs G containing M. Given the functional $P^-(M,M_1)$, one can compute a functional $P^+(M,M_1)$ as a function of w_a and w_b which is the *maximum* increment of P when replacing M with M_1 over all graphs G containing M. Specifically, we have the equality $P^+(M,M_1) = -P^-(M_1,M)$. Note that Table 2 is symmetric around the main diagonal.

Thus, to compute $P^-(M,M_1)$, we need Lemma 3 below, where M consists of different chains, denoted by 1 and 2, and M_1 consists of one corresponding chain, denoted by 3. If D is a maximal domain in a graph G *containing* M, we *denote* by G' the graph that contains all chains from M and all chains from elements in D which intersect with M. These elements themselves form a maximal domain D' in G'. Indeed, if D' is not maximal, strictly extend it to a maximal domain in G', which together with elements in $D\backslash G'$ forms a domain in G with quality strictly greater than that of D, a contradiction.

Lemma 3. *We have* $P(o(G)) - P(G) \geq P(o(G')) - P(G')$.

Proof. Recall that such a difference was called the *increment* of the corresponding quantity. Denote the left-hand side of the inequality by d, and the right-hand side, by d'. Consider $D\backslash D'$ (elements of D that do not contain chains 1 and 2), D_1, a maximal domain in $o(G')$, and D_2, the result of replacing D' with D_1 in D. Then, $d \geq P(D_2) - P(D) = P(D_1) - P(D')$, since $P(D_2) = P(D_1) + P(D_2\backslash D_1)$ and $P(D) = P(D') + P(D_2\backslash D_1)$. □

Thus, we want to find $\min_{G'} P(o(G')) - P(G')$. Let X be a set containing at most two polytypes α^* and β^* (one for each element of D'), and let Y be a set containing at most four of the types of which these polytypes are composed (types of the chains contained in the corresponding elements of D') and also containing the types of the initial chains 1 and 2. There exists a G'' for which these polytypes correspond to disjoint elements, and for such a G'' the pair $D'' = \{\alpha^*, \beta^*\}$ is a polytype of a domain in G'', not necessarily maximal. In the last case, $\sum_{\alpha*\in X} P(\alpha*)$ is smaller than the quality $P(G'')$, and such an X is inessential when searching for the minimum. Given G'', we form the corresponding $o(G'')$ and find in it an actual maximal domain and the quality $P(o(G''))$, i.e., the minuend is definitely correct, and the subtrahend may be smaller than required. Note that $P(D') = P(D'')$, as well as for other domains, since the quality of any domain does not change when changing chains but preserving their types. Through an exhaustive search over all such X, we find the desired lower bound; in this way, Table 2 is filled. Take note that the number of possible sets X is not greater than the squared number of interactions, and X contains at most eight types. Therefore, due to an exhaustive search over all X and finding a maximal domain in $o(G'')$, computing $P^-(M,M_1)$, takes constant time.

Note that if the initial chains are even, then T' increases by $1 - w_a$; otherwise, it does not change; simple observations of this kind are omitted in what follows. Let us show how seven cells in Table 2 are filled, with the other being filled similarly.

(1) Cell $1a$,$1a$. After the join, we obtain a chain of type $3a^*$. If the initial chains have hanging edges, then B reduces by 1 and S does not change. If at least one of the initial chains has an odd extremal segment, then, vice versa, B does not change and S reduces by 1, D reduces by 1, and K_b reduces by 1. Therefore, $T + P$ reduces by $w_a + w_b$. Our functionals are as follows: $P^-(1a,1a; 3a^*) = -w_a - 1$ and $P^+(1a,1a; 3a^*) = \min(1 - w_a - w_b,0)$. In other words, either P reduces by at least $w_a + w_b - 1$ but at most $w_a + 1$ (if $w_a + w_b - 1 > 0$), or P does not increase and reduces by at most $w_a + 1$ (otherwise). Therefore, we have $|\Delta T| \leq 1$.

(2) Cell $1a$,2. After the join, we obtain a chain of type 1 (it is not involved in interactions). Similarly to case (1), we obtain that $T + P$ reduces by $w_a + w_b$. We have $P^-(1a,2; 1) = -w_a - w_b$, $P^+(1a,2; 1) = \min(1 - w_a - w_b,0)$. In other words, either P reduces by at least $w_a + w_b - 1$ (if $w_a + w_b - 1 > 0$) but at most $w_a + w_b$, or P does not increase and reduces by at most $w_a + w_b$ (otherwise). Therefore, $|\Delta T| \leq 1$.

(3) Cell $2a$,$3a$. After the join, we obtain a chain of type $1a$. Then, $T + P$ reduces by $w_b - w_a$ if the type of the $3a$-chain is $3a^*$ and does not change if it is $3a'$. We have $P^-(2a,3a^*; 1a) = -w_b$, $P^-(2a,3a'; 1a) = -w_a$, $P^+(2a,3a^*; 1a) = w_a$, and $P^+(2a,3a'; 1a) = w_a$. In other words, P reduces by at most w_b in the first case and by at most w_a in the second and increases by at most w_a in both cases.

(4) Cell $2a$,$1'$. After the join, we obtain a chain of type 2. Then, B, S, and D do not change. We have $P^-(2a,1; 2) = -w_b$, $P^+(2a,1; 2) = \max(w_a + w_b - 1,0)$.

(5) Cell 1,3. After the join, we obtain a chain of type $3b$. The increment of $T + P$ is $1 - w_b$ if the chain of type 1 contains a b-vertex, and $1 - w_a$ otherwise. We have $P^-(1,3; 3b) = \min(1 - w_a - w_b,0)$ and $P^+(1,3; 3b) = w_b$. Subtracting from the increment of $T + P$ three possible increments of P, equal to $1 - w_a - w_b$, 0, or w_b, we obtain a number of absolute value no greater than 1.

(6) Cell 2,2. After the join, we obtain a chain of type $2b^*$. The increment of $T + P$ is $1 - w_a - w_b$. We have $P^-(2,2; 2b^*) = -w_a$ and $P^+(2,2; 2b^*) = 2 - w_a - w_b$. Subtracting from the increment of $T + P$ all possible increments of P, equal to $-w_a$ or $2 - w_a - w_b$, we obtain a number of absolute value no greater than 1.

(7) Cell 2,3. After the join, we obtain a chain of type $1b$. The increment of $T + P$ is $1 - w_b$. We have $P^-(2,3; 1b) = \min(1 - w_a - w_b,0)$ and $P^+(2,3; 1b) = 1$. Subtracting from the increment of $T + P$ all possible increments of P, equal to $1 - w_a - w_b$, 0 or 1, we obtain a number of absolute value no greater than 1.

Thus, the result for OM follows from the fact that, according to Table 2, T changes by at most 1.

The next simple lemma is used in what follows.

Lemma 4. *Let $X(G)$ be a function of a graph G, and let $\Delta X(G,G_1)$ be its increment when passing from G to G_1. Then,*

1. $\Delta X(G,G_1) = -\Delta X(G_1,G)$. *In particular, if o is an operation, then* $\Delta X(G,o(G)) = -\Delta X(o(G),G)$. *If o is a composition of operations o_1 and o_2 (first o_1 and then o_2), then* $\Delta X(G,o(G)) = \Delta X(G,o_1(G)) + \Delta X(o_1(G),o(G))$.
2. *For any graphs G and G_1 and any operations o and o_1, if* $\Delta X(G,G_1) = \Delta X(o(G),o_1(G_1))$, then $\Delta X(G,o(G)) = \Delta X(G_1,o_1(G_1))$.

Proof. The first claim is obvious. The second follows from the equalities $\Delta X(G,o(G)) = \Delta X(G,G_1) + \Delta X(G_1,o_1(G_1)) - \Delta X(o(G),o_1(G_1)) = \Delta X(G_1,o_1(G_1))$. □

3. *o* is a Cut. This operation is the inverse to OM; thus, the result follows from case 2 and Lemma 4.1.

4. *o* is an SM. Represent *o* as a composition of a cut and a join. By Lemma 4.1, the increment obtained in a composition of operations is the sum of increments obtained in each of them.

4.1. Reversal (i.e., rearrangement of a connected fragment of edges in reverse order at the same place) of an extremal fragment of the chain. Denote by 1 and 2 the extremities of the reversed fragment and, by 2′, the arising free endpoint of the remaining chain. Consider two joins: the first is the inverse to the cut (the first operation of the reversal), and the second is the join from the reversal. They are either both *a*- or both *b*-joins. We go over all rows of Table 2 in the role of the endpoint 2′ and over pairs of columns as the extremities 1 and 2; the latter are either of types 1*a* or 1*a*′ or they are a pair of identical extremities of types 0*a*, 2*a**, 2*a*′, 3*a**, or 3*a*′. The increment of T when making a cut is opposite to the increment of T when joining 2 and 2′, as for any mutually inverse operations. By Lemma 4.1 it suffices to check the following: when joining the extremities 2 and 2′ and extremities 1 and 2′, we obtain increments with the absolute value of their difference no greater than 1.

Let us give examples; other pairs of joins are considered similarly. Cases of *a*-joins are considered analogously to *b*-joins.

(1) The pair of cells 1*a*,1*a* and 1*a*,1*a*′ defines two joins: 2′ = 1*a*, 2 = 1*a*, 1 = 1*a*′. For the first of them, ΔT and ΔP are, respectively, $-1 + x$ and $1 - w_a - w_b - x$ with $0 \le x \le w_a + 1$; for the second, these increments are $-w_b + y$ and $0 - y$ with $0 \le y \le w_a + w_b$. The difference of increments of T is strictly greater than 1 in absolute value if $x - y > 2 - w_b$ or $x - y < -w_b$ (this region in the x,y plane is called *forbidden*). This difference is equal to the increment of T upon the reversal, and we want to show that the forbidden region is empty. The difference R of increments of P (the second minus the first) is $x - y + w_a + w_b - 1$. In the forbidden region, $R > w_a + 1$ or $R < w_a - 1$. However, R is equal to the increment of P resulting when making the reversal. The reversal result is obtained by replacing a chain of type 3*a** (the result of joining the extremities 2 and 2′) with a chain of type 1*a* (the result of joining the extremities 1 and 2′). This increment of P lies in the segment $[0,w_a]$, since $P^-(3a^*; 1a) = 0$ and $P^+(3a^*; 1a) = w_a$. Hence, the forbidden region is empty. Computation of $P^-(M;M_1)$ in the cases where $|M| = 1$ or $|M_1| = 2$ is performed as in Lemma 3.

(2) The pair of identical cells 1*a*,3*a* defines two joins: 2′ = 1*a*, 2 = 3*a*, 1 = 3*a*. For these joins, ΔT is either $-w_b + x$ and $-w_b + y$ (if the 3*a* chain is of type 3*a**) or $-w_a + x$ and $-w_a + y$ (if this chain is of type 3*a*′). Although both joins are made according to the same cell, ΔT takes different values within the same segment because of other parameters of the graph. Respectively, ΔP is $0 - x$ and $0 - y$. The forbidden region, which is shown to be empty, is as follows: $x - y > 1$ or $x - y < -1$. We have $R = x - y$. However, R is equal to the increment of P upon the reversal (replacing a chain 3*a** (the result of joining the extremities 2 and 2′) with a chain 3*a** (the result of joining the extremities 1 and 2′)), i.e., it is zero.

4.2. Circularization of an extremal fragment of a chain, i.e., the composition of a cut of the chain with closing one of the parts into a cycle or loop by a join. Denote by 1 and 2 the extremities of the circularized fragment and, by 2′, the arising free endpoint of the remaining part of the chain. Consider two joins: the first is the inverse to the cut (the first operation of the circularization), and the second is the join from the circularization. They are either both *a*- or both *b*-joins.

Let us make Table 3 for closing a chain into a cycle, analogous to Table 2, following the description given in case 2.1. In Table 3, for each odd type of a chain, we give an interval of possible increments of T and the corresponding interval of increments of P when closing this chain into a cycle by a join.

Table 3. Increments of T and P when circularizing a chain by joining extremities.

0	1a, 1b	2a, 2b	3a, 3b
$(w_a - 1)$ $[0]$	$(-1 \ldots w_a + w_b - 1)$ $[0 \ldots -w_a - w_b]$	$(-1 \ldots w_b - 1)$ $[0 \ldots -w_b]$	$(-1 \ldots w_b - 1)$ $[0 \ldots -w_{b,}]$

In Table 2 we go over all rows in the role of extremity 2′ and all columns in the role of extremity 2, the latter for all types of extremities of odd chains. By Lemma 4.1, it suffices to check the following condition:

The absolute value of the difference of ΔT for joins of 2 and 2′ and joins of 1 and 2 is at most 1. (3)

Let us give examples; other pairs of joins are considered similarly. Cases of a-joins are considered analogously to b-joins.

(1) Cell 0,1a (i.e., 2′ = 0, 2 = 1a, 1 = 1a'). Here, ΔT and ΔP are, respectively, $0 + x$ and $0 - x$, and, in Table 3, for chain 1a, these increments are $-1 + y$ and $0 - y$. The forbidden region for condition (3) is as follows: $x - y > 0$ or $x - y < -2$. The difference R of increments of P is $x - y$. In the forbidden region, we have, respectively, $R > 0$ or $R < -2$. However, R is equal to the increment of P when replacing a chain of type 3 (the result of joining the extremities 2 and 2′) with a chain of type 0 (the result of joining the extremities 1 and 2) and cannot take these values. Therefore, the forbidden region is empty.

(2) Cell 1a,1a (i.e., 2′ = 1a, 2 = 1a, 1 = 1a'). Here, ΔT and ΔP are, respectively, $-1 + x$ and $1 - w_a - w_b - x$. The forbidden region for condition (3) is as follows: $x - y > 1$ or $x - y < -1$; $R = x - y + w_a + w_b - 1$. In the forbidden region, $R > w_a + w_b$ or $R < w_a + w_b - 2$. However, R is equal to the increment of P when replacing a chain of type 3a with a chain of type 1a and cannot take these values. Therefore, the forbidden region is empty.

(3) Cell 2,3a (i.e., 2′ = 2, 2 = 3a, 1 = 3a). Here, ΔT and ΔP are, respectively, $-w_b + x$ (or $-w_a + x$) and $0 - x$, and, in Table 3, for chain 3a, these increments are $-1 + y$ and $0 - y$. The forbidden region for condition (3) is as follows: $x - y > w_b$ or $x - y < w_b - 2$ (or: $x - y > w_a$ or $x - y < w_a - 2$); $R = x - y$. In the forbidden region, $R > w_a$ or $R < w_b - 2$. However, R is equal to the increment of P when replacing a chain of type 1 with a chain of type 2 and (since $P^+(1;2) = \max(w_a + w_b - 1{,}0)$) cannot take these values. Therefore, the forbidden region is empty.

4.3. SM is applied to a cycle (or a loop) and a chain, i.e., a cycle or a loop is cut and the obtained chain is lengthened. This operation is the inverse to the preceding one (case 4.2); therefore, by Lemma 4.1, we have $|\Delta T| \leq 1$.

4.4. SM is applied to two chains: the first is cut, and the second is joined. Denote by 1 the arising free extremity in the cut chain, which is joined with an extremity of the second chain, denoted by 2′. The other arising free extremity of the cut chain is denoted by 2. Consider two joins: the first of them is the inverse to the cut (the first operation in our SM), and the second is the final OM. They are either both a- or both b-joins. We go over all rows of Table 2 in the role of extremity 1 and over all pairs of columns in the role of extremities 2 and 2′. By Lemma 4.1, it suffices to check the following: for joins of extremities 1 and 2 and extremities 1 and 2′ we obtain increments with the absolute value of their difference no greater than 1.

Let us give examples; other pairs of joins are considered similarly. Cases of a-joins are considered analogously to b-joins.

(1) The pair of cells 1a,0 and 1a,1a (i.e., 1 = 1a, 2 = 0, 2′ = 1a). When making the join according to the first cell, we have $\Delta T = 0 + x, 0 \leq x \leq 1$. When making the join according to the second cell, $\Delta T = -1 + y, 0 \leq y \leq w_a$. The difference of these expressions (the second minus the first) is $y - x - 1$. Its absolute value is greater than 1 if $y - x > 2$ or $y - x < 0$ (the forbidden region). The corresponding difference R of increments of P is $x - y + 1 - w_a - w_b$, which, for $y - x > 2$, yields $R < -1 - w_a - w_b$, and for $y - x < 0$, $R > 1 - w_a - w_b$. However, R is equal to the increment of P when making the original SM operation. The result is obtained by replacing chains of types 3 (joining the extremities 1 and 2) and 1a (the chain with extremity 2′) with chains of types 3a* (joining the extremities 1 and 2′) and 0 (the chain

with extremity 2). The equalities $P^{-}(1a,3; 3a^{*},0) = -w_a$ and $P^{+}(1a,3; 3a^{*},0) = \min\{0,1 - w_a - w_b\}$ imply that such values of R are impossible and that the forbidden region is empty.

(2) The pair $2a^{*}$,1 and $2a^{*}$,1′ (i.e., $1 = 2a^{*}$, $2 = 1$, $2' = 1'$). When making the join according to the first cell, we have $\Delta T = -w_b + x$. When making the join according to the second cell, $\Delta T = -w_b + y$. The difference of these expressions (the second minus the first) is $y - x$. The forbidden region is $x - y < -1$ or $x - y > 1$. The corresponding difference R of increments of P is $x - y + w_a$. In the forbidden region, $R < -1 + w_a$ or $R > 1 + w_a$. However, R is equal to the increment of P when replacing chains of types 1 (joining the extremities 1 and 2) and 1 (the chain with extremity 2′) with chains of types 2 (joining the extremities 1 and 2′) and 1 (the chain with extremity 2). The equalities $P^{-}(1,1; 2,1) = 0$ and $P^{+}(1,1; 2,1) = \max\{0, w_a + w_b - 1\}$ imply that the forbidden region is empty.

(3) The pair 1,1a and 1,1 (i.e., $1 = 1$, $2 = 1a$, $2' = 1$). When making the join according to the first cell, we have $\Delta T = -w_b + x$. When making the join according to the second cell, $\Delta T = 1 - 2w_b + y$. The difference of these expressions is $y - x + 1 - w_b$. The forbidden region is $x - y < -w_b$ or $x - y > 2 - w_b$. The corresponding difference R of increments of P is $x - y + w_b$. In the forbidden region, $R < 0$ or $R > 2$. Under the SM, chains of types 3 and 1 are replaced with chains of types 1a and 3b. The equalities $P^{-}(3,1; 1a,3b) = w_b$ and $P^{+}(3,1; 1a,3b) = 1$ imply that the forbidden region is empty.

(4) The pair (1,1\\$b$),1$a$ and (1,1\\$b$),3 (i.e., $1 = 1$ or $1 = 1\backslash b$, $2 = 1a$, $2' = 3$). When making the join according to the first cell, we have $\Delta T = -w_{b,a} + x$, where the index b occurs in the case $1 = 1$, and index a, in the case $1 = 1\backslash b$. When making the join according to the second cell, $\Delta T = -w_{b,a} + y$. The forbidden region is $x - y < -1$ or $x - y > 1$. The corresponding difference R of increments of P is $x - y + 1$. In the forbidden region, $R < 0$ or $R > 2$. Under the SM, chains of types 1 and 3 are replaced with chains of types 1a and 1b. The equalities $P^{-}(1,3; 1a,1b) = \min(1, w_a + w_b)$ and $P^{+}(1,3; 1a,1b) = w_a + w_b$ imply that the forbidden region is empty.

(5) The pair 1,3a^{*} and 1,1′ (i.e., $1 = 1$, $2 = 3a^{*}$, $2' = 1'$). When making the join according to the first cell, we have $\Delta T = -w_b + x$. When making the join according to the second cell, $\Delta T = 1 - w_b + w_a - y$. The forbidden region is $x + y < w_a$ or $x + y > 2 + w_a$. The corresponding difference R of increments of P is $x + y - w_a$. In the forbidden region, $R < 0$ or $R > 2$. Under the SM, chains of types 1 and 3 are replaced with chains of types 3a^{*} and 1b. The equalities $P^{-}(1,3; 3a^{*},1b) = w_b$ and $P^{+}(1,3; 3a^{*},1b) = \min(1, w_a + w_b)$ imply that the forbidden region is empty.

(6) The pair 2,1a and 2,1 (i.e., $1 = 2$, $2 = 1a$, $2' = 1$). When making the join according to the first cell, we have $\Delta T = -1 + x$. When making the join according to the second cell, $\Delta T = 1 - w_b - y$. The forbidden region is $x + y < 1 - w_b$ or $x + y > 3 - w_b$. The corresponding difference R of increments of P is $x + y - 1 + w_a + w_b$. In the forbidden region, $R < w_a$ or $R > 2 + w_a$. Under the SM, chains of types 1 and 1 are replaced with chains of types 1a and 1b. The equalities $P^{-}(1,1; 1a,1b) = w_a + w_b$ and $P^{+}(1,1; 1a,1b) = w_a + w_b$ imply that the forbidden region is empty.

(7) The pair 2,2 and 2,3 (i.e., $1 = 2$, $2 = 2$, $2' = 3$). When making the join according to the first cell, we have $\Delta T = 1 - w_b - x$. When making the join according to the second cell, $\Delta T = -w_b + y$. The forbidden region is $-x - y > 0$ or $-x - y < -2$. The corresponding difference R of increments of P is $-x - y + w_a + 1$. In the forbidden region, $R > w_a + 1$ or $R < w_a - 1$. Under the SM, chains of types 2b^{*} and 3 are replaced with chains of types 1b and 2. The equalities $P^{-}(2b^{*},3; 1b,2) = 0$ and $P^{+}(2b^{*},3; 1b,2) = w_a$ imply that the forbidden region is empty.

5. *o* is a DM. Represent o as a composition of two cuts and two joins. One can easily check that, when making any operation, $B + S$ changes by at most 1. This property of an operation is called the *B + S property*.

5.1. Reversal in a cycle or in a chain. Let both cuts be interior (i.e., on both sides of each cut there are singular vertices); then, P and D do not change, since the chain type does not change. The result follows from the $B + S$ property. Let a reversal be made in a chain where at least one cut is exterior. The extremity of the chain adjacent to an exterior cut is called *exterior;* let it be the left-hand extremity. The other cut is called *interior*, as well as the edge that it cuts. To reduce the number of cases to consider, note that, according to the upper row of Table 2, joining any chain with an odd chain of type 0, as well as the inverse cut operation, does not change T. We call this the *odd cutting property*. Therefore, if at one end (or at both ends) outside the chain there is a fragment (fragments) of oddly many conventional edges, then we can cut it (them) out, make the reversal, and then again join it (them) to the resulting chain. Thus, it remains to consider the case where exterior cuts are made at extreme edges. If the reverted segment has no singular vertices, the chain does not change. Otherwise, consider two possibilities.

(1) Two singular vertices are joined into one, i.e., B reduces by 1. Then, one cut is interior and the other exterior. Consider variants of changing the parity of the extremal segment. If it does not change, the result follows from the $B + S$ property. If an even extremal segment is replaced with an odd one, S increases by 1, since an extremal odd segment appears. The following variants of changing the chain type are possible: $3a$ changes into $1a$, 1 changes into 2, $1a$ changes into $2a$, and 3 changes into 1. In the first two variants, D does not change, and P does not reduce and increases by at most w_a. In the last two variants, D reduces by 1 and P does not increase and reduces by at most w_a. If an odd extremal segment is replaced with an even one, S does not change. Now, the following variants of changing the chain type are possible: $1a$ changes into $3a$, 2 changes into 1, $2a$ changes into $1a$, and 1 changes into 3. In the first two variants, D does not change, and P does not increase and reduces by at most w_a. In the last two variants, D increases by 1, and P does not reduce and increases by at most w_a. In all the cases, T changes by at most 1.

(2) Assume that no merging of two singular vertices into one occurs, i.e., B does not change, and let one cut be interior and the other exterior. Consider variants of changing the parity of the extremal segment. If it does not change, the result follows from the $B + S$ property. If an even extremal segment is replaced with an odd one, S either does not change or increases by 1. The following variants of changing the chain type are possible: $3a$ changes into $1a$, 1 changes into 2, $1a$ changes into $2a$, and 3 changes into 1. In the first two variants, D does not change, and P does not reduce and increases by at most w_a. In the last two variants, D reduces by 1, and P does not increase and reduces by at most w_a. If an odd extremal segment is replaced with an even one, S either does not change or reduces by 1. Now, the following variants of changing the chain type are possible: $1a$ changes into $3a$, 2 changes into 1, $2a$ changes into $1a$, and 1 changes into 3. In the first two variants, D does not change, and P does not increase and reduces by at most w_a. In the last two variants, D increases by 1, and P does not reduce and increases by at most w_a. In all the cases, $|\Delta T| \leq 1$. If both cuts are exterior, then, taking into account the odd cutting property, the chain does not change.

5.2. Breaking a cycle into two cycles or a cycle and a loop (cutting out a cycle or a loop from a cycle). In this case, T' either reduces by $1 - w_a$ or does not change; T'' either does not change, or increases by w_a, or reduces by w_a; P does not change; and $(w_b - w_a)K_b$ either does not change or increases by $w_b - w_a$. Therefore, $|\Delta T| \leq 1$.

5.3. Cutting out a fragment from a chain and circularizing it (we consider the case where a cycle is formed; the case of a loop reduces to removal of a vertex). Here, c increases by 1. If this fragment contains no singular vertices, consider two cases. Let no merging of two singular vertices into one occur. Then, one fragment splits into two, one of the new segments being cyclic and, hence, even. Then S reduces by 1, since a cyclic segment appears, and the chain type does not change; T reduces by 1. Now, let two singular vertices merge into one. Then, one odd segment is circularized with adding an additional conventional edge, S does not change, the chain type does not change, and T reduces by 1.

If the fragment contains singular vertices, we consider the same two possibilities as in case 5.1 and repeat all arguments with replacing the reversal operation by circularization. The only case that must be considered separately is the case where both cuts are exterior. One can easily see that the odd cutting property holds for the circularization operation, as well as for the reversal. Therefore, it suffices to consider the case where both cuts are made at extremal edges. This results in a 0-chain of length 1. If merging of vertices occurs, then no segment appears in the cycle, and S increases by 1 due to the appearance of the above-mentioned 0-chain. In this case, the initial chain is of type $3a$ or $3b$, such that, when it turns into a chain of type 0, D reduces by 1, and P does not increase and reduces by at most w_a. In all the cases, $|\Delta T| \leq 1$.

Let there be no merging of vertices; then, two extremal segments of the chain are transformed into other two segments with the same total length. One of them is the above-mentioned 0-chain of length 1, and the other is located inside a cycle. If both extremal segments are odd, then S reduces by 1, the initial chain is of type either $2a$ or $2b$, D does not change, and P does not increase and reduces by at most w_b. If at least one of the extremal segments is odd, then S does not change, the initial chain is of type $1a$, $1b$, $3a$, or $3b$, D reduces by 1, and P does not increase and reduces by at most $w_a + w_b$. In all the cases, $|\Delta T| \leq 1$.

5.4. Joining two cycles into one, joining a cycle and a loop, or joining two loops. In this case, T' either increases by $1 - w_a$ or does not change, T'' does not change, increases by w_a, or reduces by w_a, P does not change, and $(w_b - w_a)K_b$ either does not change or reduces by $w_b - w_a$. Therefore, $|\Delta T| \leq 1$.

5.5. One cut is made in a chain, and the other in a cycle or loop. The case of a loop is trivial; consider the case of a cycle. The cycle is cut and inserted into a chain. If no merging of vertices occurs, then this operation is the inverse to that considered in case 5.3. Otherwise, B reduces by 1. Furthermore, c reduces by 1. If the cut in the chain is interior or the parity of the extremal segment does not change, the result follows from the $B + S$ property. Next, we repeat the arguments from case 5.1, subcase (1).

5.6. Each cut is made in a separate chain.

5.6.1. Both cuts are interior. In what follows, we implicitly use the $B + S$ property without referring to it. To trace the changes of D and P, consider types of both chains. If the set of types of the two chains does not change or the change consists in replacing the «asterisk» with the «prime» in the type of one of the chains or vice versa, then the result follows from the fact that, when making such a replacement, P changes by at most $w_b - w_a$, since $P^-(2b^*; 2b') = P^-(3a^*; 3a') = w_a - w_b$ and $P^-(2b'; 2b^*) = P^-(3a'; 3a^*) = 0$. In what follows, we call this case the identical transformation and omit it from consideration. Consider other cases.

(1) Both chains are odd, and either both are a- or both are b-chains. T' does not change. If both chains are of type $1a$, the only nonidentical change of types is transformation of this pair into a pair of chains of types $2a$ and $3a$ or $2a$ and $3a'$: D reduces by 1, and $(w_b - w_a)K_b$ either reduces by $w_b - w_a$ or does not change. Since $P^-(1a,1a; 2a,3a) = -2w_a$, $P^-(1a,1a; 2a,3a') = -1$, and $P^+(1a,1a; 2a,3a) = P^+(1a,1a; 2a,3a') = 1 - w_a - w_b$, we have $|\Delta T| \leq 1$. Type $1b$ is considered similarly. Any other pair of types of the initial chains (except for $2a$ and $3a$, which give the inverse transformation) results in the identical change.

(2) Both chains are odd, one of them being an a-chain and the other a b-chain. If their pair of types is $1a,1b$, then, when making a DM, it transforms into either a 1,1 or a 2,3 pair: T' reduces by $1 - w_a$; $(w_b - w_a)K_b$ in the first case either reduces by $w_b - w_a$ or does not change and, in the second case, does not change. In the first case, D reduces by 2, and P reduces by $w_a + w_b$. In the second case, D reduces by 1, and P reduces by at most 1 and at least $\min(1, w_a + w_b)$ (since $P^+(1a,1b; 2,3) = \max(-1, -w_a - w_b)$). If the pair of initial types is $1a,2b$ or $1a,2b'$, then, when making a DM, it transforms into a pair 1,2: T' reduces by $1 - w_a$; $(w_b - w_a)K_b$ does not change, reduces by $w_b - w_a$ (in the case of $2b$), or increases by $w_b - w_a$ (in the case of $2b'$); D reduces by 1; P reduces by at most 1 and at least w_b (in the case of $2b$) or w_a (in the case of $2b'$). The pair $1a,3b$ is considered similarly. If the pair of initial types is $2a,2b$ or $2a,2b'$, then, when making a DM, it transforms into a pair 2,2: T' reduces by $1 - w_a$; $(w_b - w_a)K_b$ either does not change or increases by $w_b - w_a$; D does not change; P does not increase and reduces by at most $2 - w_a - w_b$, since $P^-(2a,2b; 2,2) = P^-(2a,2b'; 2,2) = w_a + w_b - 2$ and $P^+(2a,2b; 2,2) = P^+(2a,2b'; 2,2) = 0$. The pairs $3a,3b$ or $3a',3b$ are considered similarly. If the initial pair is $2b,3a$, $2b',3a$, $2b,3a'$, or $2b',3a'$, then, when making a DM, it transforms into a pair 1,1: $(w_b - w_a)K_b$ does not change, increases by $w_b - w_a$ (for the last three pairs), or reduces by $w_b - w_a$ (for the first pair); T' reduces by $1 - w_a$; D reduces by 1; P reduces by w_b (in the first three cases) or w_a (in the last case). Similarly, this applies to the pair $2a,3b$. In all the cases, $|\Delta T| \leq 1$.

(3) Both chains are even. The only nontrivial transformation relates to pairs 2,3 and 1,1; other transformations are either identical or inverse to those considered in case (2). T' does not

change. When a pair 2,3 transforms into 1,1, D reduces by 1, $(w_b - w_a)K_b$ either does not change or reduces by $w_b - w_a$, and $\Delta P = \min(0{,}1 - w_a - w_b)$.

(4) One chain is odd, and the other even: T' does not change. If a pair $1b$,1 transforms into $2b$,3 or $2b'$,3, then D does not change; $(w_b - w_a)K_b$ does not change, increases by $w_b - w_a$ (for the first case), or reduces by $w_b - w_a$ (for the second case); ΔP is not less than $w_b - 1$ and $w_a - 1$, respectively, and not greater than 0, whence the desired follows. Transformation into a pair $3a$,2 is similar. If a pair $1b$,2 transforms into $2b$,1 or $2b'$,1, then D reduces by 1, $(w_b - w_a)K_b$ either does not change or reduces by $w_b - w_a$, and ΔP is not less than $- w_b$ and not greater than $\min(0{,}1 - w_a - w_b)$, whence the desired follows. Transformation of a pair $1a$,3 into $3a$,1 is similar. Transformation of a pair $2a$,1 into $1a$,2 is inverse to the one considered above. If a pair $2a$,3 transforms into $3a$,2 or $3a'$,2, then D does not change, ΔP is not less than $w_a - 1$ and not greater than $1 - w_b$, and K_b does not change. Transformations $2a$,3 into $3a'$,2 and $2b'$,3 into $3b$,2 are considered similarly. Transformations $2a$,3 into $1a$,1, $3a$,1 into $1a$,3, and $3a$,2 into $1a$,1 are inverse to those considered above. For a pair $3a$,3, the identical transformation is only possible.

5.6.2. One of the cuts is exterior. We reduce this case to an SM or OM operation as follows: the extremity of the chain adjacent to an exterior cut is called exterior. The other cut is called interior, as well as the edge that it cuts. If an exterior cut is at neither an extremal edge nor the edge next to it, we remove two conventional edges from the exterior end. Then, T reduces by 1 both in the initial graph and in the resulting one, which allows us to apply Lemma 4.2. Therefore, it suffices to consider cases where the exterior cut is at an extremal edge of a chain or at the next to it.

(1) The exterior cut is at the next to extremal edge. If it is conventional, remove two extremal edges together with the exterior cut, thus replacing the DM with an SM (see Figure 5a). Then, T reduces by 1 both in the initial graph and in the resulting one, which by Lemma 4.2 reduces the problem to the SM case. If the edge with the exterior cut is singular, remove the extremal edge together with its endpoints, replacing an odd extremal segment of length 1 with a hanging edge (the singular vertex becomes an endpoint of a hanging edge). Again, the DM is changed to an SM with reducing T by 1 in both graphs (see Figure 5b).

(2) The exterior cut is at the extremal edge, and this edge is conventional. If the interior edge is conventional or its singular vertex is on the side that is joined with the exterior extremity, move the exterior cut outside the end of the chain (i.e., make it "fictitious") and move the interior cut to the next edge in the opposite direction from the potential singular vertex (and, if this edge is missing, then outside the end of the chain) (see Figure 5c). Thus, the DM is changed to an SM (or, respectively, to the identity operation, which changes nothing). The result of the operation does not change. If the singular vertex is on the other side, move the exterior cut from the end to the next edge and, accordingly, move the interior cut to the neighboring edge (see Figure 5d). The result of the operation does not change, and the problem reduces to that considered in the preceding case.

(3) The exterior cut is at the extremal edge, and this edge is singular. If the interior edge is conventional or its singular vertex is on the side that is joined with the exterior extremity, we proceed as in the preceding case (see Figure 6a). Otherwise, move the exterior cut outside the end of the chain and move the interior cut to the neighboring edge in the opposite direction from the singular vertex (and, if this edge is missing, then outside the end of the chain) (see Figure 6b–d). Thus, the DM is replaced with an SM or OM. To obtain the result of the first operation (DM) from the result of the second (SM or OM), merge the two singular vertices by deleting the conventional edge that separates them, and add a chain from one conventional edge (in the case of OM) or add two extremal conventional edges to the chain (if the interior cut was moved to a conventional edge), or replace a hanging edge with an extremal segment of length 1 (if the interior cut was moved to a singular edge). In all the cases, the transformation does not change T, which completes the analysis of the cases.

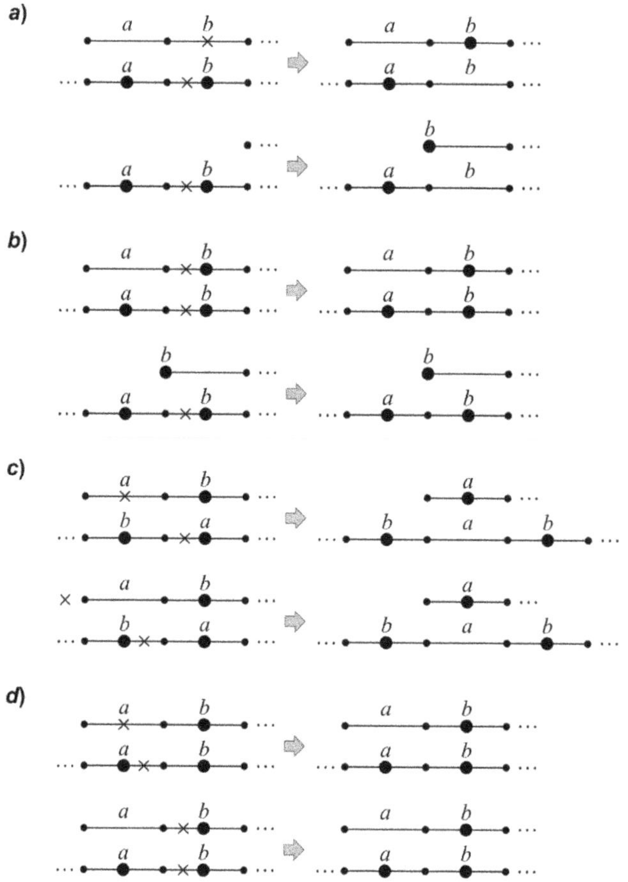

Figure 5. DM with an exterior cut at a conventional edge. (**a**) Reduction to an SM for an exterior cut at the next to extremal conventional edge. Top: initial DM. Bottom: SM obtained from the initial DM by removing two conventional edges. (**b**) Reduction to an SM for an exterior cut at the next to extremal singular edge. (**c**) Reduction to an SM for an exterior cut at an extremal conventional edge. The fictitious cut outside the end of the chain corresponds to replacement of the DM with an SM. (**d**) Reduction of a cut at an extremal conventional edge to the case considered in Figure 5b.

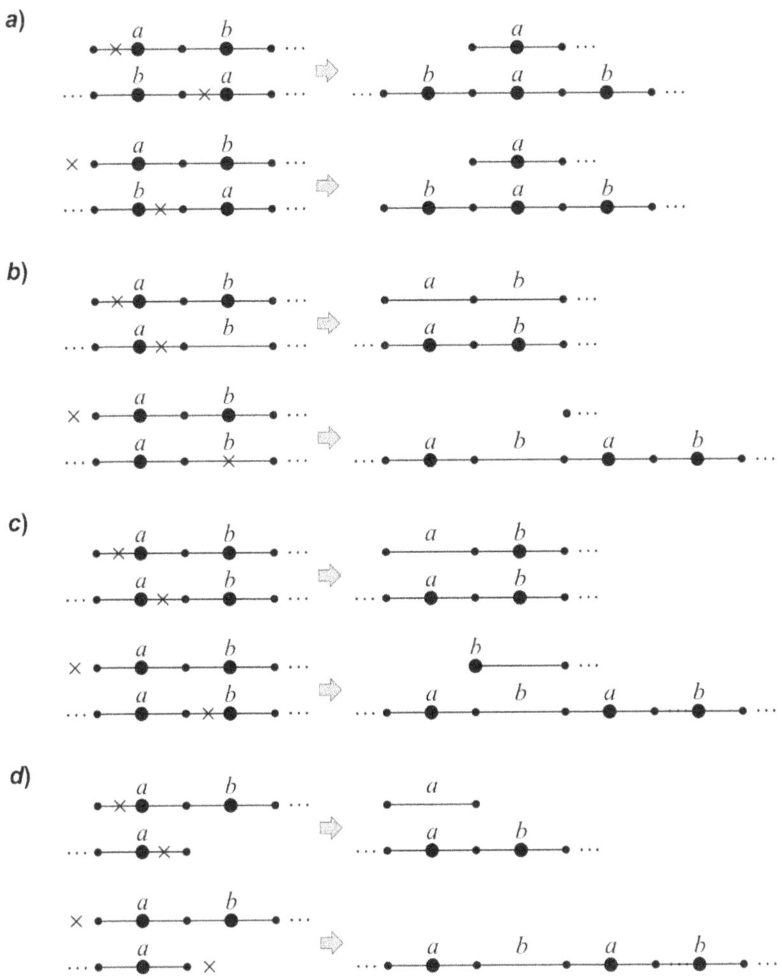

Figure 6. DM with an exterior cut at an extremal singular edge. Reduction to an SM: first case (**a**), second case (**b**), and third case (**c**); reduction to an OM (**d**).

Theorem 1 is proven.

4. Algorithm and Proof of Theorem 2 (Case III)

Theorem 2 considers Case III; it is a particular case of Theorem 1.

Theorem 2. *If DCJ operations are assigned with equal costs w and the deletion and insertion operations are assigned with any costs w_d and w_i such that one of them is strictly less and the other strictly greater than w, then the algorithm described below is of linear time complexity and outputs a sequence of operations with total cost superior to the cost of the shortest sequence by at most an additive constant that depends only on the operation costs.*

Without loss of generality, we may assume that $w = 1$. Recall that $w_d = w_a$ and $w_i = w_b$. In Theorem 2, we distinguish three subcases: (1) $w_d + w_i \leq 2$, (2) $w_d + w_i > 2$ and $\max(w_d,w_i) \leq 2$, and (3) $\max(w_d,w_i) > 2$, and we enumerate them, respectively, by 1, 2, and 3. The additive exactness constant k mentioned in the theorem is in the first subcase not greater than $w_b - 1$; in the second subcase, constant k is not greater than $4w_b + 2w_a - 6$; in the third subcase, constant k is not greater than $6w_b + 2w_a - 9$. In Section 4a, we describe the algorithm for the first subcase. Algorithms for the second and third subcases, as well as proofs of their exactness, will be given elsewhere due to natural restrictions on the size of the paper. These algorithms are akin to the algorithm for the first subcase, and the proofs of their exactness are akin to the proof for this subcase, which is given in Section 4.2. As above, we may confine ourselves with the case of $w_a \leq w_b$.

4.1. Description of the Algorithm for the First Subcase

In what follows, all comparisons are made with the Algorithm from Section 2 (Case II).

Stage 0: The same, transformation of initial graphs a and b into $a + b$.

Stage 1: The same, cutting out all conventional edges.

Let us recall what *types* of chains may occur in $a + b$. These are odd chains: $1a$, $1b$, $2a$ ($2a^*$ and $2a'$), $2b$ ($2b^*$ and $2b'$), $3a$ ($3a^*$ and $3a'$), $3b$ ($3b^*$ and $3b'$), and 0, and even chains: 1_a ($1_a{}^*$ and $1_a{}'$), 1_b ($1_b{}^*$ and $1_b{}'$), 2 (2^* and $2'$), 3, and 0. Chains of types $2a'$, $3b'$, $1'_b$ are called *problem chains*. Components of $a + b$ are divided into *types* (*a*,*b*)-, *a*-, *b*-, and 0- accordingly as they contain both *a*- and *b*-vertices, *a*- but no *b*-vertices, *b*- but no *a*-vertices, or no singular vertices at all.

Stage 2: In essence, the same: Consider former 2-interactions between different chains that were performed in Case II. Their *a-quality* (instead of the quality that was used in Case II) is shown in square brackets (its definition is given right after the algorithm description, in Section 4b): $1a + 1b = 1^*_b$ $[w_a + 1]$, $3a^* + 2b^* = 1_a$ $[1]$, $3a^* + 2b' = 1_a$ $[w_a]$, $3a' + 2b^* = 1_a$ $[w_a]$, $3a' + 2b' = 1'_a$ $[w_a]$, $3b + 2a = 1_b$ $[1]$, $3 + 2 = 1^*_b$ $[w_a]$, $(1a + 2b^*) + 3 = 1^*_b$ $[w_a + 1]$, $(1a + 2b') + 3 = 1^*_b$ $[2w_a]$, $(1b + 2a) + 3 = 1^*_b$ $[w_a + 1]$, $(3a^* + 1b) + 2 = 1^*_b$ $[w_a + 1]$, $(3a' + 1b) + 2 = 1^*_b$ $[2w_a]$, $(3b + 1a) + 2 = 1^*_b$ $[w_a + 1]$, $1a + 2 = 2a^*$ $[w_a]$, $1b + 2 = 2b^*$ $[w_a]$, $3 + 1a = 3a^*$ $[w_a]$, $3 + 1b = 3b^*$ $[w_a]$, $(3b + 1a) + (1a + 2b^*) = 1^*_b$ $[w_a + 2]$, $(3b + 1a) + (1a + 2b') = 1^*_b$ $[2w_a + 1]$, $(3a^* + 1b) + (1b + 2a) = 1^*_b$ $[w_a + 2]$, $(3a' + 1b) + (1b + 2a) = 1^*_b$ $[2w_a + 1]$, $1a + (1a + 2b^*) = 2a^*$ $[w_a + 1]$, $1a + (1a + 2b') = 2a^*$ $[2w_a]$, $1b + (1b + 2a) = 2b^*$ $[w_a + 1]$, $(3b + 1a) + 1a = 3a^*$ $[w_a + 1]$, $(3a^* + 1b) + 1b = 3b^*$ $[w_a + 1]$, $(3a' + 1b) + 1b = 3b^*$ $[2w_a]$, $1a + 1a = 3a^*$ $[w_a]$, $1b + 1b = 3b^*$ $[w_a]$, $1a + 2b^* = 2$ $[1]$, $1a + 2b' = 2$ $[w_a]$, $1b + 2a = 2$ $[1]$, $3a^* + 1b = 3$ $[1]$, $3a' + 1b = 3$ $[w_a]$, $3b + 1a = 3$ $[1]$, $3 + ((3 + 2b^*) + 2a) = 1^*_b$ $[w_a + 1]$, $3 + ((3 + 2b') + 2a) = 1^*_b$ $[2w_a]$, $(3a^* + (3b + 2)) + 2 = 1^*_b$ $[w_a + 1]$, $(3a' + (3b + 2)) + 2 = 1^*_b$ $[2w_a]$, $(3a^* + 2) + 2 = 2a^*$ $[w_a]$, $(3a' + 2) + 2 = 2a^*$ $[2w_a - 1]$, $(3b + 2) + 2 = 2b^*$ $[w_a]$, $3 + (3 + 2a) = 3a^*$ $[w_a]$, $3 + (3 + 2b^*) = 3b^*$ $[w_a]$, $3 + (3 + 2b') = 3b^*$ $[2w_a - 1]$, $(3 + 2b^*) + 2a = 2^*$ $[1]$, $(3 + 2b') + 2a = 2^*$ $[w_a]$, $3a^* + (3b + 2) = 3$ $[1]$, $3a' + (3b + 2) = 3$ $[w_a]$, $3a^* + 3b = 3$ $[1 - w_a]$, and $2a + 2b^* = 2^* + 1'_a$ $[1 - w_a]$.

Construct a maximal domain on the set of chains of $a + b$ in the same way as in Case II but with the *a*-quality instead of the quality. Having found a maximal domain M, simultaneously and independently perform all 2-interactions that correspond to it.

Notice 3. To 2-interactions, we may add the following three interactions that are applied to two different chains, with the term equalities: $1'_b + 2a' = 2a'$, $1'_b + 3b' = 3b'$, and $1'_b + 1'_b = 1'_b$. They have zero *a*-qualities and reduce the number of problem chains, which produce an additive error of the algorithm. However, they are inessential for the proof of Theorem 2. These interactions were shown in Figure 7 in [2].

Stage 3: Execute the following *3-interactions*, which reduce the number of (*a*,*b*)-components and the number of *b*-components. These interactions are shown in Figure 7 below.

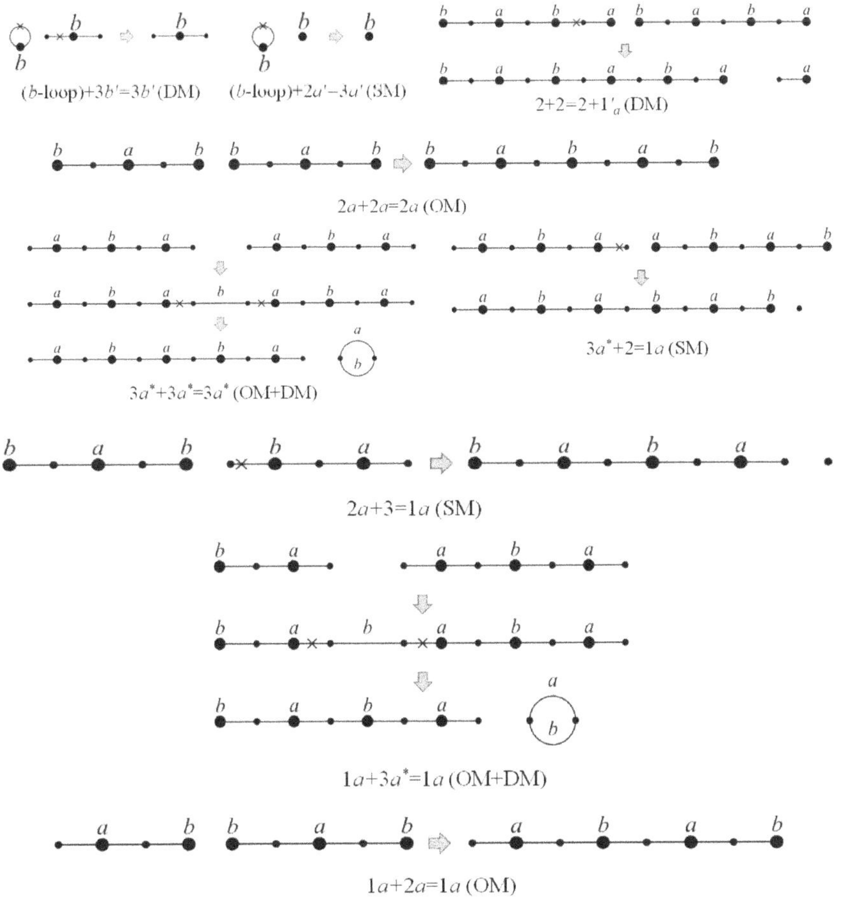

Figure 7. *Cont.*

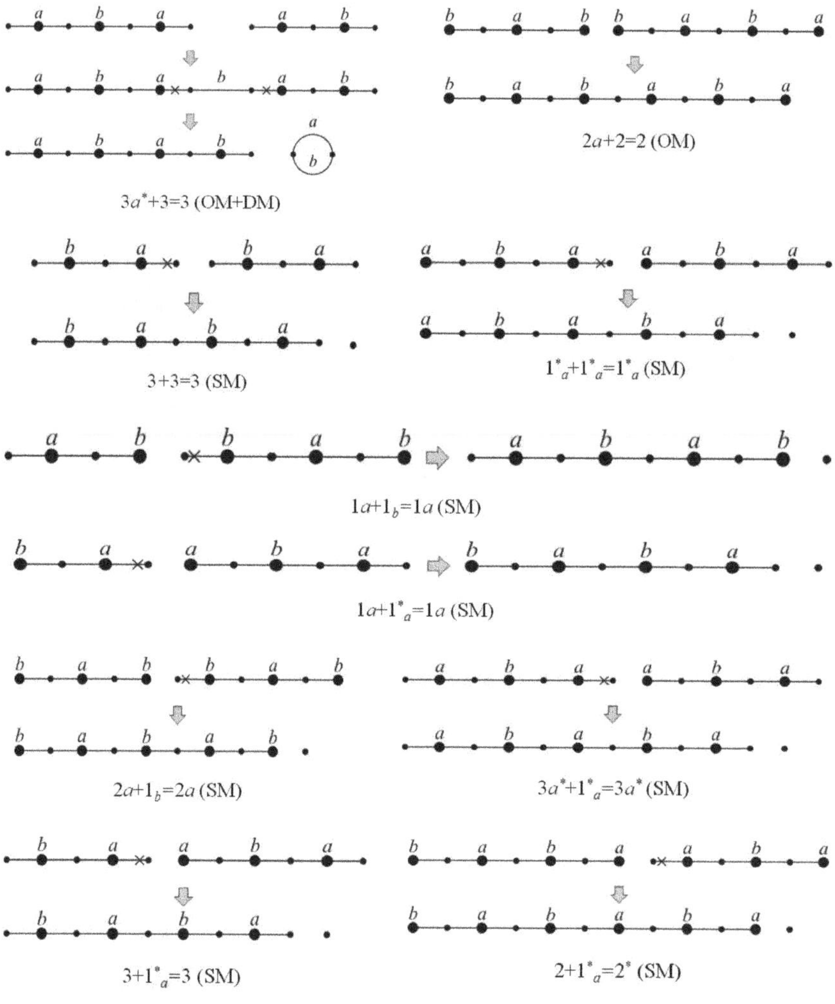

Figure 7. The 3-interactions.

1. b-loop + «component with a b-singular vertex» = the same component (DM or SM operations);
2. $2 + 2 = 2 + 1'_a$. SM operation with cutting out two vertices of a 2-chain (an extremal a-singular vertex and the neighboring conventional) and joining the obtained extremity with an extremal b-singular vertex of the other 2-chain.
3. $2a + 2a = 2a$, $2b^* + 2b^* = 2b^*$. Joining hanging vertices of two chains.
4. $3a^* + 3a^* = 3a^*$, $3b + 3b = 3b$. OM and cutting out a conventional edge.
5. $3a^* + 2 = 1a$, $3b + 2 = 1b$. SM.
6. $2a + 3 = 1a$, $2b^* + 3 = 1b$. SM.
7. $1a + 3a^* = 1a$, $1b + 3b = 1b$. OM and cutting out a conventional edge.
8. $1a + 2a = 1a$, $1b + 2b^* = 1b$. OM.
9. $3a^* + 3 = 3$, $3b + 3 = 3$. OM and cutting out a conventional edge.
10. $2a + 2 = 2$, $2b^* + 2 = 2^*$. OM.
11. $3 + 3 = 3$. SM.

12. $1^*_a + 1^*_a = 1^*_a$, $1_b + 1_b = 1_b$. SM.
13. $1a + 1_b = 1a$, $1b + 1^*_a = 1b$. SM.
14. $1a + 1^*_a = 1a$, $1b + 1_b = 1b$. SM.
15. $2a + 1_b = 2a$, $2b^* + 1^*_a = 2b^*$. SM.
16. $3a^* + 1^*_a = 3a^*$, $3b + 1_b = 3b$. SM.
17. $3 + 1^*_a = 3$, $3 + 1_b = 3$. SM.
18. $2 + 1^*_a = 2^*$, $2 + 1_b = 2$. SM.

Repeat Stage 3 until these interactions are possible.
Stage 4: Perform autonomous reduction of remaining components to a final form.
End of the algorithm description for the first subcase.

4.2. Proof of the Algorithm Exactness for the First Subcase

Define the *autonomous a-cost* $A_a(G)$ of a graph G to be the total cost of the sequence of operations given in the autonomous reduction for G after deduction of costs of all its b-special operations, i.e., those reducing the number of b-singular vertices. In other words, $A_a(G)$ accounts only for a-special (i.e., those reducing the number of a-singular vertices) and nonspecial (i.e., not reducing the number of singular vertices) operations. Together, they are called *a-operations*.

For a graph G, we use the following *notation*: B_a and B_b are the numbers of, respectively, a-singular and b-singular vertices in G; S is the sum of integral parts of halved lengths of maximal connected fragments consisting of conventional edges (referred to as *segments*) plus the number of extremal (on a chain) odd segments (i.e., those consisting of an odd number of edges) minus the number of cyclic segments; S_a is the number of segments enclosed between two singular a-edges; C_a is the number of a-cycles; D is the number of chains of types $1a$, $1b$, $3a$, $3b$, and 3; $N(t)$ is the number of chains of a certain type t. Denote by A the number of a-special DCJ operations in the autonomous reduction.

Lemma 5. *Let w_a and w_b be the removal costs for singular a- and b-vertices, and let all other operations have cost 1. Then,*

$$A_a(G) = (1 - w_a)\cdot(S_a - C_a + N(2b^*) + N(3a^*) + N(1_a^*)) + w_a\cdot B_a + S + D.$$

Proof. For every component of G, let us check the equality $A = S_a - C_a + N(2b^*) + N(3a^*) + N(1_a^*)$, and then sum up the obtained equalities. Each of them is implied by the following facts: when cutting out conventional edges from any component other than an a-cycle, the number of executed a-special operations is S_a, and, for an a-cycle, it is less by 1; the operation of circularization into a cycle is a-special if and only if this chain is of type $2b^*$, $3a^*$, or 1_a^*; breaking a cycle does not contain a-special DCJ operations. Next, the number of removals of a-singular vertices in the autonomous reduction is $B_a - A$, and, similarly to the proof of Lemma 1, we obtain that the number of nonspecial operations is $S + D$. □

Define the ***a*-quality** $P_a(s)$ of an interaction s to be

$$P_a(s) = A_a(G) - A_a(s(G)) - c_a(s),$$

where $s(G)$ is the graph obtained by applying s to G, and $c_a(s)$ is the total cost of a-operations in s. The number $P_a(S)$ shows what a-cost is saved when applying the interaction s as against the autonomous reduction of G, i.e., without using s. Take note that, in Case II, we used the *quality* of interaction, which is related to all operations instead of a-operations.

Using Lemma 4, we compute a-qualities of 2-interactions. For example, let us find the a-quality of the interaction $1a + 1b = 1_b^*$ (Figure 4), where $c_a(s) = 1$. The interaction reduces B_a by 1, reduces D by 2, and does not change all other quantities in the expression for $A_a(G)$. Therefore, $P_a(s) = w_a + 1$.

Let E be any reducing sequence for a graph G, which is arbitrarily divided into connected fragments of operations; let us for a while call the latter interactions and denote them by s. Define $T_a(G,E)$, the *total cost of a-operations* in E, and $P_a(G,E)$, the total a-quality of all interactions in E, where $P_a(s)$ is formally defined above. The following lemma is proven similarly to Lemma 2 in Section 3:

Lemma 6. *For any graph G, we have* $T_a(G,E) = A_a(G) - P_a(G,E)$.

Lemma 7. *After Stage 3, the algorithm outputs a graph G‴ with no polytype corresponding to any 2-interaction of two chains.*

Proof. It is important that the algorithms for Cases II (when $w_b = 1$) and III at Stages 0, 1, and 2 completely coincide. Maximal domains that are constructed in both cases coincide, since, through an exhaustive search over all 2-interactions, we check the following: for $w_b = 1$, the a-quality of any 2-interaction is equal to the quality if the same 2-interaction (computed in Section 2). Furthermore, through an exhaustive search, we check that all 3-interactions have zero quality. If, for G''', there are two different chains corresponding to some 2-interaction (its a-quality is automatically strictly greater than zero for $0 < w_a < 1$), then, after Stage 2 of the *algorithm for Case II* applied to G with $w_b = 1$, we use 3-interactions, followed by this 2-interaction and, then, autonomous reduction. The aggregate quality of all interactions in such a reducing sequence is strictly greater than the aggregate quality in the sequence output by the algorithm for Case II. By Lemma 2, this means that we have obtained a sequence with total cost strictly less than the absolute minimum $C(G)$, which is impossible. □

Lemma 8. *After Stage 3, the algorithm outputs a graph G‴ satisfying the following condition: it contains no more than two chains having b-singular vertices. These pairs of different chains are contained among the following pairs of types:* {2a,3a*}, {1_a*,1_b}, {3b,2b*}, {2a,1_a*}, {3a*,1_b}, {1_a*,3b}, and {1_b,2b*}. *If the initial G has no* (a,b)*-chains, then G‴ contains at most one chain with a b-singular vertex.*

Proof. First claim. Through an exhaustive search, we check that, for any three types of chains with b-singular vertices, there are two of them corresponding to a 2- or 3-interaction. The former is impossible by Lemma 7, and the latter is possible by the definition of Stage 3. □

Second claim. Exhaustively searching over all types of chains with b-singular vertices, form all pairs of them that do not correspond to any 2- or 3-interaction. We obtained precisely the pairs listed in the lemma. Any two chains in this list do not correspond to these interactions.

Third claim. If an (a,b)-chain does not occur in the initial G, it does not appear while running the algorithm, i.e., in G''', there are only a- and b-chains. If, in G''', there are two different chains with b-singular vertices, then, by the second claim, their types form one of the listed pairs. However, every chain whose type contains an asterisk is an (a,b)-chain.

Denote by $T(G)$ the *total cost of operations* executed by the described algorithm on a graph G. Let C_b and C_{ab} be the numbers of b- and (a,b)-cycles in G, respectively; let I_{pb} be the indicator function of the property «G contains a b-loop but has no components with a singular b-vertex other than loops»; similarly, let I_{cb} be the indicator function of having a chain with a singular b-vertex; let $\varepsilon_b = w_b - 1$.

Denote by $P_a(G)$ the *total a-quality* of 2-interactions when running the algorithm on a graph G.

Lemma 9. *Let w_a and w_b be removal costs for singular a- and b-vertices, and let all other operations have cost 1. Then,*

$$T(G) = A_a(G) - P_a(G) + B_b + \varepsilon_b \cdot (C_b + C_{ab} + I_{cb} + I_{pb} + E), \text{ where } E = 0 \text{ or } E = 1.$$

Proof. We declare each operation outside Stages 2 and 3 as an interaction. It is easily checked that the a-quality of any such formal interaction, as well as that of any 3-interaction, is zero. By Lemma 6, the total cost of a-operations is $A_a(G) - P_a(G)$. In a reducing sequence, the number of b-special operations is B_b. Let us show that the number of b-removals among them is $C_b + C_{ab} + I_{cb} + I_{pb} + E$. If, in G, there are no components with a b-vertex other than b-loops, the claim is trivial. Otherwise, the number of b-removals is equal to the number of components with a b-vertex that occur in G''', i.e., before Stage 4, since a removal operation was never applied in the algorithm before; at Stage 4, only autonomous reduction is used. This amounts to $C_b + C_{ab}$ original cycles, I_{cb} chains with a b-vertex, and E more chains from Lemma 8. □

We prove the additive exactness of the algorithm by induction on the cost $C(G)$ of the shortest reducing sequence for a graph G. Denote by $T'(G)$ the value of $T(G)$ for $E = 0$. Assume that, for any operation o applied to an arbitrary breakpoint graph G, the «triangle inequality»,

$$c(o) \geq T'(G) - T'(o(G)),$$

is valid, where $c(o)$ is the cost of o and $o(G)$ is the result of applying o to G. If we take for o the first operation in the shortest sequence, by the induction hypothesis, we obtain $C(o(G)) \geq T'(o(G)) \geq T'(G) - c(o)$ and $C(G) = C(o(G)) + c(o) \geq T'(G) - c(o) + c(o) = T'(G)$. Therefore, the algorithm error is at most $\varepsilon_b E \leq w_b - 1$.

The proof of the triangle inequality follows the same lines as in Section 3 (Case II) by examining all operations o. In the proof, one should use P_a, the a-quality of the maximal domain on chains in G.

5. Algorithm and Proof of Theorem 1 for Case I

It might be interesting for the reader to compare the algorithms and proofs presented above with those related to Case I, where both costs w_d and w_i are not less than w, although this case was analyzed in a preparatory work [2]. As above, it suffices to consider the case of $w_a \leq w_b$.

Recall that, in Case I, the Algorithm consists of 8 stages.

Stage 0: The same as above: Transform initial graphs a and b into $a + b$.

Stage 1: The same as above: Cut out conventional edges.

Stage 2: Perform the same 2-interactions between chains as in Section 2 (except for the two additional ones) but in the same order as they are listed, independently of their quality (i.e., we need not choose a maximal domain M). The meaning of these interactions, as always, is that they save maximally many operations as against the number of operations in the autonomous reduction of a graph.

Stage 3: Perform *3′-interactions* between chains, which are listed in [2]. The meaning of 3′-interactions is that they reduce the number of *short* chains in $a + b$ (to be precise, chains of types $2a'$, $2b'$, $3a'$, $3b'$, $1'_a$, $1'_b$, and $2'$), thereby reducing the additive error of the algorithm, which occurs precisely because of them.

Stage 4: Circularize chains (except for short ones) into cycles using OM or SM operations.

Stage 5: Join all cycles into one cycle and then detach final cycles from it.

Stage 6: Perform *6-interactions* between a cycle and short chains, which reduce the number of the latter. These are interactions defined by the following term equalities: (a,b)-cycle + $2a'$ + $2b'$ = $2'$ (two DM operations) and (a,b)-cycle + $2'$ = $2'$ (DM and cutting out a conventional edge).

Stage 7 is applied if $w_b > 2$. Perform *7-interactions* between two cycles or between a cycle and a short chain. Their meaning is to replace the removal operation for a b-singular vertex with two intermergings, which is advantageous if $w_b > 2$. These are the interactions defined by the following term equalities: (a,b)-cycle + b-cycle = (a,b)-cycle + cycle of length 2 (two DM operations), b-cycle + b-cycle = b-cycle + cycle of length 2 (two DM operations), b-cycle + $2'$-chain = $2'$-chain + cycle of length 2 (SM and cutting out an extremal conventional edge), (a,b)-cycle + $2a'$-chain = (a,b)-cycle (SM and OM), (a,b)-cycle + $1_b'$-chain = (a,b)-cycle (two SM operations), b-cycle + $2a'$-chain = $2a'$-chain + cycle of length 2 (SM and cutting out an extremal conventional edge), b-cycle + $1_b'$-chain = $1_b'$-chain + cycle of length 2 (SM and cutting out an extremal conventional edge). If $w_a > 2$, then symmetric interactions with a replaced by b are also performed.

Stage 8: Remove singular vertices.

Now, we recall the proof of exactness of this algorithm, which has an additive error of at most $2w$. As all such proofs, it is based on the triangle inequality. However, for Case I, it has a distinction: $T(G)$ takes the form $T'(G) + E(G)$, where $T'(G)$ involves only easily computable characteristics of a graph G, and $E(G) \leq 2w$. The triangle inequality for $T'(G)$ implies that $C(G) \geq T'(G)$, where, as always, $C(G)$ is the total cost of the shortest reducing sequence for G. However, for $T'(G)$, the triangle inequality is not always valid. To overcome this obstacle, all graphs G *are divided into ranks* 1, 2, and 3. Graphs of ranks 1 and 2 have a simpler structure, although graphs of a final form fall into rank 3 (from the point of view of the current proof, they are more complicated). For graphs G of ranks 1 and 2, we prove a lower bound on $C(G)$, which is stronger than $C(G) \geq T'(G)$; for graphs of rank 1, it is even stronger than for those of rank 2. When passing, by operation o, from a graph of a larger rank to that of a smaller rank, the triangle inequality is weakened by a strictly positive number Δ (depending only on these ranks), i.e., $c(o) \geq T'(G) - T'(o(G)) - \Delta$. Under the inverse transformation, it is strengthened by Δ, i.e., $c(o) \geq T'(G) - T'(o(G)) + \Delta$. Then, the proof is made by induction on the shortest reducing sequence for G; it also requires an exhaustive search over all pairs of ranks of G and $o(G)$. For example, if G is of rank 3 and $o(G)$ of rank 2, then, by the induction hypothesis, we have $C(o(G)) \geq T'(o(G)) + \Delta \geq T'(G) - c(o) - \Delta + \Delta$ and $C(G) = C(o(G)) + c(o) \geq T'(G) - c(o) + c(o) = T'(G)$. Here, we used the fact that the Δ by which the triangle inequality is weakened when passing from a graph of rank 3 to that of rank 2 is equal to the number by which the lower bound on $C(G)$ for graphs of rank 2 is strengthened. The analogous consistency of such inequalities holds for other pairs of ranks as well. All cases of the ranks are examined in [2].

6. Discussion

Usually, a small (slightly different from the original operation costs) explicitly specified additive error is not considered as a violation of the algorithm exactness. In this sense, we described an exact algorithm of linear complexity which constructs a shortest transformation of one weighted (in other words, labeled) directed graph into another under equal costs of DCJ operations and arbitrary costs of deletion and insertion operations. These graphs must be of degree 2 in the sense that each vertex is of degree 1 or 2. The labeling cannot contain repetitions of names. Note that the above can be carried over to infinite (countable) recursively enumerated sequences. The idea of the current algorithm was based on the abstract theory developed in [11,12].

The theorem proven in the paper is rather unexpected from the mathematical point of view; a problem that seems to require exhaustive search or at least to be computationally very difficult is solved by an exact algorithm whose runtime is always linear in the size of input data. In this sense, the problem is not related to any application. However, it has arisen in the context of quite various applications. Of these, the most popular concerns the description of a biological genome by a set of chains and cycles that correspond to linear and circular chromosomes. In this case, the considered operations correspond to real genome transformations in the biological evolution of the genome. It is important that different operations can be assigned with different costs, which corresponds to different frequency of their occurrence in the evolution.

For instance, the possibility to take different costs of deletion w_d and insertion w_i is essential when mitochondria of deuterostomes are considered. Their mitochondria mainly have the same set of genes, and each gene has a unique function. However, the order of these genes can vary; it is substantially different in the purple sea urchin *Strongylocentrotus purpuratus* than in vertebrates (Jacobs et al. [13]), as well as among echinoderms themselves. Thus, gene losses and acquisitions are much rarer here than DCJ operations, and rare events should be assigned higher costs. A similar pattern is observed in plastids of plants, e.g., the order of plastid genes in the red alga *Porphyridium purpureum* notably differs from that in other red algae. In the next case where one genome is smaller than the other, the loss cost should exceed that of acquisition, which surely depends on the gene and the number of its copies. For example, the nuclear genome amplifies after hybridization of two species as in the case of polyploid strawberry or wheat (Bors and Sullivan [14]). Polyploidy is much more common in plants than in animals. Among animals, it was described in nematodes including ascarids and certain amphibians. However, other full-genome duplications have been reported in chordates (Putnam et al. [15]). The genome reduction can also occur, e.g., many genes (paralogs) are lost after a full-genome duplication. In this regard, the presented algorithm has been used, e.g., in [16].

Let us give an example from another area. In some robotics and image processing problems, it is assumed that the "terrain" is bound by barriers that can be described as chains and cycles. Chains and cycles are also used to represent parts that have long or compact forms and to describe the frame of an area or human pose, among other examples. Among many such applications, which are not related to genomics, let us mention, as an example, [17].

Further research will be aimed at reducing the additive error, up to zeroing it, constructing fast algorithms in the case of unequal costs of DCJ operations, allowing name repetitions in the graph labeling, and passing to graphs of degree 3. Moreover, it would be important to extend the list of operations by adding, for example, the operation of edge duplication in the initial graph, which will allow strictly considering the duplication of biological genes [18] from the mathematical point of view.

Author Contributions: Conceptualization, V.L. and K.G.; proof, K.G. and V.L.; writing—original draft preparation, K.G.; writing—review and editing, V.L.; supervision, V.L.; project administration, V.L.; funding acquisition, V.L. All authors have read and agreed to the published version of the manuscript.

Funding: This research was funded by the Russian Foundation for Basic Research under research project No. 18-29-13037.

Acknowledgments: We are grateful to the anonymous reviewers for their thorough review, and we highly appreciate the comments and suggestions, which significantly contributed to improving the quality of the publication.

Conflicts of Interest: The authors declare no conflict of interest.

References

1. Yancopoulos, S.; Attie, O.; Friedberg, R. Efficient sorting of genomic permutations by translocation, inversion and block interchange. *Bioinformatics* **2005**, *21*, 3340–3346. [CrossRef] [PubMed]
2. Gorbunov, K.Y.; Lyubetsky, V.A. An Almost Exact Linear Complexity Algorithm of the Shortest Transformation of Chain-Cycle Graphs. *arXiv* **2020**, arXiv:2004.14351.
3. Da Silva, P.H.; Machado, R.; Dantas, S.; Braga, M.D.V. Genomic Distance with High Indel Costs. *IEEE/ACM Trans. Comput. Biol. Bioinform.* **2016**, *14*, 728–732. [CrossRef] [PubMed]
4. Compeau, P.E.C. A Generalized Cost Model for DCJ-Indel Sorting. *Lect. Notes Comput. Sci.* **2014**, *8701*, 38–51. [CrossRef]
5. Warnow, T. (Ed.) *Bioinformatics and Phylogenetics: Seminal Contributions of Bernard Moret*; Springer Nature: Aktiengesellschaft, Switzerland, 2019.
6. Yin, Z.; Tang, J.; Schaeffer, S.W.; Bader, D.A. Exemplar or matching: Modeling DCJ problems with unequal content genome data. *J. Comb. Optim.* **2015**, *32*, 1165–1181. [CrossRef]
7. Gorbunov, K.Y.; Lyubetsky, V.A. Linear algorithm of the minimal reconstruction of structures. *Probl. Inform. Transm.* **2017**, *53*, 55–72. [CrossRef]
8. Alekseyev, M.A.; Pevzner, P.A. Multi-break rearrangements and chromosomal evolution. *Theor. Comput. Sci.* **2008**, *395*, 193–202. [CrossRef]
9. Alekseyev, M.A.; Pevzner, P.A. Breakpoint graphs and ancestral genome reconstructions. *Genome Res.* **2009**, *19*, 943–957. [CrossRef] [PubMed]
10. Lenstra, H.W. Integer Programming with a Fixed Number of Variables. *Math. Oper. Res.* **1983**, *8*, 538–548. [CrossRef]
11. Kanovei, V.; Lyubetsky, V.A. Definable E0 classes at arbitrary projective levels. *Ann. Pure Appl. Log.* **2018**, *169*, 851–871. [CrossRef]
12. Kanovei, V.; Lyubetsky, V.A. Non-uniformizable sets of second projective level with countable cross-sections in the form of Vitali classes. *Izv. Math.* **2018**, *82*, 61–90. [CrossRef]

13. Jacobs, H.T.; Elliott, D.J.; Math, V.B.; Farquharson, A. Nucleotide sequence and gene organization of sea urchin mitochondrial DNA. *J. Mol. Biol.* **1988**, *202*, 185–217. [CrossRef]
14. Bors, R.H.; Sullivan, J.A. Interspecific Hybridization of Fragaria vesca subspecies with F. nilgerrensis, F. nubicola, F. pentaphylla, and F. viridis. *J. Am. Soc. Hortic. Sci.* **2005**, *130*, 418–423. [CrossRef]
15. Putnam, N.H.; Butts, T.; Ferrier, D.E.K.; Furlong, R.F.; Hellsten, U.; Kawashima, T.; Robinson-Rechavi, M.; Shoguchi, E.; Terry, A.; Yu, J.-K.; et al. The amphioxus genome and the evolution of the chordate karyotype. *Nature* **2008**, *453*, 1064–1071. [CrossRef]
16. Gershgorin, R.A.; Gorbunov, K.Y.; Zverkov, O.A.; Rubanov, L.I.; Seliverstov, A.V.; Lyubetsky, V.A. Highly Conserved Elements and Chromosome Structure Evolution in Mitochondrial Genomes in Ciliates. *Life* **2017**, *7*, 9. [CrossRef] [PubMed]
17. Kerola, T.; Inoue, N.; Shinoda, K. Cross-view human action recognition from depth maps using spectral graph sequences. *Comput. Vis. Image Underst.* **2017**, *154*, 108–126. [CrossRef]
18. Mane, A.C.; Lafond, M.; Feijao, P.C.; Chauve, C. The distance and median problems in the single-cut-or-join model with single-gene duplications. *Algorithms Mol. Biol.* **2020**, *15*, 8–14. [CrossRef] [PubMed]

mathematics

MDPI

Article

Nonstandard Analysis, Deformation Quantization and Some Logical Aspects of (Non)Commutative Algebraic Geometry

Alexei Kanel-Belov [1], Alexei Chilikov [2,3], Ilya Ivanov-Pogodaev [3], Sergey Malev [4,*], Eugeny Plotkin [1], Jie-Tai Yu [5] and Wenchao Zhang [1]

1 Department of Mathematics, Bar-Ilan University, Ramat Gan 5290002, Israel; kanelster@gmail.com (A.K.-B.); plotkin@math.biu.ac.il (E.P.); whzecomjm@gmail.com (W.Z.)
2 Department of Information Security, Bauman Moscow State Technical University, ul. Baumanskaya 2-ya, 5, 105005 Moscow, Russia; chilikov@passware.com
3 Department of Discrete Mathematics, Moscow Institute of Physics and Technology, Dolgoprudnyi, Institutskiy Pereulok, 141700 Moscow Oblast, Russia; ivanov.pogodaev@gmail.com
4 Department of Mathematics, Ariel University of Samaria, Ariel 40700, Israel;
5 College of Mathematics and Statistics, Shenzhen University, Shenzhen 518061, China; jietaiyu@163.com or jietaiyu@szu.edu.cn
* Correspondence: sergeyma@ariel.ac.il

Received: 24 August 2020; Accepted: 22 September 2020; Published: 2 October 2020

Abstract: This paper surveys results related to well-known works of B. Plotkin and V. Remeslennikov on the edge of algebra, logic and geometry. We start from a brief review of the paper and motivations. The first sections deal with model theory. In the first part of the second section we describe the geometric equivalence, the elementary equivalence, and the isotypicity of algebras. We look at these notions from the positions of universal algebraic geometry and make emphasis on the cases of the first order rigidity. In this setting Plotkin's problem on the structure of automorphisms of (auto)endomorphisms of free objects, and auto-equivalence of categories is pretty natural and important. The second part of the second section is dedicated to particular cases of Plotkin's problem. The last part of the second section is devoted to Plotkin's problem for automorphisms of the group of polynomial symplectomorphisms. This setting has applications to mathematical physics through the use of model theory (non-standard analysis) in the studying of homomorphisms between groups of symplectomorphisms and automorphisms of the Weyl algebra. The last sections deal with algorithmic problems for noncommutative and commutative algebraic geometry.The first part of it is devoted to the Gröbner basis in non-commutative situation. Despite the existence of an algorithm for checking equalities, the zero divisors and nilpotency problems are algorithmically unsolvable. The second part of the last section is connected with the problem of embedding of algebraic varieties; a sketch of the proof of its algorithmic undecidability over a field of characteristic zero is given.

Keywords: universal algebraic geometry; affine algebraic geometry; elementary equivalence; isotypic algebras; first order rigidity; Ind-group; affine spaces; automorphisms; free associative algebras; Weyl algebra automorphisms; polynomial symplectomorphisms; deformation quantization; infinite prime number; semi-inner automorphism; embeddability of varieties; undecidability; noncommutative Gröbner-Shirshov basis; finitely presented algebraic systems; algorithmic unsolvability; turing machine

Dedicated to the 70-th anniversary of A.L. Semenov
and to the 95-th anniversary of B.I. Plotkin.

Contents

1. Introduction

The connections between algebraic geometry and mathematical logic are extremely important. First of all, notice a deep connection between algebra, category theory and model theory inspired by the results of Plotkin and his school (see References [1–5]. Note that this research is related to the one of most striking examples of interaction between model theory and geometry given by solutions of the famous Tarskii's problem, see References [6,7]. Another outstanding achievement is the theory of Zariski geometries developed by B. Zilber and E. Hrushovski [8–10]. In addition, the use of non-standard analysis has allowed progress in the theory of polynomial 64 automorphisms. See the work of Belov and Kontsevich [11,12]. For a detailed bibliography see Reference [13].

The foundations of algebraic geometry had an important part of the translation of topological and arithmetical properties into a purely algebraic language ([14]). Translation of the algebraic properties of a variety into the language of mathematical logic can be considered somehow in the spirit of this program.

This short survey is related to ideas contained in the works of B. Plotkin and V. Remeslennikov and their followers. We assume that lots of questions still require further illumination.

In particular, the following question is of interest: given two algebraic sets. Is there an algorithm for checking isomorphism? Similarly for birational equivalence. Is there a solution to the nesting problem of two varieties? For characteristics 0, the answer is not Reference [15], but for a positive characteristic the answer is unknown.

On should mention a number of conjectures related to the theory of models and polynomial automorphisms expressed in the paper by Belov-Kontsevich [11]. The investigations of the Plotkin school are far from completion, thus the relationship between the theory of medallions and the theory of categories is relevant.

One of the goals of this paper is to narrow the gap and to draw attention to this topic. We deal with commutative and non-commutative algebraic geometry. The latter notion can be understood in several ways. There are many points of view on the subject. We touch universal algebraic geometry, some of its relations with reformational quantization and Gröbner basis in non-commutative situation.

The paper is organized as follows—Section 2 is devoted to various model-theoretical aspects and their applications. More precisely, Section 2.1 deals with universal algebraic geometry and is focused around the interaction between algebra, logic, model theory and geometry. All these subjects are collected under the roof of the different kinds of logical rigidity of algebras. Under logical rigidity we mean some logical invariants of algebras whose coincidence gives rise to structural closeness of algebras in question. If such an invariant is strong enough then there is a solid ground to look for isomorphism of algebras whose logical invariants coincide.

We are comparing three types of logical description of algebras. Namely, we describe geometric equivalence of algebras, elementary equivalence of algebras and isotypicity of algebras. We look at these notions from the positions of universal algebraic geometry and logical geometry. This approach was developed by B. Plotkin and resulted in the consistent series of papers where algebraic logic, model theory, geometry and categories come together. In particular, an important role plays the study of automorphisms of categories of free algebras of the varieties. This question is highly related to description of such objects as $\mathrm{Aut}(\mathrm{Aut})(A)$ and $\mathrm{Aut}(\mathrm{End})(A)$, where A is a free algebra in a variety.

We formulate the principal problems in this area and make a survey of the known results. Some of them are very recent while the others are quite classical. In any case we shall emphasize that we attract attention to the widely open important problem whether the finitely generated isotypic groups are isomorphic.

The line started in Section 2.1 is continued in Section 2.2. Problems related to universal algebraic geometry (i.e., algebraic geometry over algebraic systems) and logical foundations of category theory gave rise to natural questions on automorphisms of categories and their auto-equivalences. The latter ones stimulate a new motivation to investigation of semigroups of endomorphisms and groups of automorphisms of universal algebras (Plotkin's problem) (see Reference [16]).

Let Θ be a variety of linear algebras over a commutative-associative ring K and $W = W(X)$ be a free algebra from Θ generated by a finite set X. Let H be an algebra from Θ and $AG_\Theta(H)$ be the category of algebraic sets over H. Throughout the work, we refer to References [13,17] for definitions of the Universal Algebraic Geometry (UAG).

The category $AG_\Theta(H)$ is considered as the logical invariant of an algebra H. By Definition 1, two algebras H_1 and H_2 are geometrically similar if the categories $AG_\Theta(H_1)$ and $AG_\Theta(H_2)$ are isomorphic. It has been shown in Reference [17], (cf., Proposition 3) that geometrical similarity of algebras is determined by the structure of the group $\mathrm{Aut}(Theta^0)$, where Θ^0 is the category of free finitely generated algebras of Θ. The latter problem is treated by means of Reduction Theorem (see References [17–20]). This theorem reduces investigation of automorphisms of the whole category Θ^0 of free algebras in Θ to studying the group $\mathrm{Aut}(\mathrm{End}(W(X)))$ associated with $W(X)$ in Θ^0.

In Section 2.2 we provide the reader with the results, describing Aut(End(A)), where A is finitely generated free commutative or associative algebra, over a field K.

We prove that the group Aut(End(A)) is generated by semi-inner and mirror automorphisms of End(A) and the group Aut($\mathcal{A}^\circ$) is generated by semi-inner and mirror automorphisms of the category of free algebras $\mathcal{A}^\circ$.

Earlier, the description of Aut($\mathcal{A}^\circ$) for the variety $\mathcal{A}$ of associative algebras over algebraically closed fields has been given in Reference [21] and, over infinite fields, in Reference [22]. Also in the same works, the description of Aut(End($W(x_1, x_2)$)) has been obtained.

Note that a description of the groups Aut(End($W(X)$)) and Aut(Θ°) for some other varieties Θ has been given in References [18,19,21–29].

A group of automorphisms of ind-schemes was computed in Reference [30]. In investigating the Jacobian conjecture and automorphisms of the Weyl algebra, Plotkin's problem for symplectomorphisms is also extremely important. Such problems are associated with mathematical physics and the theory of $\mathcal{D}$-modules.

Section 2.3 is devoted to mathematical physics and model theory. This relation deals with nonstandard analysis. We refer the reader to the review in Reference [31].

The Belov–Kontsevich conjecture [11], sometimes Kanel-Belov–Kontsevich conjecture, dubbed $B{-}KKC_n$ for positive integer n, seeks to establish a canonical isomorphism between automorphism groups of algebras

$$\mathrm{Aut}(A_{n,\mathbb{C}}) \simeq \mathrm{Aut}(P_{n,\mathbb{C}}).$$

Here $A_{n,\mathbb{C}}$ is the n-th Weyl algebra over the complex field,

$$A_{n,\mathbb{C}} = \mathbb{C}\langle x_1, \ldots, x_n, y_1, \ldots, y_n\rangle / (x_i x_j - x_j x_i,\ y_i y_j - y_j y_i,\ y_i x_j - x_j y_i - \delta_{ij}),$$

and $P_{n,\mathbb{C}} \simeq \mathbb{C}[z_1, \ldots, z_{2n}]$ is the commutative polynomial ring viewed as a $\mathbb{C}$-algebra and equipped with the standard Poisson bracket:

$$\{z_i, z_j\} = \omega_{ij} \equiv \delta_{i,n+j} - \delta_{i+n,j}.$$

The automorphisms from Aut($P_{n,\mathbb{C}}$) preserve the Poisson bracket.

Let ζ_i, $i = 1, \ldots, 2n$ denote the standard generators of the Weyl algebra (the images of x_j, y_i under the canonical projection). The filtration by total degree on $A_{n,\mathbb{C}}$ induces a filtration on the automorphism group:

$$\mathrm{Aut}^{\leq N}(A_{n,\mathbb{C}}) := \{f \in \mathrm{Aut}(A_{n,\mathbb{C}}) \mid \deg f(\zeta_i),\ \deg f^{-1}(\zeta_i) \leq N, \forall i = 1, \ldots, 2n\}.$$

The obvious maps

$$\mathrm{Aut}^{\leq N}(A_{n,\mathbb{C}}) \to \mathrm{Aut}^{\leq N+1}(A_{n,\mathbb{C}})$$

are Zariski-closed embeddings, the entire group Aut($A_{n,\mathbb{C}}$) is a direct limit of the inductive system formed by $\mathrm{Aut}^{\leq N}$ together with these maps. The same can be said for the symplectomorphism group Aut($P_{n,\mathbb{C}}$).

The Belov–Kontsevich conjecture admits a stronger form, with $\mathbb{C}$ being replaced by the rational numbers. The latter conjecture will not be treated here in any way.

Since Makar-Limanov [32,33], Jung [34] and van der Kulk [35], the B-KK conjecture is known to be true for $n = 1$. The proof is essentially a direct description of the automorphism groups. Such a direct approach however seems to be completely out of reach for all $n > 1$. Nevertheless, at least one known candidate for isomorphism may be constructed in a rather straightforward fashion. The idea is to start with an arbitrary Weyl algebra automorphism, lift it after a shift by a certain automorphism of $\mathbb{C}$ to an automorphism of a larger algebra (of formal power series with powers taking values in the ring ${}^*\mathbb{Z}$ of hyperintegers) and then restrict to a subset of its center isomorphic to $\mathbb{C}[z_1, \ldots, z_{2n}]$.

This construction goes back to Tsuchimoto [36], who devised a morphism $\mathrm{Aut}(A_{n,\mathbb{C}}) \to \mathrm{Aut}(P_{n,\mathbb{C}})$ in order to prove the stable equivalence between the Jacobian and the Dixmier conjectures. It was independently considered by Kontsevich and Kanel-Belov [12], who offered a shorter proof of the Poisson structure preservation which does not employ p-curvatures. It should be noted, however, that Tsuchimoto's thorough inquiry into p-curvatures has exposed a multitude of problems of independent interest, in which certain statements from the present paper might appear.

The construction we describe in detail in the following sections differs from that of Tsuchimoto in one aspect: an automorphism f of the Weyl algebra may in effect undergo a shift by an automorphism of the base field $\gamma : \mathbb{C} \to \mathbb{C}$ prior to being lifted, and this extra procedure is homomorphic. Taking γ to be the inverse nonstandard Frobenius automorphism (see below), we manage to get rid of the coefficients of the form $a^{[p]}$, with $[p]$ an infinite prime, in the resulting symplectomorphism. The key result here is that for a large subgroup of automorphisms, the so-called tame automorphisms, one can completely eliminate the dependence of the whole construction on the choice of the infinite prime $[p]$. Also, the resulting ind-group morphism $\varphi_{[p]}$ is an isomorphism of the tame subgroups. In particular, for $n = 1$ all automorphisms of $A_{1,\mathbb{C}}$ are tame (Makar-Limanov's theorem), and the map $\varphi_{[p]}$ is the conjectured canonical isomorphism.

These observations motivate the question whether for any n the group homomorphism $\varphi_{[p]}$ is independent of infinite prime.

The next Section makes emphasis on algorithmic questions. First we dwell on Non-Commutative Gröbner basis. Questions of algorithmic decidability in algebraic structures have been studied since the 1940s. In 1947 Markov [37] and independently Post [38] proved that the word equality problem in finitely presented semigroups (and in algebras) cannot be algorithmically solved. In 1952 Novikov constructed the first example of the group with unsolvable problem of word equality (see References [39,40]). In 1962 Shirshov proved solvability of the equality problem for Lie algebras with one relation and raised a question about finitely defined Lie algebras [41]. In 1972 Bokut settled this problem. In particular, he showed the existence of a finitely defined Lie algebra over an arbitrary field with algorithmically unsolvable identity problem [42].

Nevertheless, some problems become decidable if a finite Gröbner basis defines a relations ideal. In this case it is easy to determine whether two elements of the algebra are equal or not (see Reference [43]). In his work, D. Piontkovsky extended the concept of obstruction, introduced by V. Latyshev (see References [44–47]). V.N. Latyshev raised the question concerning the existence of an algorithm that can find out if a given element is either a zero divisor or a nilpotent element when the ideal of relations in the algebra is defined by a finite Gröbner basis.

Similar questions for monomial automaton algebras can be solved. In this case the existence of an algorithm for nilpotent element or a zero divisor was proved by Kanel-Belov, Borisenko and Latyshev [48]. Note that these algebras are not Noetherian and not weak Noetherian. Iyudu showed that the element property of being one-sided zero divisor is recognizable in the class of algebras with a one-sided limited processing (see References [49,50]). It also follows from a solvability of a linear recurrence relations system on a tree (see Reference [51]).

An example of an algebra with a finite Gröbner basis and algorithmically unsolvable problem of zero divisor is constructed in Reference [52].

A notion of *Gröbner basis* (better to say *Gröbner-Shirshov basis*) first appeared in the context of noncommutative (and not Noetherian) algebra. Note also that Poincaré-Birkhoff-Witt theorem can be canonically proved using Gröbner bases. More detailed discussions of these questions see in References [42,48,53].

To solve these two problems we simulate a universal Turing machine, each step of which corresponds to a multiplication from the left by a chosen letter.

The problem of the algorithmic decidability of the existence of an isomorphism between two algebraic varieties is extremely interesting and fundamental. A closely related problem is the embeddability problem. In the general form, it is formulated as follows.

Embeddability problem. *Let $\mathscr{A}$ and $\mathscr{B}$ be two algebraic varieties. Determine whether or not there exists an embedding of $\mathscr{A}$ in $\mathscr{B}$. Find an algorithm or prove its nonexistence.*

In this paper, a negative solution to this problem is given even for affine varieties over an arbitrary field of characteristic zero whose coordinate rings are given by generators and defining relations.

Questions related to the Gröbner basis were investigated in [54–58]. For details of nonstandard analysis see [59–61].

2. Model-Theoretical Aspects

Algebraic geometry over algebraic systems was investigated by B.I. Plotkin and his school. The Section 2.1 is devoted to this approach. In connection with this approach, Plotkin's problem about the automorphism of semigroups of endomorphisms of free algebra and categories (and also of groups of automorphisms) arose. The Section 2.2 is devoted to Plotkin's problem of endomorphisms and automorphisms. The problem of describing automorphisms for groups of polynomial symplectomorphisms and automorphisms of the Weyl algebra is extremely important, both from the point of view of mathematical physics and from the point of view of the Jacobian conjecture. Section 2.3 is dedicated to this problem.

2.1. Algebraic Geometry over Algebraic Systems

2.1.1. Three Versions of Logical Rigidity

Questions we are going to illuminate in this section are concentrated around the interaction between algebra, logic, model theory and geometry.

The main question behind further considerations is as follows. Suppose we have two algebras equipped with a sort of logical description.

Problem 1. *When the coincidence of logical descriptions provides an isomorphism between algebras in question?*

With this aim we consider different kinds of logical equivalences between algebras. Some of the notions we are dealing with are not formally defined in the text. For precise definitions and references use References [1,2,17,62–65].

2.1.2. Between Syntax and Semantics

By syntax we will mean a language intended to describe a certain subject area. In syntax we ask questions, express hypotheses and formulate the results. In syntax we also build chains of formal consequences. For our goals we use first-order languages or their fragments. Each language is based on some finite set of variables that serve as the alphabet, and a number of rules that allow us to build words based on this alphabet. In general, its signature includes Boolean operations, quantifiers, constants, and also functional symbols and predicate symbols. The latter ones are included in atomic formulas and, in fact, determine the face of a particular language. Atomic formulas will be called words. Words together with logical operations between them will be called formulas.

By semantics we understand the world of models, or in other words, the subject area of our knowledge. This world exists by itself, and develops according to its laws.

Fix a variety of algebras Θ. Let $W(X)$, $X = \{x_1, \ldots, x_n\}$ denote the finitely generated free algebra in Θ. By equations in Θ we mean expressions of the form $w \equiv w'$, where w, w' are words in $W(X)$ for some X. This is our first syntactic object. Next, let $\tilde{\Phi} = (\Phi(X), X \in \Gamma)$ be the multi-sorted Halmos algebra of first order logical formulas based on atoms $w \equiv w'$, w, w' in $W(X)$, see References [17,65,66]. There is a special procedure to construct such an algebraic object which plays the same role with respect to First Order Logic as Boolean algebras do with respect to Propositional calculus. One can view elements of $\tilde{\Phi} = (\Phi(X), X \in \Gamma)$ just as first order formulas over $w \equiv w'$.

Let $X = \{x_1, \ldots, x_n\}$ and let H be an algebra in the variety Θ. We have an affine space H^X of points $\mu : X \to H$. For every μ we have also the n-tuple $(a_1, \ldots, a_n) = \bar{a}$ with $a_i = \mu(x_i)$. For the given Θ we have the homomorphism

$$\mu : W(X) \to H$$

and, hence, the affine space is viewed as the set of homomorphisms

$$\mathrm{Hom}(W(X), H).$$

The classical kernel $\mathrm{Ker}(\mu)$ corresponds to each point $\mu : W(X) \to H$. This is exactly the set of equations for which the point μ is a solution. Every point μ has also the logical kernel $\mathrm{LKer}(\mu)$, see References [3,64,66]. Logical kernel $\mathrm{LKer}(\mu)$ consists of all formulas $u \in \Phi(X)$ valid on the point μ. This is always an ultrafilter in $\Phi(X)$.

So we define syntactic and semantic areas where logic and geometry operate, respectively. Connect them by a sort of Galois correspondence.

Let T be a system of equations in $W(X)$. The set A in the affine space $\mathrm{Hom}(W(X), H)$ consisting of all solutions of the system T corresponds to T. Sets of such kind are called *algebraic sets*. Vice versa, given a set A of points in the affine space consider all equations T having A as the set of solutions. Sets T of such kind are called *closed congruences* over W.

We can do the same correspondence with respect to arbitrary sets of formulas. Given a set T of formulas in algebra of formulas (set of elements) $\Phi(X)$, consider the set A in the affine space, such that every point of A satisfies every formula of Φ. Sets of such kind are called *definable sets*. Points of A are called solutions of the set of formulas T. Conversely, given a set A of points in the affine space consider all formulas (elements) T having A as the set of solutions. Sets T of such kind are *closed filters* in $\Phi(X)$.

Let us formalize the Galois correspondence described above.

2.1.3. Galois Correspondence in the Logical Geometry

Let us start with a particular case when the set of formulas T in $\Phi(X)$ is a set of equations of the form $w = w'$, $w, w' \in W(X)$, $X \in \Gamma$.

We set

$$A = T'_H = \{\mu : W(X) \to H \mid T \subset Ker(\mu)\}.$$

Here A is an *algebraic set* in $\mathrm{Hom}(W(X), H)$, determined by the set T.

Let, further, A be a subset in $\mathrm{Hom}(W(X), H)$. We set

$$T - A'_H - \bigcap_{\mu \in A} Ker(\mu).$$

Congruences T of such kind are called H-closed in $W(X)$. We have also Galois-closures T''_H and A''_H.

Let us pass to the general case of logical geometry. Let now T be a set of arbitrary formulas in $\Phi(X)$. We set

$$A = T^L_H = \{\mu : W(X) \to H \mid T \subset \mathrm{LKer}(\mu)\}.$$

We have also

$$A = \bigcap_{u \in T} \mathrm{Val}^X_H(u).$$

Here A is called a *definable set* in $\mathrm{Hom}(W(X), H)$, determined by the set T. We use the term "definable" for A of such kind, meaning that A is defined by some set of formulas T.

For the set of points A in $\mathrm{Hom}(W(X), H)$ we set

$$T = A^L_H = \bigcap_{\mu \in A} \mathrm{LKer}(\mu).$$

We have also

$$T = A_H^L = \{u \in \Phi(X) \mid A \subset \mathrm{Val}_H^X(u)\}.$$

Here T is a Boolean filter in $\Phi(X)$ determined by the set of points A. Filters of such kind are Galois-closed and we can define the Galois-closures of arbitrary sets T in $\Phi(X)$ and A in $\mathrm{Hom}(W(X), H)$ as T^{LL} and A^{LL}.

Remark 1. *The principal role in all considerations plays the value homomorphism* $\mathrm{Val} : \tilde{\Phi} \to \mathrm{Hal}_\Theta$, *where* Hal_Θ *is a special Halmos algebra associated with the vector space* $\mathrm{Hom}(W(X), H)$, *see References [65,66]. Its meaning is to make the procedure of verification whether a point satisfies the formula a homomorphism.*

2.1.4. Logical Similarities of Algebras

Now we are in a position to introduce several logical equivalences between algebras. Since the Galois correspondence yields the duality between syntactic and semantic objects, every definition of equivalence between algebras formulated in terms of formulas, that is logically, has its semantical counterpart, that is a geometric formulation, and vice versa.

All algebraic sets constitute a category with special rational maps as morphisms [65]. The same is true with respect to definable sets [65]. So, we can formulate logical closeness of algebras geometrically.

Definition 1. *We call algebras* H_1 *and* H_2 *geometrically similar if the categories of algebraic sets* $AG_\Theta(H_1)$ *and* $AG_\Theta(H_2)$ *are isomorphic.*

By Galois duality between closed congruences and algebraic sets, H_1 and H_2 are geometrically similar if and only if the corresponding categories $C_\Theta(H_1)$ and $C_\Theta(H_2)$ of closed congruences over $W(X)$ are isomorphic.

Definition 2. *We call algebras* H_1 *and* H_2 *logically similar, if the categories of definable sets* $LG_\Theta(H_1)$ *and* $LG_\Theta(H_2)$ *are isomorphic.*

By Galois duality between closed filters in $\Phi(X)$ and definable sets, H_1 and H_2 are logically similar if and only if the corresponding categories $F_\Theta(H_1)$ and $F_\Theta(H_2)$ of closed filters in $F(X)$ are isomorphic.

We will be looking for conditions $\mathcal{A}$ on algebras H_1 and H_2 that provide geometrical or logical similarity.

Let two algebras H_1 and H_2 subject to some condition $\mathcal{A}$ be given. Here $\mathcal{A}$ is any condition of logical or, dually, geometrical character, formulated in terms of closed sets of formulas or definable sets.

Definition 3. *We call the condition* $\mathcal{A}$ *rigid (or* $\mathcal{A}$*-rigid) if two algebras* H_1 *and* H_2 *subject to* $\mathcal{A}$ *are isomorphic.*

2.1.5. Geometric Equivalence of Algebras

Definition 4. *Algebras* H_1 *and* H_2 *are called AG-equivalent, if for every* X *and every system of equations* T *holds* $T''_{H_1} = T''_{H_2}$.

AG-equivalent algebras are called also *geometrically equivalent* algebras, see References [3,64,65]. The closure T''_H is called, sometimes, a *radical* of T with respect to H. This is a normal subgroup and an ideal in cases of groups and associative (Lie) algebras, respectively.

The meaning of Definition 4 is as follows. Two algebras H_1 and H_2 are AG-equivalent if they have the same solution sets with respect to any system of equations T. We have the following criterion, see Reference [65].

Proposition 1. *If algeras* H_1 *and* H_2 *are AG-equivalent, then they are AG-similar.*

So, geometric equivalence of algebras provides their geometrical similarity. The next statement describes geometrically equivalent algebras. Assume, for simplicity, that our algebras are geometrically noetherian (see References [3,63]), which means that every system of equations T is equivalent to a finite subsystem T' of T. Then, see Reference [67],

Proposition 2. *Geometrically noetherian algebras H_1 and H_2 are AG-equivalent if and only if they generate the same quasi-variety.*

Hence, two algebras H_1 and H_2 are AG-equivalent if and only if they have the same quasi-identities. If we drop the condition of geometrical noetherianity, then algebras H_1 and H_2 are AG-equivalent if they have the same infinitary quasi-identities.

Let Θ be the variety of all groups. Now the question of AG-rigidity for groups reduces to the question when two groups generating one and the same quasi-variety are isomorphic. Of course the condition on groups to have one and the same quasi-identities is very weak and the rigidity of such kind can happen if both groups belong to a very narrow class of groups. In general, such a condition does not seem sensible.

Geometrical equivalence of algebras gives a sufficient condition for AG-similarity. It turns out that for some varieties Θ this condition is also sufficient.

Theorem 1. *Let $Var(H_1) = Var(H_2) = \Theta$. Let Θ be one of the following varieties*

- $\Theta = Grp$, *the variety of groups,*
- $\Theta = Jord$, *the variety of Jordan algebras,*
- $\Theta = Semi$, *the variety of semigroups,*
- $\Theta = Inv$, *the variety of inverse semigroups,*
- $\Theta = \mathfrak{N}_d$, *the variety of nilpotent groups of class d.*

Categories $AG_\Theta(H_1)$ and $AG_\Theta(H_2)$ are isomorphic if and only if the algebras H_1 and H_2 are geometrically equivalent (see References [68–71]).

Let Θ^0 be the category of all free algebras of the variety Θ. The following proposition is the main tool in the proof of Theorem 1.

Proposition 3 ([5]). *If for the variety Θ every automorphism of the category Θ^0 is inner, then two algebras H_1 and H_2 are geometrically similar if and only if they are geometrically equivalent.*

So, studying automorphisms of Θ^0 plays a crucial role. The latter problem is treated by means of Reduction Theorem (see References [17–20]). This theorem reduces investigation of automorphisms of the whole category Θ^0 of free in Θ algebras to studying the group $\mathrm{Aut}(\mathrm{End}(W(X)))$ associated with a single object $W(X)$ in Θ^0. Here, $W(X)$ is a finitely generated free in Θ hopfian algebra, which generates the whole variety Θ. In fact, if all automorphisms of the endomorphism semigroup of a free algebra $W(X)$ are close to being inner, then all automorphisms of Θ^0 possess the same property. More precisely, denote by $Inn(\mathrm{End}(W(X)))$ the group of inner automorphisms of $\mathrm{Aut}(\mathrm{End}(W(X)))$. Then the group of outer automorphisms $\mathrm{Aut}(\mathrm{End}(W(X)))/Inn(\mathrm{End}(W(X)))$ measures, in some sense, the difference between the notions of geometric similarity and geometric equivalence.

2.1.6. Elementary Equivalence of Algebras

As we saw in the previous section AG-equivalence of algebraic sets reduces to coincidence of quasi-identities of algebras. This is a weak invariant, a small part of elementary theory, and, of course, coincidence of quasi-identities does not imply isomorphism of algebras. Hence AG-equivalence does not make much sense from the point of view of rigidity. Now we recall a more powerful logical invariant of algebras.

Given algebra H, its *elementary theory* $Th(H)$ is the set of all sentences (closed formulas) valid on H. We modify a bit this definition and adjust it to the Galois correspondence. Fix $X = \{x_1, \ldots, x_n\}$. Define X-elementary theory $Th^X(H)$ to be the set of all formulas $u \in \Phi(X)$ valid in every point of the affine space $\mathrm{Hom}(W(X), H)$. In general we have a multi-sorted representation of the elementary theory

$$Th(H) = (Th^X(H), X \in \Gamma),$$

where Γ is a certain system of sets.

Definition 5. *Two algebras H_1 and H_2 are said to be elementarily equivalent if their elementary theories coincide.*

Remark 2. *From the geometric point of view this definition does not make difference between different points of the affine space. Given algebras H_1 and H_2, we collect all together formulas valid in every point of the affine spaces* $\mathrm{Hom}(W(X), H_1)$ *and* $\mathrm{Hom}(W(X), H_2)$, *and declare algebras H_1 and H_2 elementarily equivalent if these sets coincide.*

Importance of the elementary classification of algebraic structures goes back to the famous works of A.Tarski and A.Malcev. The main problem is to figure out *what are the algebras elementarily equivalent to a given one.* Very often we fix a class of algebras C and ask what are the algebras elementarily equivalent to a given algebra inside the class C. So, the rigidity question with respect to elementary equivalence looks as follows.

Problem 2. *Let a class of algebras C and an algebra $H \in C$ be given. Suppose that the elementary theories of algebras H and $A \in C$ coincide. Are they elementarily rigid, that is, are H and A isomorphic?*

Remark 3. *What we call elementary rigidity has different names. This notion appeared in the papers by A. Nies [72] under the name of quasi definability of groups. The corresponding name used in Reference [73] is first order rigidity. For some reasons which will be clear in the next section we use another term.*

In other words we ask for which algebras their logical characterization by means of the elementary theory is strong enough and define the algebra in the unique, up to an isomorphism, way?

We restrict our attention to the case of groups, and, moreover, assume quite often that our groups are finitely generated. Elementary rigidity of groups occurs not very often. Usually various extra conditions are needed. Here is the incomplete list of some known cases:

Theorem 2. *We will consider the following cases*

- *Finitely generated abelian groups are elementarily rigid in the class of such groups, see References [74,75].*
- *Finitely generated torsion-free class 2 nilpotent groups are elementarily rigid in the class of finitely generated groups, see References [76,77] (this is wrong for such groups of class 3 and for torsion groups of class 2, see Reference [78]).*
- *If two finitely generated free nilpotent groups are elementarily equivalent, then they are isomorphic, that is a free finitely generated nilpotent group is elementarily rigid in the class of such groups, see References [79,80].*
- *If two finitely generated free solvable groups are elementarily equivalent, then they are isomorphic, that is a free finitely generated solvable group is elementarily rigid in the class of such groups, see References [79,80].*
- *Baumslag-Solitar group $BS(1, n)$ is elementarily rigid in the class of countable groups, see Reference [81]. General Baumslag-Solitar groups $BS(m, n)$ are elementarily rigid in the class of all Baumslag-Solitar groups, see Reference [81].*
- *Right-angled Coxeter group is elementarily rigid in the class of all right-angled Coxeter groups, see Reference [82].*

- *A good rigidity example is provided by profinite groups: if two finitely generated profinite groups are elementarily equivalent (as abstract groups), then they are isomorphic [83].*

Consider, separately, examples of elementary rigidity for linear groups. First of all, a group which is elementarily equivalent to a finitely generated linear group is a residually finite linear group [84]. The rigidity cases are collected in the following theorem.

Theorem 3. *We will consider the following cases.*

- *Historically, the first result was obtained by Malcev [85]. If two linear groups $GL_n(K)$ and $GL_m(F)$, where K and F are fields, are elementarily equivalent, then $n = m$ and the fields K and F are elementarily equivalent.*
- *This result was generalized to the wide class of Chevalley groups. Let $G_1 = G_\pi(\Phi, R)$ and $G_2 = G_\mu(\Psi, S)$ be two elementarily equivalent Chevalley groups. Here Φ, Ψ denote the root systems of rank ≥ 1, R and S are local rings, and π, μ are weight lattices. Then root systems and weight lattices of G_1 and G_2 coincide, while the rings are elementarily equivalent. In other words Chevalley groups over local rings are elementarily rigid in the class of such groups modulo rigidity of the ground rings [86].*
- *Let $G_\pi(\Phi, K)$ be a simple Chevalley group over the algebraically closed field K. Then $G_\pi(\Phi, K)$ is elementarily rigid in the class of all groups (cardinality is fixed). This result can be deduced from Reference [87]. In fact, this is true for a much wider class of algebraic groups over algebraically closed fields and, modulo elementary equivalence of fields, over arbitrary fields [87].*
- *Any irreducible non-uniform higher-rank characteristic zero arithmetic lattice is elementarily rigid in the class of all groups, see Reference [73]. In particular, $SL_n(\mathbb{Z})$, $n > 2$ is elementarily rigid.*
- *Recently, the results of Reference [73] have been extended to a much more wide class of lattices, see Reference [88].*
- *Let $\mathcal{O}$ be the ring of integers of a number field, and let $n \geqslant 3$. Then every group G which is elementarily equivalent to $SL_n(\mathcal{O})$ is isomorphic to $SL_n(\mathcal{R})$, where the rings $\mathcal{O}$ and $\mathcal{R}$ are elementarily equivalent. In other words $SL_n(\mathcal{O})$ is elementarily rigid in the class of all groups modulo elementary equivalence of rings. The similar results are valid with respect to $GL_n(\mathcal{O})$ and to the triangular group $T_n(\mathcal{O})$ [89]. These results intersect in part with the previous items, since the ring $R = \mathbb{Z}$ is elementarily rigid in the class of all finitely generated rings [72], and thus $SL_n(\mathbb{Z})$ is elementarily rigid in the class of all finitely generated groups.*
- *For the case of arbitrary Chevalley groups the results similar to above cited are obtained in Reference [90] by different machinery for a wide class of ground rings. Suppose the Chevalley group $G = G(\Phi, R)$ of rank $\geqslant 2$ over the ring R is given. Suppose that the ring R is elementarily rigid in the class C of rings. Then $G = G(\Phi, R)$ is elementarily rigid in the corresponding class C_1 of groups if R is a field, R is a local ring and G is simply connected, R is a Dedekind ring of arithmetic type, that is the ring of S-integers of a number field, R is Dedekind ring with at least 4 units and G is adjoint. In particular, if a ring of such kind is finitely generated then it gives rise to elementary rigidity of $G = G(\Phi, R)$ in the class of all finitely generated groups. If R of such kind is not elementarily rigid then $G = G(\Phi, R)$ is elementarily rigid in the class of all groups modulo elementary equivalence of rings.*

Absolutely free groups lie on the other side of the scale of groups. It was Tarski who asked whether one can distinguish between finitely generated free groups by means of their elementary theories. This formidable problem has been solved in affirmative, that is all free groups have one and the same elementary theory [6,7]. In fact, the variety of all groups is the only known variety of groups, such that a free in this variety finitely generated group is not rigid in the class of all such groups.

Problem 3. *Construct a variety of groups different from the variety of all groups such that all free finitely generated groups in this variety have one and the same elementary theory.*

2.1.7. Logical Equivalence of Algebras

In this Section we introduce the notion of logical equivalence of algebras which can be viewed as first order equivalence. We proceed following exactly the same scheme which was applied in Section 2.1.5 with respect to the definition of geometric equivalence of algebras.

Let H_1 and H_2 be two algebras. We will be looking for semantic logical invariant of these algebras, that is compare the definable sets over H_1 and H_2. Recall that according to Definition 2 two algebras H_1 and H_2 are *logically similar*, if the categories of definable sets $LG_\Theta(H_1)$ and $LG_\Theta(H_2)$ are isomorphic.

Using the duality provided by Galois correspondence from Section 2.1.3 we will raise logical similarity to the level of syntax. The principal Definition 6 is the first order counterpart of Definition 4.

Definition 6. *Algebras H_1 and H_2 are called LG-equivalent (aka logically equivalent), if for every X and every set of formulas T in $\Phi(X)$ the equality $T^{LL}_{H_1} = T^{LL}_{H_2}$ holds .*

It is easy to see that

Proposition 4. *If algebras H_1 and H_2 are LG-equivalent then they are elementarily equivalent.*

Now we want to understand what is the meaning of logical equivalence.

Definition 7. *Two algebras H_1 and H_2 are called LG-isotypic if for every point $\mu : W(X) \to H_1$ there exists a point $\nu : W(X) \to H_2$ such that $\mathrm{LKer}(\mu) = \mathrm{LKer}(\nu)$ and, conversely, for every point $\nu : W(X) \to H_2$ there exists a point $\mu : W(X) \to H_1$ such that $\mathrm{LKer}(\nu) = \mathrm{LKer}(\mu)$.*

The meaning of Definition 7 is the following. Two algebras are isotypic if the sets of realizable types over H_1 and H_2 coincide. So, by some abuse of language these algebras have the same logic of types. Some references for the notion of isotypic algebras are contained in References [64,65,67,91–93]. Note that the notion was introduced in Reference [91,92] while Reference [65] gives the most updated survey.

The main theorem is as follows, see Reference [93].

Theorem 4. *Algebras H_1 and H_2 are LG-equivalent if and only if they are LG-isotypic.*

Now we are in a position to study rigidity of algebras with respect to isotypicity property. It is clear, that since isotypicity is stronger than elementary equivalence, this phenomenon can occur quite often. Let us state this problem explicitly.

Problem 4. *Let a class of algebras C and an algebra $H \in C$ be given. Suppose that algebras $H \in C$ and $A \in C$ are isotypic. Are they isotypically rigid, that is are H and A isomorphic?*

Remark 4. *In many papers from the list above isotypically rigid algebras are called logically separable [65,67], or type definable [94].*

Theorem 5. *We will consider the following cases of rigidity:*

- *Finitely generated free abelian groups are isotypically rigid in the class of all groups, see Reference [93].*
- *Finitely generated free nilpotent groups of class at most n are isotypically rigid in the class of all groups [93].*
- *Finitely generated metabelian groups are isotypically rigid in the class of all groups [94].*
- *Finitely generated virtually polycyclic groups are isotypically rigid in the class of all groups [94].*
- *Finitely generated free solvable groups of derived length $d > 1$ are isotypically rigid in the class of all groups [94].*
- *All surface groups, which are not non-orientable surface groups of genus 1,2 or 3 are isotypically rigid in the class of all groups [94].*

- *Finitely generated absolutely free groups are isotypically rigid in the class of all groups, see Reference [95] based on Reference [96] (also follows from References [97,98]).*
- *Finitely generated free semigroups are isotypically rigid in the class of semigroups, see Reference [93].*
- *Finitely generated free inverse semigroups are isotypically rigid in the class of inverse semigroups, see Reference [93].*
- *Finitely generated free associative algebras are isotypically rigid in the class of such algebras.*

The number of examples can be continued to co-Hopf groups, some Burnside groups, and so forth.

In fact, using either logical equivalence of algebras, or what is the same, the isotypicity of algebras, we compare the possibilities of individual points in the affine space to define the sets of formulas (in fact ultrafilters in $\Phi(X)$) which are valid in these points. Given a point μ in the affine space, the collection of formulas valid on the point μ is *a type* of μ. If these individual types are, roughly speaking, the same for both algebras, then these algebras are declared isotypic. Thus, for isotypic algebras we compare types of formulas realizable on these algebras. Of course, this is significantly stronger than elementary equivalence, where the individuality of points disappeared and we compare only formulas valid in all points of the affine space.

The following principal problem was stated in Reference [65] and is widely open.

Problem 5 (Rigidity problem). *Is it true that every two isotypic finitely generated groups are isomorphic?*

We will finish with the one more tempting problem of the same spirit.

Problem 6. *What are the isotypicity classes of fields? When two isotypic fields are isomorphic?*

The elementary equivalence of fields was one of motivating engines for Tarski to develop the whole model-theoretic staff related to elementary equivalence. Problem 6, in a sense, takes us back to the origins of the theory.

2.2. Plotkin's Problem: Automorphisms of Endomorphism Semigroups and Groups of Polynomial Automorphisms

In the light of B.I. Plotkin's activity on creation of algebraic geometry over algebraic systems, he drew a special attention to studying the groups of their automorphisms, see Reference [16]. Later on he emphasized that automorphisms of categories of free algebras of the varieties play here a role of exceptional importance. This role was underlined in Proposition 3 of Section 2.1. The meaning of Reduction Theorem (see References [17–20]) was explained just after this proposition. Reduction Theorem reduces investigation of automorphisms of the whole category Θ^0 of free in the variety Θ algebras to studying the group $\mathrm{Aut}(\mathrm{End}(W(X)))$ associated with a single object $W(X)$ in Θ^0. Here, $W(X)$ is a finitely generated free in Θ algebra. In fact, if all automorphisms of the endomorphism semigroup of a free algebra $W(X)$ are close to being inner, then all automorphisms of Θ^0 possess the same property.

This philosophy forms a clear basis for investigation of automorphisms of the semigroup of polynomial endomorphisms and the group of polynomial automorphisms. The automorphisms of the endomorphism semigroup of a free associative algebra A were given by Belov, Berzins and Lipyanski, (see Reference [99] for details and definitions of semi-inner and mirror automorphisms):

Theorem 6. *The group* $\mathrm{Aut}(\mathrm{End}(A))$ *is generated by semi-inner and mirror automorphisms of* $\mathrm{End}(A)$. *Correspondingly, the group of automorphisms of the category of free associative algebras is generated by semi-inner and mirror automorphisms of this category.*

In the same spirit, the description of an endomorphism semigroup of the ring of commutative polynomials A is given by Belov and Lipyanski in Reference [100]:

Theorem 7. *Every automorphism of the group* $\mathrm{Aut}(\mathrm{End}(A))$ *is semi-inner.*

The automorphisms of the group of polynomial automorphisms on free associative algebras and commutative algebras at the level of *Ind*-schemes were obtained by Belov, Elishev and J.-T.Yu in Reference [30]. Let $K[x_1, \ldots, x_n]$ and $K\langle x_1, \ldots, x_n\rangle$ be the free commutative polynomial algebra and the free associative algebra with n generators, respectively. Denote by NAut the group of *nice* automorphisms, that is, the group of automorphisms which can be approximated by tame ones. One can prove that in characteristic zero case every automorphism is nice.

Theorem 8. *Any* Ind-*scheme automorphism* φ *of* $\mathrm{NAut}(K[x_1, \ldots, x_n])$ *for* $n \geqslant 3$ *is inner, that is, it is a conjugation via some automorphism of* $K[x_1, \ldots, x_n]$. *Any* Ind-*scheme automorphism* φ *of* $\mathrm{NAut}(K\langle x_1, \ldots, x_n\rangle)$ *for* $n \geqslant 3$ *is semi-inner (see Reference [30] for the precise definition).*

Here, the Ind-scheme is defined as follows:

Definition 8. *An* Ind-*variety* M *is the direct limit of algebraic varieties* $M = \varinjlim\{M_1 \subseteq M_2 \cdots\}$. *An* Ind-*scheme is an* Ind-*variety which is a group such that the group inversion is a morphism* $M_i \to M_{j(i)}$ *of algebraic varieties, and the group multiplication induces a morphism from* $M_i \times M_j$ *to* $M_{k(i,j)}$. *A map* φ *is a morphism of an* Ind-*variety* M *to an* Ind-*variety* N, *if* $\varphi(M_i) \subseteq N_{j(i)}$ *and the restriction* φ *to* M_i *is a morphism for all* i. *Monomorphisms, epimorphisms and isomorphisms are defined similarly in a natural way.*

2.3. On the Independence of the B-KK Isomorphism of Infinite Prime and Plotkin Conjecture for Symplectomorphisms

2.3.1. Plotkin's Problem for Symplectomorphism and the Kontsevich Conjecture

Observe that the study of automorphisms of the group of polynomial symplectomorphisms, as well as automorphisms of the Weyl algebra (Plotkin's problem) is extremely important in course of the Kontsevich conjecture, as well as the Jacobian conjecture.

2.3.2. Ultrafilters and Infinite Primes

Let $\mathcal{U} \subset 2^{\mathbb{N}}$ be an arbitrary non-principal ultrafilter on the set of all positive numbers (in this note $\mathbb{N}$ will almost always be regarded as the index set). Let $\mathbb{P}$ be the set of all prime numbers, and let $\mathbb{P}^{\mathbb{N}}$ denote the set of all sequences $p = (p_m)_{m \in \mathbb{N}}$ of prime numbers. We refer to a generic set $A \in \mathcal{U}$ as an index subset in situations involving the restriction $p_{|A} : A \to \mathbb{P}$. We will call a sequence p of prime numbers $\mathcal{U}$-stationary if there is an index subset $A \in \mathcal{U}$ such that its image $p(A)$ consists of one point.

A sequence $p : \mathbb{N} \to \mathbb{P}$ is bounded if the image $p(\mathbb{N})$ is a finite set. Thanks to the ultrafilter finite intersection property, bounded sequences are necessarily $\mathcal{U}$-stationary.

Any non-principal ultrafilter $\mathcal{U}$ generates a congruence

$$\sim_{\mathcal{U}} \subseteq \mathbb{P}^{\mathbb{N}} \times \mathbb{P}^{\mathbb{N}}$$

in the following way. Two sequences p^1 and p^2 are $\mathcal{U}$-congruent iff there is an index subset $A \in \mathcal{U}$ such that for all $m \in A$ the following equality holds:

$$p^1_m = p^2_m.$$

The corresponding quotient

$$^*\mathbb{P} \equiv \mathbb{P}^{\mathbb{N}} / \sim_{\mathcal{U}}$$

contains as a proper subset the set of all primes $\mathbb{P}$ (naturally identified with classes of $\mathcal{U}$-stationary sequences), as well as classes of unbounded sequences. The latter are referred to as nonstandard, or infinitely large, primes. We will use both names and normally denote such elements by $[p]$, mirroring the convention for equivalence classes. The terminology is justified, as the set of nonstandard primes is in one-to-one correspondence with the set of prime elements in the ring ${}^*\mathbb{Z}$ of nonstandard integers in the sense of Robinson [101].

Indeed, one may utilize the following construction, which was thoroughly studied (also cf. Reference [102]) in Reference [103]. Consider the ring $\mathbb{Z}^\omega = \prod_{m\in\mathbb{N}} \mathbb{Z}$—the product of countably many copies of $\mathbb{Z}$ indexed by $\mathbb{N}$. The minimal prime ideals of $\mathbb{Z}^\omega$ are in bijection with the set of all ultrafilters on $\mathbb{N}$ (perhaps it is opportune to remind that the latter is precisely the Stone-Cech compactification $\beta\mathbb{N}$ of $\mathbb{N}$ as a discrete space). Explicitly, if for every $a = (a_m) \in \mathbb{Z}^\omega$ one defines the support complement as

$$\theta(a) = \{m \in \mathbb{N} \mid a_m = 0\}$$

and for an arbitrary ultrafilter $\mathcal{U} \in 2^{\mathbb{N}}$ sets

$$(\mathcal{U}) = \{a \in \mathbb{Z}^\omega \mid \theta(a) \in \mathcal{U}\},$$

then one obtains a minimal prime ideal of $\mathbb{Z}^\omega$. It is easily shown that every minimal prime ideal is of such a form. Of course, the index set $\mathbb{N}$ may be replaced by any set I, after which one easily gets the description of minimal primes of $\mathbb{Z}^I$ (since those correspond to ultrafilters, there are exactly $2^{2^{|I|}}$ of them if I is infinite and $|I|$ when I is a finite set). Note that in the case of finite index set all ultrafilters are principal, and the corresponding $(\mathcal{U})$ are of the form $\mathbb{Z} \times \cdots \times (0) \times \cdots \times \mathbb{Z}$—a textbook example.

Similarly, one may replace each copy of $\mathbb{Z}$ by an arbitrary integral domain and repeat the construction above. If for instance all the rings in the product happen to be fields, then, since the product of any number of fields is von Neumann regular, the ideal $(\mathcal{U})$ will also be maximal.

The ring of nonstandard integers may be viewed as a quotient (an ultrapower)

$$\mathbb{Z}^\omega/(\mathcal{U}) = {}^*\mathbb{Z}.$$

The class of $\mathcal{U}$-congruent sequences $[p]$ corresponds to an element (also an equivalence class) in ${}^*\mathbb{Z}$, which may as well as $[p]$ be represented by a prime number sequence $p = (p_m)$, only in the latter case some but not too many of the primes p_m may be replaced by arbitrary integers. For all intents and purposes, this difference is insignificant.

Also, observe that $[p]$ indeed generates a maximal prime ideal in ${}^*\mathbb{Z}$: if one for (any) $p \in [p]$ defines an ideal in $\mathbb{Z}^\omega$ as

$$(p, \mathcal{U}) = \{a \in \mathbb{Z}^\omega \mid \{m \mid a_m \in p_m\mathbb{Z}\} \in \mathcal{U}\},$$

then, taking the quotient $\mathbb{Z}^\omega/(p, \mathcal{U})$ in two different ways, one arrives at an isomorphism

$${}^*\mathbb{Z}/([p]) \simeq \left(\prod_m \mathbb{Z}_{p_m}\right)/(\mathcal{U}),$$

and the right-hand side is a field by the preceding remark. For a fixed non-principal $\mathcal{U}$ and an infinite prime $[p]$, we will call the quotient

$$\mathbb{Z}_{[p]} \equiv {}^*\mathbb{Z}/([p])$$

the nonstandard residue field of $[p]$. Under our assumptions this field has characteristic zero.

2.3.3. Algebraic Closure of Nonstandard Residue Field

We have seen that the objects $[p]$—the infinite prime—behaves similarly to the usual prime number in the sense that a version of a residue field corresponding to this object may be constructed. Note that the standard residue fields are contained as a degenerate case in this construction, namely if we drop the condition of unboundedness and instead consider $\mathcal{U}$-stationary sequences, we will arrive at a residue field isomorphic to $\mathbb{Z}_p$, with p being the image of the stationary sequence in the chosen class. The fields of the form $\mathbb{Z}_{[p]}$ are a realization of what is known as pseudofinite field, cf. Reference [104].

The nonstandard case is surely more interesting. While the algebraic closure of a standard residue field is countable, the nonstandard one itself has the cardinality of the continuum. Its algebraic closure is also of that cardinality and has characteristic zero, which implies that it is isomorphic to the field of complex numbers. We proceed by demonstrating these facts.

Proposition 5. *For any infinite prime $[p]$ the residue field $\mathbb{Z}_{[p]}$ has the cardinality of the continuum (There is a general statement on cardinality of ultraproduct due to Frayne, Morel, and Scott [105]. We believe the proof of this particular instance may serve as a neat example of what we are dealing with in the present paper.).*

Proof. It suffices to show there is a surjection

$$h^* : \mathbb{Z}_{[p]} \to \mathfrak{P},$$

where $\mathfrak{P} = \{0,1\}^\omega$ is the Cantor set given as the set of all countable strings of bits with the 2-adic metric

$$d_2(x,y) = 1/k, \quad k = \min\{m \mid x_m \neq y_m\}.$$

The map h^* is constructed as follows. If $\mathfrak{Z} \subset \mathfrak{P}$ is the subset of all strings with finite number of ones in them, and

$$e : \mathbb{Z}_+ \to \mathfrak{Z}, \quad e\left(\sum_{k<m} f_k 2^k\right) = (f_1, \ldots, f_{m-1}, 0, \ldots)$$

is the bijection that sends a nonnegative integer to its binary decomposition, then for a class representative $a = (a_m) \in [a] \in \mathbb{Z}_{[p]}$ set $h^*(a)$ to be the (unique) ultralimit of the sequence of points $\{x_m = e(a_m)\}$. The correctness of this map rests on the property of the Cantor set being Hausdorff quasi-compact. Surjectivity is then established directly: consider an arbitrary $x \in \mathfrak{P}$. For each $m \in \mathbb{N}$ the set

$$\mathfrak{P}_m = \{e(0), e(1), \ldots, e(p_m - 1)\}$$

consists of p_m distinct points. Let x_m be the nearest to x point from this set with respect to the 2-adic metric. The sequence (p_m) is unbounded, so that for every $m \in \mathbb{N}$ the index subset

$$A_m = \{k \in \mathbb{N} \mid p_k > 2^m\}$$

belongs to the ultrafilter $\mathcal{U}$. It is easily seen that for every $k \in A_m$ one has:

$$d_2(x, x_k) < 1/m$$

But that effectively means that the sequence (x_m) has the ultralimit x, after which $a_m = e^{-1}(x_m)$ yields the desired preimage. □

As an immediate corollary of this proposition and the well-known Steinitz theorem, one has

Theorem 9. *The algebraic closure $\overline{\mathbb{Z}_{[p]}}$ of $\mathbb{Z}_{[p]}$ is isomorphic to the field of complex numbers.*

We now fix the notation for the aforementioned isomorphisms in order to employ it in the next section.

For any nonstandard prime $[p] \in {}^*\mathbb{P}$ fix an isomorphism $\alpha_{[p]} : \mathbb{C} \to \overline{\mathbb{Z}_{[p]}}$ coming from the preceding theorem. Denote by $\Theta_{[p]} : \overline{\mathbb{Z}_{[p]}} \to \overline{\mathbb{Z}_{[p]}}$ the nonstandard Frobenius automorphism—that is, a well-defined field automorphism that sends a sequence of elements to a sequence of their p_m-th powers:

$$(x_m) \mapsto (x_m^{p_m}).$$

The automorphism $\Theta_{[p]}$ is identical on $\mathbb{Z}_{[p]}$; conjugated by $\alpha_{[p]}$, it yields a wild automorphism of complex numbers, as by assumption no finite power of it (as always, in the sense of index subsets $A \in \mathcal{U}$) is the identity homomorphism.

2.3.4. Extension of the Weyl Algebra

The n-th Weyl algebra $A_{n,\mathbb{C}} \simeq A_{n,\overline{\mathbb{Z}_{[p]}}}$ can be realized as a proper subalgebra of the following ultraproduct of algebras

$$\mathcal{A}_n(\mathcal{U},[p]) = \left(\prod_{m\in\mathbb{N}} A_{n,\mathbb{F}_{p_m}}\right)/\mathcal{U}.$$

Here for any m the field $\mathbb{F}_{p_m} = \overline{\mathbb{Z}_{p_m}}$ is the algebraic closure of the residue field $\mathbb{Z}_{p_m}$. This larger algebra contains elements of the form $(\zeta^{I_m})_{m\in\mathbb{N}}$ with unbounded $|I_m|$—something which is not present in $A_{n,\overline{\mathbb{Z}_{[p]}}}$, hence the proper embedding. Note that for the exact same reason (with degrees $|I_m|$ of differential operators having been replaced by degrees of minimal polynomials of algebraic elements) the inclusion

$$\overline{\mathbb{Z}_{[p]}} \subseteq \left(\prod_{m\in\mathbb{N}} F_{p_m}\right)/\mathcal{U}$$

is also proper.

It turns out that, unlike its standard counterpart $A_{n,\mathbb{C}}$, the algebra $\mathcal{A}_n(\mathcal{U},[p])$ has a huge center described in this proposition:

Proposition 6. *The center of the ultraproduct of Weyl algebras over the sequence of algebraically closed fields $\{\mathbb{F}_{p_m}\}$ coincides with the ultraproduct of centers of $A_{n,\mathbb{F}_{p_m}}$:*

$$C(\mathcal{A}_n(\mathcal{U},[p])) = \left(\prod_m C(A_{n,\mathbb{F}_{p_m}})\right)/\mathcal{U}.$$

The proof is elementary and is left to the reader. As in positive characteristic the center $C(A_{n,\mathbb{F}_p})$ is given by the polynomial algebra

$$\mathbb{F}_p[x_1^p,\ldots,x_n^p,y_1^p,\ldots,y_n^p] \simeq \mathbb{F}_p[\xi_1,\ldots,\xi_{2n}],$$

There is an injective $\mathbb{C}$-algebra homomorphism

$$\mathbb{C}[\xi_1,\ldots\xi_{2n}] \to \left(\prod_m \mathbb{F}_{p_m}[\xi_1^{(m)},\ldots\xi_{2n}^{(m)}]\right)/\mathcal{U}$$

From the algebra of regular functions on $\mathbb{A}^{2n}_{\mathbb{C}}$ to the center of $\mathcal{A}_n(\mathcal{U},[p])$, evaluated on the generators in a straightforward way:

$$\xi_i \mapsto [(\xi_i^{(m)})_{m\in\mathbb{N}}].$$

Just as before, this injection is proper.

Furthermore, the image of this monomorphism (the set which we will simply refer to as the polynomial algebra) may be endowed with the canonical Poisson bracket. Recall that in positive characteristic case for any $a,b \in \mathbb{Z}_p[\xi_1,\ldots,\xi_{2n}]$ one can define

$$\{a, b\} = -\pi\left(\frac{[a_0, b_0]}{p}\right).$$

Here $\pi : A_{n,\mathbb{Z}} \to A_{n,\mathbb{Z}_p}$ is the modulo p reduction of the Weyl algebra, and a_0, b_0 are arbitrary lifts of a, b with respect to π. The operation is well defined, takes values in the center and satisfies the Leibnitz rule and the Jacobi identity. On the generators one has

$$\{\xi_i, \xi_j\} = \omega_{ij}.$$

The Poisson bracket is trivially extended to the entire center $\mathbb{F}_p[\xi_1, \ldots, \xi_{2n}]$ and then to the ultraproduct of centers. Observe that the Poisson bracket of two elements of bounded degree is again of bounded degree, hence one has the bracket on the polynomial algebra.

2.3.5. Endomorphisms and Symplectomorphisms

The point of this construction lies in the fact that thus defined Poisson structure on the (injective image of) polynomial algebra is preserved under all endomorphisms of $\mathcal{A}_n(\mathcal{U}, [p])$ of bounded degree. Every endomorphism of the standard Weyl algebra is specified by an array of coefficients $(a_{i,I})$ (which form the images of the generators in the standard basis); these coefficients are algebraically dependent, but with only a finite number of bounded-order constraints. Hence the endomorphism of the standard Weyl algebra can be extended to the larger algebra $\mathcal{A}_n(\mathcal{U}, [p])$. The restriction of any such obtained endomorphism on the polynomial algebra $\mathbb{C}[\xi_1, \ldots, \xi_{2n}]$ preserves the Poisson structure. In this setup the automorphisms of the Weyl algebra correspond to symplectomorphisms of $\mathbb{A}^{2n}_{\mathbb{C}}$.

Example 1. *If x_i and y_i are standard generators, then one may perform a linear symplectic change of variables:*

$$f(x_i) = \sum_{j=1}^{n} a_{ij}x_j + \sum_{j=1}^{n} a_{i,n+j}y_j, \quad i = 1, \ldots, n,$$
$$f(d_i) = \sum_{j=1}^{n} a_{i+n,j}x_j + \sum_{j=1}^{n} a_{i+n,n+j}y_j, \quad a_{ij} \in \mathbb{C}.$$

In this case the corresponding polynomial automorphism f^c of

$$\mathbb{C}[\xi_1, \ldots, \xi_{2n}] \simeq \mathbb{C}[x_1^{[p]}, \ldots, x_n^{[p]}, y_1^{[p]}, \ldots, y_n^{[p]}]$$

acts on the generators ξ as

$$f^c(\xi_i) = \sum_{j=1}^{2n} (a_{ij})^{[p]}\xi_j,$$

where the notation $(a_{ij})^{[p]}$ means taking the base field automorphism that is conjugate to the nonstandard Frobenius via the Steinitz isomorphism.

Let $\gamma : \mathbb{C} \to \mathbb{C}$ be an arbitrary automorphism of the field of complex numbers. Then, given an automorphism f of the Weyl algebra $A_{n,\mathbb{C}}$ with coordinates $(a_{i,I})$, one can build another algebra automorphism using the map γ. Namely, the coefficients $\gamma(a_{i,I})$ define a new automorphism $\gamma_*(f)$ of the Weyl algebra, which is of the same degree as the original one. In other words, every automorphism of the base field induces a map $\gamma_* : A_{n,\mathbb{C}} \to A_{n,\mathbb{C}}$ which preserves the structure of the ind-object. It obviously is a group homomorphism.

Now, if $P_{n,\mathbb{C}}$ denotes the commutative polynomial algebra with Poisson bracket, we may define an ind-group homomorphism $\varphi : \mathrm{Aut}(A_{n,\mathbb{C}}) \to \mathrm{Aut}(P_{n\mathbb{C}})$ as follows. Previously we had a morphism $f \mapsto f^c$, however as the example has shown it explicitly depends on the choice of the infinite prime $[p]$. We may eliminate this dependence by pushing the whole domain $\mathrm{Aut}(A_{n,\mathbb{C}})$ forward with a specific base field automorphism γ, namely $\gamma = \Theta_{[p]}^{-1}$—the field automorphism which is Steinitz-conjugate with the inverse nonstandard Frobenius, and only then constructing the symplectomorphism f_{Θ}^{c} as the restriction to the (nonstandard) center. For the subgroup of tame automorphisms such as linear changes of variables this procedure has a simple meaning: just take the $[p]$-th root of all coefficients $(a_{i,I})$ first. We thus obtain a group homomorphism which preserves the filtration by degree and is in fact well-behaved with respect to the Zariski topology on Aut (indeed, the filtration $\mathrm{Aut}^N \subset \mathrm{Aut}^{N+1}$ is given by Zariski-closed embeddings). Formally, we have a proposition:

Proposition 7. *There is a system of morphisms*

$$\varphi_{[p],N} : \mathrm{Aut}^{\leq N}(A_{n,\mathbb{C}}) \to \mathrm{Aut}^{\leq N}(P_{n,\mathbb{C}}).$$

such that the following diagram commutes for all $N \leq N'$*:*

$$\begin{array}{ccc} \mathrm{Aut}^{\leq N}(A_{n,\mathbb{C}}) & \xrightarrow{\varphi_{[p],N}} & \mathrm{Aut}^{\leq N}(P_{n,\mathbb{C}}) \\ \big\downarrow{\scriptstyle \mu_{NN'}} & & \big\downarrow{\scriptstyle \nu_{NN'}} \\ \mathrm{Aut}^{\leq N'}(A_{n,\mathbb{C}}) & \xrightarrow{\varphi_{[p],N'}} & \mathrm{Aut}^{\leq N'}(P_{n,\mathbb{C}}). \end{array}$$

The corresponding direct limit of this system is given by $\varphi_{[p]}$*, which maps a Weyl algebra automorphism* f *to a symplectomorphism* f_{Θ}^{c}.

The Belov–Kontsevich conjecture then states:

Conjecture 1. $\varphi_{[p]}$ *is a group isomorphism.*

Injectivity may be established right away.

Theorem 10. $\varphi_{[p]}$ *is an injective homomorphism.*

(See Reference [11] for the fairly elementary proof).

2.3.6. On the Loops Related to Infinite Primes

Let us at first assume that the Belov–Kontsevich conjecture holds, with $\varphi_{[p]}$ furnishing the isomorphism between the automorphism groups. This would be the case if all automorphisms in $\mathrm{Aut}(A_{n,\mathbb{C}})$ were tame, which is unknown at the moment for $n > 1$.

The main result of the paper is as follows:

Theorem 11. *If one assumes that* $\varphi_{[p],N}$ *is surjective for any infinite prime* $[p]$*, then* Φ_N *is quasifinitedimensional and its eigenvalues are roots of unity.*

Let $[p]$ and $[p']$ be two distinct classes of $\mathcal{U}$-congruent prime number sequences—that is, two distinct infinite primes. We then have the following diagram:

$$\begin{array}{ccc} \mathrm{Aut}(A_{n,\mathbb{C}}) & \xrightarrow{\varphi_{[p]}} & \mathrm{Aut}(P_{n,\mathbb{C}}) \\ \big\downarrow{\scriptstyle isom} & & \big\downarrow{\scriptstyle isom} \\ \mathrm{Aut}(A_{n,\mathbb{C}}) & \xrightarrow{\varphi_{[p']}} & \mathrm{Aut}(P_{n,\mathbb{C}}) \end{array}$$

with all arrows being isomorphisms. Vertical isomorphisms answer to different presentations of $\mathbb{C}$ as $\overline{\mathbb{Z}_{[p]}}$ and $\overline{\mathbb{Z}_{[p']}}$. The corresponding automorphism $\mathbb{C} \to \overline{\mathbb{Z}_{[p]}}$ is denoted by $\alpha_{[p]}$ for any $[p]$.

The fact that all the arrows in the diagram are isomorphisms allows one instead to consider a loop of the form

$$\Phi : \mathrm{Aut}(A_{n,\mathbb{C}}) \to \mathrm{Aut}(A_{n,\mathbb{C}}).$$

Furthermore, as it was noted in the previous section, the morphism Φ belongs to $\mathrm{Aut}(\mathrm{Aut}(A_{n,\mathbb{C}}))$. We need to prove that Φ is a trivial automorphism. The first observation is as follows.

Proposition 8. *The map Φ is a morphism of algebraic varieties.*

Proof. Basically, this is a property of $\varphi_{[p]}$ (or rather its unshifted version, $f_p \mapsto f_p^c$). More precisely, it suffices to show that, given an automorphism f_p of the Weyl algebra in positive characteristic p with coordinates $(a_{i,I})$, its restriction to the center (a symplectomorphism) f_p^c has coordinates which are polynomials in $(a_{i,I}^p)$.

The switch to positive characteristic and back is performed for a fixed $f \in \mathrm{Aut}(A_{n,\mathbb{C}})$ on an index subset $A_f \in \mathcal{U}$.

Let f be an automorphism of $A_{n,\mathbb{C}}$ and let $N = \deg f$ be its degree. The automorphism f is given by its coordinates $a_{i,I} \in \mathbb{C}$, $i = 1, \dots, 2n$, $I = \{i_1, \dots, i_{2n}\}$, obtained from the decomposition of algebra generators ζ_i in the standard basis of the free module:

$$f(\zeta_i) = \sum_{i,I} a_{i,I} \zeta^I, \quad \zeta^I = \zeta_1^{i_1} \cdots \zeta_{2n}^{i_{2n}}.$$

Let $(a_{i,I,p})$ denote the class $\alpha_{[p]}(a_{i,I})$, $p = (p_m)$, and let $\{R_k(a_{i,I} \mid i, I) = 0\}_{k=1,\dots,M}$ be a finite set of algebraic constraints for coefficients $a_{i,I}$. Let us denote by $A_1, \dots, A_M$ the index subsets from the ultrafilter $\mathcal{U}$, such that A_k is precisely the subset, on whose indices the constraint R_k is valid for $(a_{i,I,p})$. Take $A_f = A_1 \cap \dots \cap A_M \in \mathcal{U}$ and for p_m, $m \in A_f$, define an automorphism f_{p_m} of the Weyl algebra in positive characteristic $A_{n,\mathbb{F}_{p_m}}$ by setting

$$f_{p_m}(\zeta_i) = \sum_{i,I} a_{i,I,p_m} \zeta^I.$$

All of the constraints are valid on A_f, so that f corresponds to a class $[f_p]$ modulo ultrafilter $\mathcal{U}$ of automorphisms in positive characteristic. The degree of every f_{p_m} ($m \in A_f$) is obviously less than or equal to $N = \deg f$.

Now consider $f \in \mathrm{Aut}^{\leq N}(A_{n,\mathbb{C}})$ with the index subset A_f over which its defining constraints are valid. The automorphisms $f_{p_m} = f_p : A_{n,\mathbb{F}_p} \to A_{n,\mathbb{F}_p}$ defined for $m \in A_f \in \mathcal{U}$ provide arrays of coordinates $a_{i,I,p}$. Let us fix any valid $p_m = p$ denote by F_{p^k} a finite subfield of $\mathbb{F}_p$ which contains the respective coordinates $a_{i,I,p}$ (one may take k to be equal to the maximum degree of all minimal polynomials of elements $a_{i,I,p}$ which are algebraic over $\mathbb{Z}_p$).

Let $a_1, \dots, a_s$ be the transcendence basis of the set of coordinates $a_{i,I,p}$ and let $t_1, \dots, t_s$ denote s independent (commuting) variables. Consider the field of rational functions:

$$F_{p^k}(t_1, \dots, t_s).$$

The vector space

$$\mathrm{Der}_{\mathbb{Z}_p}(F_{p^k}(t_1, \dots, t_s), F_{p^k}(t_1, \dots, t_s))$$

of all $\mathbb{Z}_p$-linear derivations of the field $F_{p^k}(t_1, \dots, t_s)$ is finite-dimensional with $\mathbb{Z}_p$-dimension equal to ks; a basis of this vector space is given by elements

$$\{e_a D_{t_b} \mid a = 1, \dots, k,\ b = 1, \dots, s\}$$

where e_a are basis vectors of the $\mathbb{Z}_p$-vector space F_{p^k}, and D_{t_b} is the partial derivative with respect to the variable t_b.

Set $a_1, \ldots, a_s = t_1, \ldots, t_s$ (i.e., consider an s-parametric family of automorphisms), so that the rest of the coefficients $a_{i,I,p}$ are algebraic functions of s variables $t_1, \ldots, t_s$. We need to show that the coordinates of the corresponding symplectomorphism f_p^c are annihilated by all derivations $e_a D_{t_b}$.

Let δ denote a derivation of the Weyl algebra induced by an arbitrary basis derivation $e_a D_{t_b}$ of the field. For a given i, let us introduce the short-hand notation

$$a = f_p(\zeta_i), \quad b = \delta(a).$$

We need to prove that

$$\delta(f^c(\xi_i)) = \delta(f_p(\zeta_i^p)) = 0.$$

In our notation $\delta(f_p(\zeta_i^p)) = \delta(a^p)$, so by Leibnitz rule we have:

$$\delta(f_p(\zeta_i^p)) = ba^{p-1} + aba^{p-2} + \cdots + a^{p-1}b.$$

Let $\mathrm{ad}_x : A_{n,\mathbb{F}_p} \to A_{n,\mathbb{F}_p}$ denote a $\mathbb{Z}_p$-derivation of the Weyl algebra corresponding to the adjoint action (all Weyl algebra derivations are inner!):

$$\mathrm{ad}_x(y) = [x, y].$$

We will call an element $x \in A_{n,\mathbb{F}_p}$ locally ad-nilpotent if for any $y \in A_{n,\mathbb{F}_p}$ there is an integer $D = D(y)$ such that

$$\mathrm{ad}_x^D(y) = 0.$$

All algebra generators ζ_i are locally ad-nilpotent. Indeed, one could take $D(y) = \deg y + 1$ for every ζ_i.

If f is an automorphism of the Weyl algebra, then $f(\zeta_i)$ is also a locally ad-nilpotent element for all $i = 1, \ldots, 2n$. That means that for any $i = 1, \ldots, 2n$ there is an integer $D \geq N + 1$ such that

$$\mathrm{ad}_{f_p(\zeta_i)}^D(\delta(f_p(\zeta_i))) = \mathrm{ad}_a^D(b) = 0.$$

Now, for $p \geq D + 1$ the previous expression may be rewritten as

$$0 = \mathrm{ad}_a^{p-1}(b) = \sum_{l=0}^{p-1}(-1)^l \binom{p-1}{l} a^l ba^{p-1-l} \equiv \sum_{l=0}^{p-1} a^l ba^{p-1-l} \pmod{p},$$

and this is exactly what we wanted.

We have thus demonstrated that for an arbitrary automorphism f_p of the Weyl algebra in characteristic p the coordinates of the corresponding symplectomorphism f_p^c are polynomial in p-th powers of the coordinates of f_p, provided that p is greater than $\deg f_p + 1$. As the sequence $(\deg f_{p_m})$ is bounded from above by N for all $m \in A_f$, we see that there is an index subset $A_f^* \in \mathcal{U}$ such that the coordinates of the symplectomorphism $f_{p_m}^c$ for $m \in A_f^*$ are polynomial in p_m-th powers of a_{i,I,p_m}. This implies that f^c in characteristic zero is given by coefficients polynomial in $\alpha_{[p]}(a_{i,I})^{[p]}$ as desired.

It follows, after shifting by the inverse nonstandard Frobenius, that Φ is an endomorphism of the algebraic variety $\mathrm{Aut}(A_{n,\mathbb{C}})$. $\square$

The automorphism Φ acting on elements $f \in \mathrm{Aut}(A_{n,\mathbb{C}})$, takes the set of coordinates $(a_{i,I})$ and returns a set $(G_{i,I}(a_{k,K}))$ of the same size. All functions $G_{i,I}$ are algebraic by the above proposition. It is convenient to introduce a partial order on the set of coordinates. We say that $a_{i,I'}$ is higher than $a_{i,I}$ (for the same generator i) if $|I| < |I'|$ and we leave pairs with $i \neq j$ or with $|I| = |I'|$ unconnected. We define the dominant elements $a_{i,I}$ (or rather, dominant places (i, I)) to be the maximal elements with respect to this partial order, and subdominant elements to be the elements covered by maximal ones (in other words, for fixed i, subdominant places are the ones with $|I| = |I_{\max}| - 1$).

The next observation follows from the fact that the morphisms in question are algebra automorphisms.

Lemma 1. *Functions $G_{i,I}$ corresponding to dominant places (i, I) are identities:*

$$G_{i,I}(a_{k,K}) = a_{i,I}.$$

Proof. Indeed, it follows from the commutation relations that for any $i = 1, \dots, 2n$ and $f_p, p = p_m, m \in A_f \in \mathcal{U}$, the highest-order term in $f_p^c(\xi_i) = f_p(\zeta_i^p) = f_p(\zeta_i)^p$ has the coefficient $a_{i,I,p}^p$. The shift by the inverse Frobenius then acts as the p-th root on the dominant place, so that we deduce that the latter is independent of the choice of $[p]$. □

Let us now fix $N \geq 1$ and consider

$$\Phi_N : \mathrm{Aut}^{\leq N} A_{n,\mathbb{C}} \to \mathrm{Aut}^{\leq N} A_{n,\mathbb{C}}$$

– the restriction of Φ to the subvariety $\mathrm{Aut}^{\leq N} A_{n,\mathbb{C}}$, which is well defined by the above lemma. The morphism corresponds to an endomorphism of the ring of functions

$$\Phi_N^* : \mathcal{O}(\mathrm{Aut}^{\leq N} A_{n,\mathbb{C}}) \to \mathcal{O}(\mathrm{Aut}^{\leq N} A_{n,\mathbb{C}})$$

Let us take a closer look at the behavior of Φ_N (and of Φ_N^*, which is essentially the same up to an inversion), specifically at how Φ_N affects one-dimensional subvarieties of automorphisms. Let $\mathcal{X}_N$ be the set of all algebraic curves of automorphisms in $\mathrm{Aut}^{\leq N} A_{n,\mathbb{C}}$; by virtue of Lemma 2 we may without loss of generality consider the subset of all curves with fixed dominant places—we denote such a subset by $\mathcal{X}'_N$, and, for that same matter, the subsets $\mathcal{X}_N^{(k)}$ of curves with fixed places of the form (i, I'), which are away from a dominant place by a path of length at most $(k-1)$. In particular one has $\mathcal{X}'_N = \mathcal{X}_N^{(1)}$.

The morphism Φ_N yields a map

$$\tilde{\Phi}_N : \mathcal{X}_N \to \mathcal{X}_N$$

and its restrictions

$$\tilde{\Phi}_N^{(k)} : \mathcal{X}_N^{(k)} \to \mathcal{X}_N.$$

Our immediate goal is to prove that for all attainable k we have

$$\tilde{\Phi}_N^{(k)} : \mathcal{X}_N^{(k)} \to \mathcal{X}_N^{(k)},$$

that is, the map Φ_N preserves the terms corresponding to non-trivial differential monomials.

In spite of minor abuse of language, we will call the highest non-constant terms of a curve in $\mathcal{X}_N^{(k)}$ dominant, although they cease to be so when that same curve is regarded as an element of $\mathcal{X}_N$.

Let $\mathcal{A} \in \mathcal{X}_N$ be an algebraic curve in general position. Coordinate-wise $\mathcal{A}$ answers to a set $(a_{i,I}(\tau))$ of coefficients parameterized by an indeterminate. By Lemma 3.3, Φ_N leaves the (coefficients corresponding to) dominant places of this curve unchanged, so we may well set $\mathcal{A} \in \mathcal{X}_N^{(1)}$. In fact, it is easily seen that the subdominant terms are not affected by Φ_N either, thanks to the commutation relations that define the Weyl algebra: for every p participating in the ultraproduct decomposition, after one raises to the p-th power one should perform a reordering within the monomials—a procedure which degrades the cardinality $|I|$ by an even number. Therefore, nothing contributes to the image of any subdominant term other than that subdominant term itself, which therefore is fixed under Φ_N. We are then to consider the image

$$\tilde{\Phi}_N^{(2)}(\mathcal{A}) \in \mathcal{X}_N^{(2)}.$$

Again, given a positive characteristic p within the ultraproduct decomposition, suppose the curve $\mathcal{A}$ (or rather its component answering to the chosen element p) has a number of poles attained on dominant (With respect to $\mathcal{X}_N^{(2)}$, that is, the highest terms that actually change—see above where we specify this convention.) terms. Let us pick among these poles the one of the highest order k, and let (i_0, I_0) be its place. By definition of an automorphism of Weyl algebra as a set of coefficients, the number i_0 does not actually carry any meaningful data, so that we are left with a pair $(k, |I_0|)$. As we can see, this pair is maximal from two different viewpoints; in fact, the pair represents a vertex of a Newton polygon taken over the appropriate field, with the discrete valuation given by $|I|$. The coordinate function a_{i_0,I_0} corresponding to this pole admits a decomposition

$$a_{i_0,I_0} = \frac{a_{-k}}{t^k} + \cdots,$$

with t a local parameter. Acting upon this curve by the morphism Φ_N amounts to two steps: first, we raise everything to the p-th power and then assemble the components within the ultraproduct decomposition, then we take the preimage, which is essentially the same as taking the p'-root, with respect to a different ultraproduct decomposition. The order of the maximal pole is then multiplied by an integer during the first step and divided *by the same integer* during the second one. By maximality, there are no other terms that might contribute to the resulting place in $\tilde{\Phi}_N^{(2)}(\mathcal{A})$. It therefore does not change under Φ_N.

We may process the rest of the dominant (with respect to $\mathcal{X}_N^{(2)}$) terms similarly: indeed, it suffices to pick a different curve in general position. We then move down to $\mathcal{X}_N^{(k)}$ with higher k and argue similarly.

After we have exhausted the possibilities with non-constant terms, we arrive at the conclusion that all that Φ_N does is permute the irreducible components of $\mathrm{Aut}^{\leq N} A_{n,\mathbb{C}}$. That in turn implies the existence of a positive integer l such that

$$\Phi_N^l = \mathrm{Id}.$$

In fact, the preceding argument gives us more than just the observation that Φ_N is unipotent. Let $\Phi_{N,M}^*$ denote the linear map of finite-dimensional vector spaces obtained by restricting Φ_N^* to regular functions of total degree less than or equal to M. Then the following proposition holds.

Proposition 9. *If λ is an eigenvalue of $\Phi_{N,M}^*$, then $\lambda^k = 1$ for some integer k.*

Proof. Indeed, should there exist $\lambda_0 \neq 1$, we may find an exceptional curve whose singularity changes under Φ_N, note that coefficients are products of normalization coordinates. □

2.3.7. Discussion

The investigation of decomposition of polynomial algebra-related objects into ultraproducts over the prime numbers $\mathbb{P}$ leads to a problem of independence of the choice of infinite prime. In the case of the Tsuchimoto–Belov–Kontsevich homomorphism the answer turns out to be affirmative, although there are other constructions, which are of algebraic or even polynomial nature but for which the independence fails. The reason for such arbitrary behavior has a lot to do with growth functions (in which case the situation is similar to the one described in Reference [106], and in fact in Reference [107], where one has a non-injective endomorphism $f_p : A_{n,\mathbb{F}_p} \to A_{n,\mathbb{F}_p}$, whose degree grows with p, which disallows for the construction of a naive counterexample to the Dixmier Conjecture in the ultralimit). It is, in our view, worthwhile to study such behavior in greater detail.

3. Algorithmic Aspects of Algebraic Geometry

The section contains two subsections: the first one is devoted to noncommutative Finite Gröbner basis issues and the second one is devoted to algorithmic inclusion undecidability.

3.1. *Finite Gröbner Basis Algebras with Unsolvable Nilpotency Problem and Zero Divisors Problem*

3.1.1. The Sketch of Construction

Let A be an algebra over a field K.

The set of all words in the alphabet $\{a_1, \ldots, a_N\}$ is a semigroup. The main idea of the construction is a realization of a universal Turing machine in this semigroup. We use the universal Turing machine constructed by Marvin Minsky in Reference [108]. This machine has 7 states and 4-color tape. The machine can be completely defined by 28 instructions.

Note that 27 of them have a form

$$(i, j) \to (L, q(i, j), p(i, j)) \text{ or } (i, j) \to (R, q(i, j), p(i, j)),$$

where $0 \le i \le 6$ is the current machine state, $0 \le j \le 3$ is the current cell color, L or R (left or right) is the direction of a head moving after execution of the current instruction, $q(i, j)$ is the state after current instruction, $p(i, j)$ is the new color of the current cell.

Thus, the instruction $(2, 3) \to (L, 3, 1)$ means the following: "If the color of the current cell is 3 and the state is 2, then the cell changes the color to 1, the head moves one cell to the left, the machine changes the state to 3.

The last instruction is $(4, 3) \to$ STOP. Hence, if the machine is in state 4 and the current cell has color 3, then the machine halts.

Letters

By Q_i, $0 \le i \le 6$ denote the current state of the machine. By P_j, $0 \le j \le 3$ denote the color of the current cell.

The action of the machine depends on the current state Q_i and current cell color P_j. Thus every pair Q_i and P_j corresponds to one instruction of the machine.

The instructions moving the head to the left (right) are called *left* (*right*) ones. Therefore there are *left pairs* (i, j) for the left instructions, *right pairs* for the right ones and instruction STOP for the pair $(4, 3)$.

All cells with nonzero color are said to be *non-empty cells*. We shall use letters a_1, a_2, a_3 for nonzero colors and letter a_0 for color zero. Also, we use R for edges of colored area. Hence, the word $Ra_{u_1}a_{u_2}\ldots a_{u_k}Q_iP_ja_{v_1}a_{v_2}\ldots a_{v_l}R$ presents a full state of Turing machine.

We model head moving and cell painting using computations with powers of a_i (cells) and P_i and Q_i (current cell and state of the machine's head).

We use the universal Turing machine constructed by Minsky. This machine is defined by the following instructions:

$(0,0) \to (L,4,1)$ $(0,1) \to (L,1,3)$ $(0,2) \to (R,0,0)$ $(0,3) \to (R,0,1)$
$(1,0) \to (L,1,2)$ $(1,1) \to (L,1,3)$ $(1,2) \to (R,0,0)$ $(1,3) \to (L,1,3)$
$(2,0) \to (R,2,2)$ $(2,1) \to (R,2,1)$ $(2,2) \to (R,2,0)$ $(2,3) \to (L,4,1)$
$(3,0) \to (R,3,2)$ $(3,1) \to (R,3,1)$ $(3,2) \to (R,3,0)$ $(3,3) \to (L,4,0)$
$(4,0) \to (L,5,2)$ $(4,1) \to (L,4,1)$ $(4,2) \to (L,4,0)$ $(4,3) \to$ STOP
$(5,0) \to (L,5,2)$ $(5,1) \to (L,5,1)$ $(5,2) \to (L,6,2)$ $(5,3) \to (R,2,1)$
$(6,0) \to (R,0,3)$ $(6,1) \to (R,6,3)$ $(6,2) \to (R,6,2)$ $(6,3) \to (R,3,1)$

We use the following alphabet:

$$\{t, a_0, \ldots a_3, Q_0, \ldots Q_6, P_0 \ldots P_3, R\}$$

For every pair except $(4, 3)$ the following functions are defined: $q(i, j)$ is a new state, $p(i, j)$ is a new color of the current cell (the head leaves it).

3.1.2. Defining Relations for the Nilpotency Question

Consider the following defining relations:

$$tRa_l = Rta_l; \quad 0 \leq l \leq 3 \tag{1}$$
$$ta_lR = a_lRt; \quad 0 \leq l \leq 3 \tag{2}$$
$$ta_ka_j = a_kta_j; \quad 0 \leq k, j \leq 3 \tag{3}$$
$$ta_kQ_iP_j = Q_{q(i,j)}P_kta_{p(i,j)}; \text{for left pairs } (i,j) \text{ and } 0 \leq k \leq 3 \tag{4}$$
$$tRQ_iP_j = RQ_{q(i,j)}P_0ta_{p(i,j)}; \text{for left pairs } (i,j) \text{ and } 0 \leq k \leq 3 \tag{5}$$
$$ta_lQ_iP_ja_ka_n = a_la_{p(i,j)}Q_{q(i,j)}P_kta_n; \text{for right pairs } (i,j) \text{ and } 0 \leq k \leq 3 \tag{6}$$
$$ta_lQ_iP_ja_kR = a_la_{p(i,j)}Q_{q(i,j)}P_kRt; \text{for right pairs } (i,j) \text{ and } 0 \leq k \leq 3 \tag{7}$$
$$tRQ_iP_ja_ka_n = Ra_{p(i,j)}Q_{q(i,j)}P_kta_n; \text{for right pairs } (i,j) \text{ and } 0 \leq k \leq 3 \tag{8}$$
$$tRQ_iP_ja_kR = Ra_{p(i,j)}Q_{q(i,j)}P_kRt; \text{for right pairs } (i,j) \text{ and } 0 \leq k \leq 3 \tag{9}$$
$$ta_lQ_iP_jR = a_la_{p(i,j)}Q_{q(i,j)}P_0Rt; \text{for right pairs } (i,j) \text{ and } 0 \leq l \leq 3 \tag{10}$$
$$tRQ_iP_jR = Ra_{p(i,j)}Q_{q(i,j)}P_0Rt; \text{for right pairs } (i,j) \tag{11}$$
$$Q_4P_3 = 0. \tag{12}$$

The relations (1) and (3) are used to move t from the left edge to the last letter a_l standing before Q_iP_j which represent the head of the machine. The relations (4)–(11) represent the computation process. The relation (2) is used to move t through the finishing letter R.

Finally, the relation (12) halts the machine.

Let us call $tRa_{u_1}a_{u_2}\dots a_{u_k}Q_iP_ja_{v_1}a_{v_2}\dots a_{v_l}R$ *the main word.*

Theorem 12. *Consider an algebra A presented by the defining relations* (1)–(12)*. The word $tRUQ_iP_jVR$ is nilpotent in A if and only if machine $M(i,j,U,V)$ halts.*

Actually we can prove that multiplication on the left by an element t leads to the transition to the next state of the machine.

3.1.3. Defining Relations for a Zero Divisors Question

We use the following alphabet:

$$\Psi = \{t,\ s,\ a_0, \dots a_3,\ Q_0, \dots Q_6,\ P_0 \dots P_3,\ L,\ R\}.$$

For every pair except $(4,3)$ the following functions are defined: $q(i,j)$ is a new state, $p(i,j)$ is a new color of the current cell (the head leaves it).

Consider the following defining relations:

$$tLa_k = Lta_k; \ 0 \leq k \leq 3 \tag{13}$$
$$ta_ka_l = a_kta_l; \quad 0 \leq k, l \leq 3 \tag{14}$$
$$sR = Rs; \tag{15}$$
$$sa_k = a_ks; \ 0 \leq k \leq 3 \tag{16}$$
$$ta_kQ_iP_j = Q_{q(i,j)}P_ka_{p(i,j)}s; \text{for left pairs } (i,j) \text{ and } 0 \leq k \leq 3 \tag{17}$$
$$tLQ_iP_j = LQ_{q(i,j)}P_0a_{p(i,j)}s; \text{for left pairs } (i,j) \tag{18}$$
$$ta_lQ_iP_ja_k = a_la_{p(i,j)}Q_{q(i,j)}P_ks; \text{for right pairs } (i,j) \text{ and } 0 \leq k, l \leq 3 \tag{19}$$

$$tLQ_iP_ja_k = La_{p(i,j)}Q_{q(i,j)}P_ks; \text{for right pairs } (i,j) \text{ and } 0 \le k \le 3 \tag{20}$$
$$ta_lQ_iP_jR = a_la_{p(i,j)}Q_{q(i,j)}P_0Rs; \text{for right pairs } (i,j) \text{ and } 0 \le l \le 3 \tag{21}$$
$$tLQ_iP_jR = La_{p(i,j)}Q_{q(i,j)}P_0Rs; \text{for right pairs } (i,j) \tag{22}$$
$$Q_4P_3 = 0; \tag{23}$$

The relations (13)–(14) are used to move t from the left edge to the letters Q_i, P_j which present the head of the machine. The relations (15)–(16) are used to move s from the letter Q_i, P_j to the right edge. The relations (17)–(21) represent the computation process. Here we use relations of the form $tU = Vs$.

Finally, the relation (23) halts the machine.

3.1.4. Zero Divisors and Machine Halt

Let us call $La_{u_1}a_{u_2}\dots a_{u_k}Q_iP_ja_{v_1}a_{v_2}\dots a_{v_l}R$ *the main word.*

Theorem 13. *The machine halts if and only if the main word is a zero divisor in the algebra presented by the defining relations* (13)–(23).

Remark. *We can consider two semigroups corresponding to our algebras: in both algebras each relation is written as an equality of two monomials. Therefore the same alphabets together with the same sets of relations define semigroups. In both semigroups the equality problem is algorithmically solvable, since it is solvable in algebras. However in the first semigroup a nilpotency problem is algorithmically unsolvable, and in the second semigroup a zero divisor problem is algorithmically unsolvable.*

The entire proofs can be found at Reference [109].

3.2. On the Algorithmic Undecidability of the Embeddability Problem for Algebraic Varieties over a Field of Characteristic Zero

3.2.1. The Case of Real Numbers

By a *Matiyasevich family of polynomials* we mean a family of polynomials

$$Q(\sigma_1, \dots, \sigma_T, x_1, \dots, x_s)$$

for which the existence of a solution for a given set of parameters of the polynomial is undecidable. As was established in Reference [110], such a polynomial exists.

Consider the affine space of dimension $5d + 1$. We denote coordinates in this space by $X_i, Y_i, Z_i, U_i, W_i, 1 \le i \le d$, and T. Consider the variety $B_{(d)}$ given by the following system of generators and relations:

$$\begin{cases} X_i^2 - (T^2 - 1)\, Y_i^2 = 1, \\ Y_i - (T-1)Z_i = V_i, \\ V_iU_i = 1, \end{cases} \tag{24}$$

where $1 \le i \le d$. For fixed i, the admissible values of the coordinates X_i, Y_i, Z_i, U_i, and W_i are determined by the same value of T. Consider the "short" subsystem

$$\begin{cases} X^2 - (T^2 - 1)\, Y^2 = 1, \\ Y - (T-1)Z = V, \\ VU = 1, \end{cases} \tag{25}$$

Lemma 2. *The following assertions hold for every solution of system* (25)*:*

(1) *U and V are nonzero constants in $\mathbb{F}[t]$* ($\deg U = \deg V = 0$);

(2) *either $T = \pm 1$ and $X = \pm 1$ or*

$$Y = \sum_{k=0}^{[N/2]} \binom{N}{2k+1} \left(T^2 - 1\right)^k T^{N-1-2k}$$

for some integer N.

Let R denote a root of the equation $R^2 = T^2 - 1$ such that R belongs to the algebraic extension $\mathbb{F}[t]$. Then the element $(T + R)^n$ can be uniquely represented in the form $X_n + RY_n$, where X_n and Y_n are polynomials in $\mathbb{F}[t]$. All solutions of the equation

$$X^2 - (T^2 - 1)Y^2 = 1 \tag{26}$$

are of the form $X = \pm X_n$, $Y = \pm Y_n$ (see Reference [111]).

The structure of this set depends on T. In the case $T = \pm 1$, the first equation of the system imposes no conditions at all on Y. In turn, the other equation implies $Y = (T - 1)Z + V$. For every choice of $V \in \mathbb{F} \setminus \{0\}$ and $Z \in \mathbb{F}[t]$, the corresponding solution exists and is unique.

Lemma 3. *If $\deg T > 0$, then $V = Y \bmod (T - 1) = N$ for an integer N and $Z = (Y - N)/(T - 1)$. If $T = const \neq \pm 1$, then Y and Z are constants in $\mathbb{F}[t]$.*

Thus, the following three cases are possible:

(1) for $\deg T > 0$, to every set of integers N_i there correspond polynomial solutions Y_i and X_i determined up to sign, as well as the constants $V_i = N_i$ and $U_i = 1/V_i$, and $Z_i = (Y_i - V_i)/(T - 1)$;
(2) for $\deg T = 0$ and $T \neq \pm 1$, there are constant solutions for Y_i chosen from a given sequence; the values X_i, Z_i, V_i and U_i are also constants, and they are determined by the chosen values of Y_i;
(3) for $T = \pm 1$, we obtain $X_i = \pm 1$; for arbitrarily chosen constants V_i and polynomials Z_i, we set $U_1 = 1/V_i$ and $Y_i = (T - 1)Z_i + V_i$.

So far, these considerations are valid for an arbitrary ground field $\mathbb{F}$ of characteristic zero. In the case $\mathbb{F} = \mathbb{R}$, we introduce a new coordinate S by completing the main system of equations by the equation

$$T = S^2 + 2, \tag{27}$$

which ensures the impossibility of $T = \pm 1$. All common solutions of systems (24) and (27) either are constants (if $\deg T = 0$, $T \neq \pm 1$) or correspond to some set of integer parameters $(N_1, \ldots, N_d)$. We refer to solutions of the first kind as "bad" and to those of the second kind as "good".

Consider a Matiyasevich family of polynomials $Q(\sigma_1, \ldots, \sigma_\tau, x_1, \ldots, x_s)$. Let $d \leq s$. Then, adding the new equation $Q(\sigma, V_1, \ldots, V_s) = 0$ to systems (24) and (27), we obtain a system defining a new variety. We denote this variety by $\mathscr{B}'_{(d),\sigma}$.

If $Q = 0$ has no integer solutions, then the original system has no good solutions. In this case, the variety $\mathscr{B}'_{(d),\sigma}$ is zero-dimensional, and there are no embeddings of A in $\mathscr{B}'_{(d),\sigma}$.

Otherwise, for every solution $N_1, \ldots, N_s$, we can explicitly construct functions $Y_i(S), X_i(S)$, and $Z_i(S)$ which are solutions. They define an embedding of the line in the variety $\mathscr{B}'_{(d),\sigma}$.

Since the existence of integer solutions for Q is undecidable, it follows that so is the embeddability of A in $\mathscr{B}'_{(d),\sigma}$ (in particular, in $\mathscr{B}'_{(s),\sigma}$). Here the input data is the equations defining $\mathscr{B}'_{(d),\sigma}$. We have proved the following theorem.

Theorem 14. *The problem of the embeddability of the affine line (and, therefore, the general embedding problem for an arbitrary variety) over $\mathbb{R}$ in an arbitrary algebraic variety $\mathscr{B}$ (defined by generators and relations) is undecidable.*

3.2.2. The Complex Case

In this case, the situation is more complicated: it is hard to eliminate the case in which $T = \pm 1$ and $X_i = \pm 1$, since no constraints on Y_i arise in this case. Therefore, we consider the problem of the embeddability of an affine space A^m in a given variety $\mathscr{B}$ and construct a class of varieties such that it is impossible to decide whether a desired embedding exists from the defining relations for representatives of this class (for a certain suitable integer m). We define the coordinate ring of the variety $\mathscr{B}_{(d,e)}$ by the following system of generators and relations:

$$\begin{cases} X_{ij}^2 - \left(T_j^2 - 1\right) Y_{ij}^2 = 1, \\ Y_{ij} - (T_j - 1)\, Z_{ij} = V_{ij}, \\ V_{ij} U_{ij} = 1, \\ T_{j+1} = \prod_{k=1}^{j} \left((T_k^2 - 1)\, W_k \right) W_{j+1}, \end{cases} \tag{28}$$

where $1 \leq i \leq d$ and $1 \leq j \leq e$. In fact, we compose a system of many "clones" of the main system of the previous subsection and augment it by the "linking" relations between the parameters T_j. Let us study the solutions of the resulting system in $\mathbb{C}[t]$.

The relations for $X_{ij}, Y_{ij}, Z_{ij}, U_{ij}$, and V_{ij} for each fixed T_j are similar to those considered above. For a fixed set of T_j, the set of solutions is the direct sum of the sets $\mathscr{B}_{(d)}$, which have already been studied above.

As above, for each j, the following cases can occur: $T_j = \pm 1$ and $\deg T_j = 0$; $T_j \neq \pm 1$, and $\deg T_j > 0$.

The case most important from the point of view of "elimination" is the case where $T_{\hat{j}} = \pm 1$ for some $\hat{j}$. In this case, $T_{\hat{j}}^2 - 1 = 0$, and for all $j < \hat{j}$, we obtain

$$T_j = \prod_{k=1}^{j-1} ((T_k^2 - 1) W_k) W_j.$$

Lemma 4. *If $T_N = C_N \neq 0$ for some N, then all W_k with $k \leq N$ and all T_k with $k \leq N - 1$ are constants.*

By Lemma 4, we have $T_j = C_j$ for $j < \hat{j}$. Here $C_j \neq \pm 1$ (otherwise $C_{j+1} = 0$). Thus, if $T_{\hat{j}} = \pm 1$ for some $\hat{j}$, then the corresponding component has dimension d. However, in this case, all other components are zero-dimensional, and the total dimension of the variety does not exceed d.

In the second case, we have $T_{\hat{j}} = C_{\hat{j}} \neq \pm 1$ for some $\hat{j}$. The corresponding component of the variety has dimension 0. Moreover, Lemma 4 implies $T_j = C_j$ for $j < \hat{j}$. The corresponding $\hat{j} - 1$ components of the variety are zero-dimensional as well.

The case $\deg T_j > 0$ was considered in Section 3.2.1. Each component of the variety is parametrized by a set of integers $N_{1j}, \ldots, N_{dj}$,for which the corresponding solutions for $X_{ij}, Y_{ij}, Z_{ij}, U_{ij}$, and V_{ij} are constructed explicitly. The corresponding component has dimension 1.

Consider a Matiyasevich family of polynomials $Q(\sigma_1, \ldots, \sigma_\tau, x_1, \ldots, x_s)$. The solvability problem of the Diophantine equation $Q(\sigma_1, \ldots, \sigma_\tau, V_{1j}, \ldots, V_{sj}) = 0$ is algorithmically undecidable. Let $d \leq s$. Adding the new equations $Q(\sigma, V_{i1}, \ldots, V_{is}) = 0$ to system (28), we obtain a system defining a new variety. We denote it by $\mathscr{B}'_{(d,e),\sigma}$.

If $Q = 0$ has no integer solutions, then the original system has no solutions for which $\deg T_0 > 0$. In this case, the possible solutions correspond either to the case where $T_j = \pm 1$ for some j (and the set of solutions has dimension d) or to the case $T_j = C_j \neq \pm 1$. In the latter case, assuming that j is the maximum index for which $T_j = C_j \neq \pm 1$, we see that all the succeeding $e - j$ components are one-dimensional and the total dimension of the set equals precisely $e - j \leq e - 1$. Setting $e = s$ and $d = s - 1$, we obtain

$$\dim \mathscr{B}'_{(d,e)} \leq \max(e - 1, d) = s - 1 < s.$$

Obviously, in this case, for $m \geq s$, there is no embedding of $\mathscr{A} = A^m$ in $\mathscr{B}'_{(d,e),\sigma} = \mathscr{B}'_{(s-1,s),\sigma}$. In particular, A^s cannot be embedded in $\mathscr{B}'_{(s-1,s),\sigma}$.

If Q has integer solutions, then, for every such solution $N_1, \ldots, N_s$, we can explicitly construct functions $Y_{ij}(T), X_{ij}(T)$, and $Z_{ij}(T)$ which are solutions. These functions define an embedding of $A = A^s$ in the variety $\mathscr{B}'_{(d,e),\sigma}$.

Since the existence of integer solutions for Q is undecidable, it follows that the embeddability of A^s in $\mathscr{B}'_{(s-1,s),\sigma}$ is undecidable as well (the input data is the equations defining $\mathscr{B}'_{(s-1,s),\sigma}$). The proof is valid for any field K of of characteristic zero. The following theorem holds.

Theorem 15. *There is a positive integer s for which the embeddability of an affine space A^s over K in an arbitrary algebraic variety $\mathscr{B}$ (defined by generators and relations) is undecidable. Thus, the general embeddability problem for an arbitrary algebraic variety is undecidable as well.*

Author Contributions: Conceptualization, A.K.-B. and E.P.; methodology, A.K.-B., S.M., E.P. and J.-T.Y.; software, S.M. and W.Z.; validation, A.K.-B. and E.P.; formal analysis, A.C., I.I.-P. and S.M.; investigation, S.M. and W.Z.; resources, A.K.-B. and E.P.; data curation, E.P.; writing—original draft preparation, A.C. and I.I.-P.; writing—review and editing, S.M. and E.P.; visualization, E.P.; supervision, A.K.-B.; project administration, E.P.; funding acquisition, A.K.-B. and E.P. All authors have read and agreed to the published version of the manuscript.

Funding: The first three authors were supported by the Russian Science Foundation grant No. 17-11-01377, fourth, fifth and seventh named authors were supported by the ISF (Israel Science Foundation) grant 1994/20.

Acknowledgments: We would like to thank the anonymous referee for suggestions regarding this paper.

Conflicts of Interest: The authors declare no conflict of interest.

References

1. Plotkin, B. *Universal Algebra, Algebraic Logic and Databases*; Kluwer Academic Publishers: Dordrecht, The Netherlands, 1994.
2. Plotkin, B. Algebra, categories, and databases. In *Handbook of Algebra*; Elsevier: Amsterdam, The Netherlands, 2000; Volume 2, pp. 81–148.
3. Plotkin, B. Seven lectures on the universal algebraic geometry. In *Groups, Algebras, and Identities, Contemporary Mathematics, Israel Mathematical Conferences Proceedings*; AMS: Providence, RI, USA, 2019; Volume 726, pp. 143–217. [CrossRef]
4. Plotkin, B. Varieties of algebras and algebraic varieties. *Israel J. Math.* **1996**, *96*, 511–522. [CrossRef]
5. Plotkin, B. Varieties of algebras and algebraic varieties. Categories of algebraic varieties. *Sib. Adv. Math.* **1997**, *7*, 64–97.
6. Kharlampovich, O.; Myasnikov, A. Elementary theory of free nonabelian groups. *J. Algebra* **2006**, *302*, 451–552. [CrossRef]
7. Sela, Z. Diophantine geometry over groups VI: The elementary theory of a free group. *Geometr. Funct. Anal. GAFA* **2006**, *16*, 707–730. [CrossRef]
8. Hrushovski, E.; Zilber, B. Zariski Geometries. *J. Am. Math. Soc.* **1996**, *9*, 1–56. [CrossRef]
9. Zilber, B. *Zariski Geometries: Geometry from the Logician's Point of View*; London Mathematical Society Lecture Note Series Book 360; Cambridge University Press: Cambridge, UK, 2010.
10. Zilber, B. Zariski geometries. Geometry from the logician's point of view. *Bull. Am. Math. Soc.* **2013**, *50*, 175–180.
11. Kanel-Belov, A.; Kontsevich, M. Automorphisms of Weyl algebras. *Lett. Math. Phys.* **2005**, *74*, 181–199. [CrossRef]
12. Kanel-Belov, A.; Kontsevich, M. The Jacobian Conjecture is stably equivalent to the Dixmier Conjecture. *arXiv* **2005**, arXiv:math/0512171v2.
13. Plotkin, B. Seven lectures in universal algebraic geometry. *arXiv* **2002**, arXiv:math.RA/0502212.
14. Grothendieck, A. Éléments de géométrie algébrique: I. Le langage des schémas. *Publ. Math. De l'IHÉS* **1960**, *4*, 5–228. [CrossRef]
15. Kanel-Belov, A.Y.; Chilikov, A.A. On the Algorithmic Undecidability of the Embeddability Problem for Algebraic Varieties over a Field of Characteristic Zero. *Math. Notes* **2019**, *106*, 299–302. [CrossRef]

16. Plotkin, B. *Groups of Automorphisms of Algebraic Systems*; Wolters Noordhoff Publishing: Groningen, The Netherlands, 1972.
17. Plotkin, B. Algebras with the same algebraic geometry. *Proc. Steklov Inst. Math.* **2003**, *242*, 165–196.
18. Katsov, Y.; Lipyanski, R.; Plotkin, B. Automorphisms of the categories of free modules, free semimodules and free Lie algebras. *Comm. Algebra* **2007**, *35*, 931–952. [CrossRef]
19. Mashevitzky, G.; Plotkin, B.; Plotkin, E. Automorphisms of categories of free algebras of varieties. *Electron. Res. Announc. AMS* **2002**, *8*, 1–10. [CrossRef]
20. Plotkin, B.; Zhitomirski, G. On automorphisms of categories of universal algebras. *J. Algebra Comput.* **2007**, *17*, 1115–1132. [CrossRef]
21. Mashevitzky, G. Automorphisms of the semigroup of endomorphisms of free ring and free associative algebras. Preprint.
22. Berzins, A. The group of automorphisms of the category of free associative algebra. *arXiv* **2005**, arXiv:math/0502426.
23. Berzins, A.; Plotkin, B.; Plotkin, E. Algebraic geometry in varieties of algebras with the given algebra of constants. *J. Math. Sci.* **2000**, *102*, 4039–4070. [CrossRef]
24. Berzins, A. The group of automorphisms of semigroup of endomorphisms of free commutative and free associative algebra. *arXiv* **2005**, arXiv:math/0504015.
25. Plotkin, B.; Zhitomirski, G. Automorphisms of categories of free algebras of some varieties. *J. Algebra* **2006**, *306*, 344–367. [CrossRef]
26. Lipyanski, R. Automorphisms of the semigroup of endomorphisms of free algebras of homogeneous varieties. *arXiv* **2005**, arXiv:math.RA//0511654v1. Available online: https://www.tandfonline.com/doi/abs/10.1080/00927870601115856 (accessed on 28 March 2007)
27. Mashevitzky, G.; Schein, B. Automorphisms of the endomorphism semigroup of a free monoid or a free semigroup. *Proc. Am. math. Soc.* **2002**, *8*, 1–10.
28. Mashevitzky, G.; Plotkin, B.; Plotkin, E. Automorphisms of the category of free Lie algebras. *J. Algebra* **2004**, *282*, 490–512. [CrossRef]
29. Zhitomirski, G. Automorphisms of the semigroup of all endomorphisms of free algebras. *arXiv* **2005**, arXiv:math.GM/0510230v1.
30. Kanel-Belov, A.; Yu, J.-T.; Elishev, A. On the Augmentation Topology of Automorphism Groups of Affine Spaces and Algebras. *Int. J. Algebra Comput.* **2018**, *28*, 1449–1485. [CrossRef]
31. Kanovei, V.G.; Lyubetskii, V.A. Problems of set-theoretic non-standard analysis. *Russ. Math. Surv.* **2007**, *62*, 45–111. [CrossRef]
32. Makar-Limanov, L. On automorphisms of Weyl algebra. *Bull. Soc. Math. Fr.* **1984**, *112*, 359–363. [CrossRef]
33. Makar-Limanov, L. Automorphisms of a free algebra with two generators. *Funct. Anal. Appl.* **1970**, *4*, 107–108. [CrossRef]
34. Jung, H.W.E. Uber ganze birationale Transformationen der Eben. *J. Reine Angew. Math.* **1942**, *184*, 161–174.
35. Van der Kulk, W. On polynomial rings in two variables. *Nieuw Arch. Wisk.* **1953**, *1*, 33–41.
36. Tsuchimoto, Y. Endomorphisms of Weyl algebra and p-curvatures. *Osaka J. Math.* **2005**, *42*, 435–452.
37. Markov, A. The impossibility of certain algorithms in the theory of associative systems. *Doklady Akad. Nauk SSSR* **1947**, *55*, 587–590. (In Russian)
38. Post, E. Recursive unsolvability of a problem of Thue. *J. Symb. Logic* **1947**, *12*, 1–11. [CrossRef]
39. Novikov, P. On algorithmic unsolvability of the problem of identity. *Doklady Akad. Nauk SSSR* **1952**, *85*, 709–712. (In Russian)
40. Novikov, P. On the algorithmic unsolvability of the word problem in group theory. *Trudy Mat. Inst. im. Steklov.* **1955**, *44*, 3–143.
41. Shirshov, A. Some algorithmic problems for Lie algebras. *Sib. Mat. J.* **1962**, *3*, 292–296.
42. Bokut, L. Unsolvability of the equality problem and subalgebras of finitely presented lie algebras. *Izv. Akad. Nauk SSSR* **1972**, *36*, 1173–1219. [CrossRef]
43. Bergman, G. The diamond lemma for ring theory. *Adv. Math.* **1978**, *29*, 178–218. [CrossRef]
44. Piontkovsky, D. Gröbner base and coherence of monomial associative algebra. *Pure Appl. Math.* **1996**, *2*, 501–509.
45. Piontkovsky, D. Noncommutative Gröbner bases, coherence of monomial algebras and divisibility in semigroups, *Pure Appl. Math.* **2001**, *7*, 495–513.

46. Piontkovsky, D. On the Kurosh problem in varieties of algebras. *J. Math. Sci.* **2009**, *163*, 743–750. [CrossRef]
47. Piontkovsky, D. Graded algebras and their differential graded extensions. *J. Math. Sci.* **2007**, *142*, 2267–2301. [CrossRef]
48. Kanel-Belov, A.; Borisenko, V.; Latysev, V. *Monomial Algebras*; Plenum: New York, NY, USA, 1997.
49. Iyudu, N. Algorithmic solvability of zero divisors problem in one class of algebras. *Pure Appl. Math.* **1995**, *2*, 541–544.
50. Iyudu, N. Standart Bases and Property Solvability in the Algebras Defined by Relations. Ph.D. Thesis, Moscow State University, Moscow, Russia, 1996.
51. Belov, A.Y. Linear recurrence relations on tree. *Math Notes*. **2005**, *78*, 603–609. [CrossRef]
52. Ivanov-Pogodaev, I. An algebra with a finite Gröbner basis and an unsolvable problem of zero divisors. *J. Math. Sci.* **2008**, *152*, 191–202. [CrossRef]
53. Ufnarovsky, V. Combinatorial and asymototic methods in algebra, Itogi nauki i tehniki. *Mod. Probl. Pure Math.* **1990**, *57*, 5–177.
54. Bokut, L.; Chen, Y. Gröbner-Shirshov bases and PBW theorems, *J. Sib. Fed. Univ. Math. Phys.* **2013**, *6*, 417–427.
55. Bokut, L.; Kukin, G. Undecidable algorithmic problems for semigroups, groups and rings. *J. Soviet Math.* **1989**, *45*, 871–911. [CrossRef]
56. Gateva-Ivanova, T.; Latyshev, V. On the recognizable properties of associative algebras. *J. Symb. Comp.* **1989**, *6*, 371–398. [CrossRef]
57. Kanel-Belov, A. Classification of weakly Noetherian monomial algebras. (Russian. English summary) *Fundam. Prikl. Mat.* **1995**, *1*, 1085–1089.
58. Latyshev, V. *Combinatorial Ring Theory, Standard Bases (Russian)*; Izdatelstvo Moskovskogo Gosudarstvennogo Universiteta: Moskva, Russia, 1988; 68p.
59. Kanel-Belov, A.; Elishev, A. On Planar Algebraic Curves and Holonomic $\mathcal{D}$-modules in Positive Characteristic. *arXiv* **2014**, arXiv:1412.6836.
60. Keisler, H.J. On Cardinalities of Ultraproducts. *Bull. Am. Math. Soc.* **1964**, *70*, 644–647. [CrossRef]
61. Kontsevich, M. Holonomic $\mathcal{D}$-modules and positive characteristic. *arXiv* **2010**, arXiv:1010.2908v1.
62. Halmos, P.R. *Algebraic Logic*; Courier Dover Publications: Chelsea, NY, USA, 1969.
63. Myasnikov, A.; Remeslennikov, V. Algebraic geometry over groups II, Logical foundations. *J. Algebra* **2000**, *234*, 225–276. [CrossRef]
64. Plotkin, B.; Aladova, E.; Plotkin, E. Algebraic logic and logically-geometric types in varieties of algebras. *J. Algebra Appl.* **2012**, *12*, 1250146. [CrossRef]
65. Plotkin, B.; Plotkin, E. Multi-sorted logic and logical geometry: Some problems. *Demonstr. Math.* **2015**, *48*, 577–618. [CrossRef]
66. Plotkin, B. Algebraic geometry in First Order Logic. *J. Math. Sci.* **2006**, *137*, 5049–5097. [CrossRef]
67. Plotkin, B. Algebraic logic and logical geometry in arbitrary varieties of algebras. In Proceedings of the Conference on Group Theory, Combinatorics and Computing, Boca Raton, FL, USA, 3–8 October 2012; Volume 611; pp. 151–169.
68. Formanek, E. A question of B. Plotkin about the semigroup of endomorphisms of a free group. *Proc. Am. Math. Soc.* **2002**, *30*, 935–937.
69. Mashevitzky, G.; Schein, B.M.; Zhitomirski, G.I. Automorphisms of the semigroup of endomorphisms of free inverse semigroups. *Commun. Algebra* **2006**, *34*, 3569–3584. [CrossRef]
70. Tsurkov, A. Automorphisms of the category of the free nilpotent groups of the fixed class of nilpotency. *Int. J. Algebra Comput.* **2007**, *17*, 1273–1281. [CrossRef]
71. Tsurkov, A. Automorphic equivalence in the classical varieties of linear algebras. *Int. J. Algebra Comput.* **2017**, *27*, 979–979. [CrossRef]
72. Nies, A. Describing Groups. *Bull. Symb. Log.* **2007**, *13*, 305–339. [CrossRef]
73. Avni, N.; Lubotzky, A.; Meiri, C. First order rigidity of non-uniform higher-rank arithmetic groups. *Invent. Math.* **2019**, *217*, 219–240. [CrossRef]
74. Szmielew, W. Elementary properties of Abelian groups. *Fund. Math.* **1955**, *41*, 203–271. [CrossRef]
75. Eklof, P.C.; Fischer, R.F. The elementary theory of abelian groups. *Ann. Math. Log.* **1972**, *4*, 115–171. [CrossRef]
76. Hirshon, R. Some cancellation theorems with applications to nilpotent groups. *J. Aust. Math. Soc. Ser. A* **1977**, *23*, 147–165. [CrossRef]

77. Oger, F. Cancellation and elementary equivalence of finitely generated finite by nilpotent groups. *J. Lond. Math. Soc.* **1991**, *30*, 293–299. [CrossRef]
78. Zilber, B. An example of two elimentarily quivalent, but not isomorphic finitely generated nilpotent groups of class 2. *Algebra Log.* **1971**, *10*, 173–188
79. Malcev, A.I. On free solvable groups. *Dokl. AN SSSR* **1960**, *130*, 495–498.
80. Rogers, P.; Smith, H.; Solitar, D. Tarski's Problem for Solvable Groups. *Proc. Am. Math. Soc.* **1986**, *96*, 668–672.
81. Casals-Ruiz, M.; Kazachkov, I.; Remeslennikov, V. Elementary equivalence of right-angled Coxeter groups and graph products of finite abelian groups. *Bull. Lond. Math. Soc.* **2010**, *42*, 130–136. [CrossRef]
82. Casals-Ruiz, M.; Kazachkov, I. Two remarks on First Order Theories of Baumslag-Solitar groups. *arXiv* **2020**, arXiv:1002.2658v3.
83. Jarden, M.; Lubotzky, A. Elementary equivalence of profinite groups. *Bull. Lond. Math. Soc.* **2008**, *40*, 887–896. [CrossRef]
84. Malcev, A.I. On isomorphic matrix representations of infinite groups (Russian. English summary). *Rec. Math. Mat. Sbornik N.S.* **1940**, *8*, 405–422.
85. Malcev, A.I. The elementary properties of linear groups, Certain Problems in Math. and Mech., Sibirsk. Otdelenie Akad. Nauk SSSR, Novosibirsk, 1961. *Metamath. Algebr. Syst.* **1936**, *1967*, 110–132.
86. Bunina, E.I. Isomorphisms and elementary equivalence of Chevalley groups over commutative rings. *Sb. Math.* **2019**, *210*, 1067–1091. [CrossRef]
87. Zilber, B. Some model theory of simple algebraic groups over algebraically closed fields. *Colloq. Math.* **1984**, *48*, 173–180. [CrossRef]
88. Avni, N.; Meiri, C. On the model theory of higher rank arithmetic groups. *arXiv* **2020**, arXiv:2008.01793v1.
89. Sohrabi, M.; Myasnikov, A. Bi-interpretability with Z and models of the complete elementary theories of $SL_n(\mathcal{O})$, $T_n(\mathcal{O})$ and $GL_n(\mathcal{O})$, $n \geq 3$. *arXiv* **2020**, arXiv:2004.03585.
90. Segal, D.; Tent, K. Defining R and $G(R)$. *arXiv* **2020**, arXiv:2004.13407.
91. Plotkin, B. Isotyped algebras. *Proc. Steklov Inst. Math.* **2012**, *278*, 91–115. [CrossRef]
92. Plotkin, B.; Zhitomirski, G. Some logical invariants of algebras and logical relations between algebras. *St. Petersburg Math. J.* **2008**, *19*, 859–879. [CrossRef]
93. Zhitomirski, G. On types of points and algebras. *Int. J. Algebra Comput.* **2018**, *28*, 1717–1730. [CrossRef]
94. Myasnikov, A.G.; Romanovskii, N.S. Characterization of finitely generated groups by types. *Int. J. Algebra Comput.* **2018**, *28*, 1613–1632. [CrossRef]
95. Sklinos, R. Private correspondence. Unpublished Work.
96. Pillay, A. On genericity and weight in the free group. *Proc. Am. Math. Soc.* **2009**, *137*, 3911–3917. [CrossRef]
97. Perin, C.; Sklinos, R. Homogenety in the free group. *Duke Math. J.* **2012**, *161*, 2635–2658. [CrossRef]
98. Houcine, A.O. Homogenity and prime models in torsion-free hyperbolic groups. *Conflu. Math.* **2011**, *3*, 121–155. [CrossRef]
99. Kanel-Belov, A.; Berzins, A.; Lipyanski, R. Automorphisms of the semigroup of endomorphisms of free associative algebras. *arXiv* **2005**, arXiv:math/0512273v3.
100. Belov-Kanel, A.; Lipyanski, R. Automorphisms of the endomorphism semigroup of a free commutative algebra. *arXiv* **2009**, arXiv:0903.4839.
101. Robinson, A. Non-standard Analysis. *Nederl. Akad. Wetensch. Proc. Ser. A Indag. Math.* **1961**, 64, 432–440. [CrossRef]
102. Cherlin, G. Ideals of integers in non-standard number fields. In *Model Theory and Algebra*; Lecture Notes in Math; Springer: Berlin/Heidelberg, Germany, 1975; Volume 498, pp. 60–90.
103. Levy, R.; Loustaunau, P.; Shapiro, J. The prime spectrum of an infinite product of copies of $\mathbb{Z}$. *Fundam. Math.* **1991**, 138, 155–164. [CrossRef]
104. Beyarslan, O.; Hrushovski, E. On Algebraic Closure in Pseudofinite Fields. *J. Symb. Log.* **2012**, *77*, 1057–1066. [CrossRef]
105. Frayne, T.; Morel, A.C.; Scott, D. Reduced Direct Products. *Fund. Math.* **1962**, *51*, 195–228. [CrossRef]
106. Makar-Limanov, L. A conjecture of Bavula on homomorphisms of the Weyl algebras. *Linear Multilinear Algebra* **2012**, *60*, 787–796. [CrossRef]
107. Tsuchimoto, Y. Preliminaries on Dixmier conjecture. *Mem. Fac. Sci. Kochi Univ. Ser. A Math.* **2003**, *24*, 43–59.
108. Minsky, M. Computation: Finite and Infinite Machines. In *Prentice-Hall Series in Automatic Computation*; Prentice-Hall Inc.: Englewood Cliffs, NJ, USA, 1967; xvii+317p.

109. Ivanov-Pogodaev, I.; Malev, S. Finite Gröbner basis algebras with unsolvable nilpotency problem and zero divisors problem. *J. Algebra* **2018**, *508*, 575–588. [CrossRef]
110. Matiyasevich, Y.V. *Hilbert's Tenth Problem*; MIT Press: Cambridge, MA, USA, 1993.
111. Denef, J. The Diophantine problem for polynomial rings and fields of rational functions. *Trans. Am. Math. Soc.* **1978**, *242*, 391–399. [CrossRef]

mathematics

MDPI

Communication

Physiological Balance of the Body: Theory, Algorithms, and Results

Irina Alchinova and Mikhail Karganov *

Research Institute of General Pathology and Pathophysiology, Baltiyskaya Str. 8, 125315 Moscow, Russia; niiopp@mail.ru
* Correspondence: mkarganov@mail.ru; Tel.: +79-03-610-9132

Abstract: Aim: To confirm algorithm of determination of risk groups with physiological imbalance in the population exposed to unfavorable anthropogenic influences. Methods: The testing included such functional systems as constitution, myocardial contractility, autonomic regulation of the heart rate, regulation of peripheral circulation, psychomotor regulation, respiratory regulation and metabolism. Monitoring is carried out using computerized measurement instrumentation and data processing systems. Results: A risk group with pronounced shifts in the physiological balance was identified, which made up 38% of the surveyed population. The greatest contribution to the imbalance was made by the psychomotor system. Conclusion: We analyzed two different components of organism's adaptation: resistance and resilience. Physiological systems experiencing increasing load attain a tipping points, where even a weak disturbing influence can induce transition to a qualitatively different state. This transition can result in either recovery of the regulatory stability of the system, or its transition to a lower level (dysregulation) with further development of a pathology. In this regard, of paramount importance is early detection of the signals about approaching the tipping points, one of these is the slowing down phenomenon during functional tests. In view of intricate interaction of physiological systems, recording of as much indicators as possible is advisable. The method of partial correlations is effective for evaluation of adaptive interaction of systems.

Keywords: adaptive resource; resilience; resistance; tipping point; theory of catastrophes

Citation: Alchinova, I.; Karganov, M. Physiological Balance of the Body: Theory, Algorithms, and Results. *Mathematics* **2021**, *9*, 209. https://doi.org/10.3390/math9030209

Academic Editor: Vassily Lyubetsky
Received: 2 November 2020
Accepted: 18 January 2021
Published: 20 January 2021

The concepts of health and disease are fundamental in medicine at all stages of its development. The WHO definition of health works well to differentiate between health and disease. However, the transition from one state to another is not instantaneous. Consequently the definition proposed by G. N. Kryzhanovsky [1]—"health is the state ensuring optimal performance of body functions to the extent necessary for productive interaction with the environment"—seems to be the most promising to assess the state of the body under the influence of environmental factors. To achieve this state, specialized systems in the body analyze the changing situation inside and outside the body and adjust the latter to the optimal level of interaction with the environment. Thus, the state of health is primarily provided by the resource of functions spent for adaptation. There are practically no functional states of the whole organism, the sufficiency of which solely depends on only a single system. An important task is to determine the limits of resilience (i.e., ability of a system to perform its functions under the influence of variable external factors) [2]. This gave rise to the "tipping point" concept (i.e., a point corresponding to a transition of a self-maintaining system to a different state, just as slowly cooling water turns into ice) [3]. Here, different stages of adaptation should be distinguished: before and after the tipping point (Figure 1). The resistance mechanisms prevent the system from reaching the tipping point, while resilience mechanisms come into action after the system passed this critical point [4–6].

In experimental studies, it is quite difficult to distinguish between the resistance and resilience stages due to activation of repair mechanisms in response to damage. For separate analysis of these phenomena, special conditions should be created. For instance, the resistance of mammalian genetic apparatus to cosmic radiation can be evaluated only in dried cells [6]. The stability of DNA exposed to irradiation with fast heavy ions was also evaluated in dry specimens to exclude the influence of free radicals formed from water molecules [7]. The resilience stage can be studied under more natural conditions—in living cells with preserved reparative mechanisms.

A relatively simple approach is evaluation of the radioadaptive response of human lymphocytes. It is known that human lymphocytes irradiated in a low dose become more resistant to higher doses of radiation or to other damaging factors. We have previously demonstrated that individuals with or without this response that indirectly characterizes reparative capacity of the body differ by functional parameters of the cardiorespiratory system under conditions of mild stress [8]. This attests to either a genetically determined degree of organism's resilience at different levels of its organization, or the presence of poorly studied mechanisms of interaction between the systems.

It should be noted that resilience is an interdisciplinary concept that attracts steadily growing interest [2]. To describe abrupt changes in the dynamical systems, the theory of catastrophes (i.e., a program for predicting system instability) was developed [9]. In this case, the catastrophe means loss of resilience, even if it does not lead to system death or destruction, but only causes transition to a different trajectory of its development. Dynamic systems are characterized by a number of features: current state of the system depends on the way how it came to it; the system maintains its state as long as possible (the principle of maximum delay); trajectories of the system can be irreversible. Physiological systems are complex, that is, their properties are not confined to characteristics of their components and some new properties can emerge. Thus, fluctuations of individual elements and their combinations under the influence of gradual changes in conditions affecting these elements can lead to a qualitative change of the system in the point of bifurcation, where it switches from one to another more stable mode of functioning.

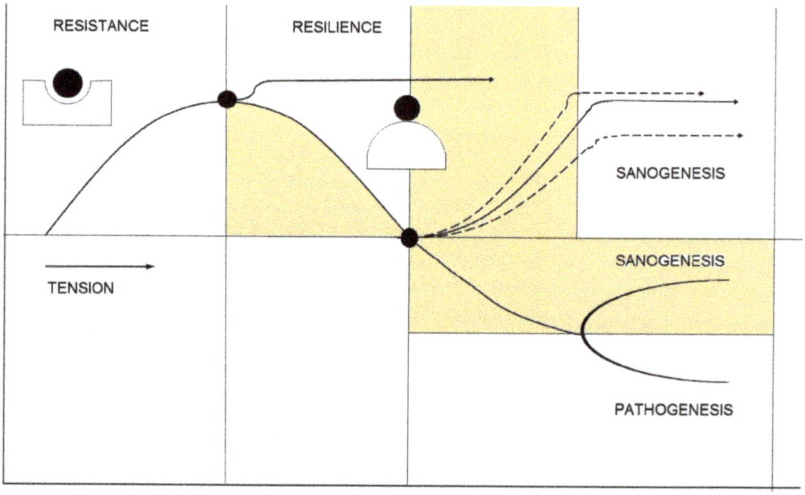

Figure 1. Scheme of the response of a physiological system to stressful exposure/load.

This brings us to the task of quantifying the proximity of the tipping point, which allows us to take measures to avoid transition or, on the contrary, to encourage it in case of initially unfavorable state of the system.

One of the early markers of approaching the tipping point is the critical slowing down phenomenon [10,11]. In physiology and medicine, functional tests are widely used for this studying of the adaptive capacities of the body. These tests are based on the use of "perturbing influences", such as physical activity, changes in spatial body position, breath holding, administration of various substances, etc., while the analyzed parameters are recorded before (at rest), during, and after the test (which is most important in this case). Changes in the pattern of system fluctuations manifesting itself in an increase in temporal autocorrelation, describing the status of the system in successive time points and an increase in variance of its parameters, can be a warning signal of critical slowing down (Figure 2).

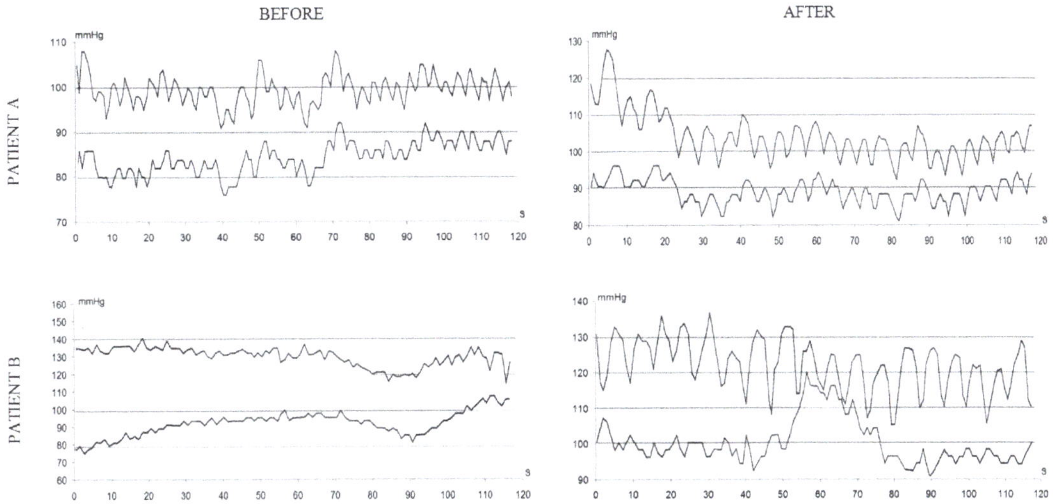

Figure 2. Changes in peripheral blood pressure during functional test (controlled breathing, six cycles/min). Upper panel—high resistance, fast recovery after exposure, lower panel—low resistance, slow recovery, increased variance. Abscissa—time, s. Ordinate—blood pressure, mmHg.

Based on these indicators, so-called dynamic indicators of resilience (DIORs), an algorithm for assessing the risk of catastrophe for ecological systems has been proposed [11]. In clinical practice, an anticipatory care system for patients approaching a tipping point stage has been successfully tested [12]. To determine the risk groups in the population exposed to unfavorable anthropogenic influences, we have to identify groups demonstrating dramatically increased frequency of fixation of certain pathological processes. Biological aftereffects of the detected shifts can reflect either favorable (development of resistance) or unfavorable (fixation of certain pathologies) outcomes. In most cases, epidemiological approaches allow assessing the shifts caused by relatively high doses/concentrations of anthropogenic factors and are unapplicable for mild influences. The most probable pathological outcomes can be predicted by polysystemic monitoring that detects signs of dysregulation in various systems of the body (cardiorespiratory, psychomotor, and metabolism systems). The decrease in health reserve impairs adaptive capacities of the organism, which indicates the first (prenosological) stage of transition from health to disease, when functional parameters remain within the normal range, but at the expense of the strained work of the adaptation mechanisms aimed at the maintenance of health parameters at the required level.

The organism is a complex of interacting and mutually influencing systems and subsystems that form a network, the components of which depend on each other. Impaired resilience of some components can lead to strengthening of cross-correlation relationships with peaks in activity of other parts of the system [13]. For collecting data array on the functional state of the main body systems (cardiorespiratory, psychomotor, and metabolic), a software-and-hardware complex suitable for mass surveys was developed. The correlations between the parameters of different systems revealed by the methods employed by us are not always obvious. The revealed correlation between the content of toxic metals and shifts in the subfractional composition of biological fluids seems to be quite reasonable [14].

Due to individual variability of physiological balance processes, an anthropogenic factor of the same strength (dose, concentration, etc.) can induce shifts in some organisms sensitive to it, but not causes in others, while the thirds can become resistant to this factor.

At the population level, low-dose and low-concentration exposure leads to the formation of three subpopulations: sensitive, neutral, and super resistant. The proportions between these subpopulations reflect the population risk from this exposure.

The monitoring of physiological balance of the body is carried out using computerized measurement instrumentation and data processing systems. The following three major instruments adapted to non-invasive screening survey are used:

– A spiro-arterio-cardio-rhythmo graph with a highly sensitive ultrasonic transducer for continuous non-invasive recording of blood pressure, expiration and inspiration air flows, and electrocardiogram;
– A computer-aided device for express-evaluation of psychomotor activity during motor tests;
– A laser correlation spectrometer intended for identification of the pattern of regulation of metabolic and immune processes.

In physiological systems, around fifty parameters are recorded during screening studies and, for analysis of these systems, unified ranking methods for the recorded values should be developed [15]. The need in this stage is dictated by the concept of dysregulation diagnosis based on discordance between the functions of individual systems, the parameters of which can be compared only if they are presented in universal units of deviation. The sizes of age subgroups are chosen with consideration of sample representativeness for providing proper 5th and 95th centiles and peculiarities of the age-related dynamics of the parameter [16]. The centile table is corrected as the number of measurements increases. The parameters are ranked using the hypo-hyperfunction scale and the centile boundaries correspond to certain scores. The measured parameter is compared with the previously determined (reference) values for persons of different age/gender and then, its correspondence to a particular centile and, thereby, its score is determined. In the expert system, the scores k_0–k_5 take on the following values: −2.5, −1.5, −0.5, 0.5, 1.5, 2.5; all parameters will be scored in the range from −2.5 to 2.5.

Thus, the scores calculated using the above-described method immediately show whether the recorded value of a certain parameter corresponds to the most common range of the same parameter in the population of the same age and gender or deviate towards the of area hypo- or hyperfunction (Figure 3a).

For example, we examined 130 workers (76% men and 24% women) of a shipyard; 77% workers were at the age of 25–54 years; 52% had continuous service length from three to 15 years. By the degree of contact with potential hazard sources, the sample population was divided into three groups. Group 1 (n = 24) included storekeepers, cloakroom attendants, cleaners, engineers, and other technical workers, who had minimum contact with hazardous factors. Group 2 (n = 53) consisted of crane operators, painters, electricians, strapper, maintenance men, and vessel trolley-transporter men who indirectly contacted with hazardous factors. Group 3 (n = 53) comprised welders, burners, riveters, fitters, and other workers on the vessel hull who are directly exposed to hazardous factors.

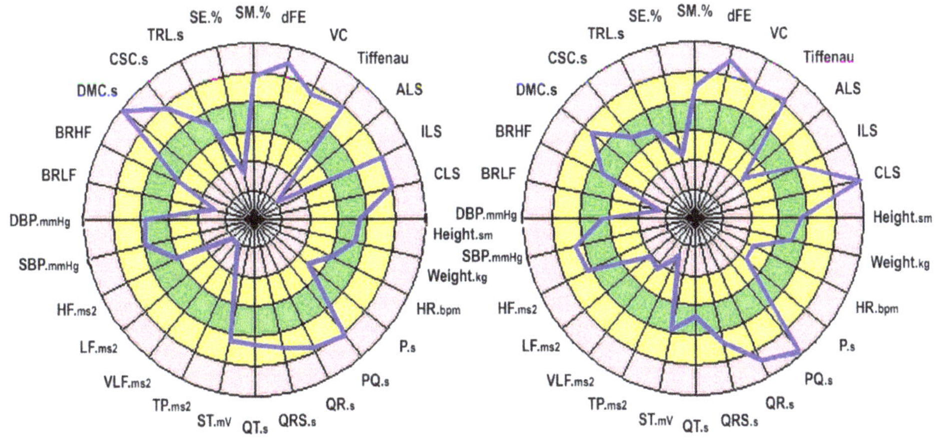

Figure 3. Evaluation of the physiological balance of the body using age-specific centile distributions (**a**) and with computation of partial correlations (**b**). The rings in the diagram: Green—optimal level (from −0.5 to 0.5), yellow—sufficient level (from −0.5 to −1.5 and from 0.5 to 1.5), red—stressed level (from −1.5 to −2.5 and 1.5 to 2.5). Abbreviations: HR—mean heart rate over the recording interval, bpm; P—P wave duration, s; PQ—duration of the interval from the beginning of P wave to the beginning of Q wave (or to the beginning of R wave in case of the absence of P wave), s; QR—duration of the interval from the peak of Q wave (or from the beginning of R wave in case of the absence of Q wave) to the peak of R wave, s; QRS—duration of the interval from the beginning of Q wave (or from the beginning of R wave in case of the absence of Q wave) to the end of S wave, s; QT—duration of the interval from the beginning of the Q wave (or from the beginning of R wave in case of the absence of Q wave) to the end of the T wave, s; ST—depression or elevation of the ST segment (from the end of the S wave to the beginning of the T wave) relative to the isoline, mV; TP—area under RR interval distribution curve, range of heart rate variability; VLF—total power in very low frequency range, ms^2; LF—total power in low frequency range, ms^2; HF—total power in high frequency range, ms^2; SBP—mean systolic BP over the recording interval, mmHg; DBP—mean diastolic BP over the recording interval, mmHg; BRLF, BRHF—baroreflex—change in RR interval duration in response blood pressure changes with consideration for LF- (low frequency) and HF- (high frequency) components; DMC—duration of movement cycle—the mean time (s) of lever movement from one marker (LED) to the other and back; CSC—central settings changing—number of cycles required to achieve the necessary movement accuracy upon change in marker-to-marker distance; TRL—time of response to light stimulus-latency of a simple motor responce time from the beginning of stimulation to the beginning of lever movement from the start point; CE—correction error is the ratio of the mean deviation from the boundaries of the preset movement range to the total amplitude of lever movement over the entire cycle, %; SM—smoothness of movement—assessed by the percentage of the main harmonics of the Fourier spectrum (%); the higher is the contribution of the main frequency, the higher is smoothness of movement; dEF—predominance of extensor/flexor tone—the examinee is asked to move the lever side to side with a certain amplitude with eyes closed, and displacement of the extreme positions of the lever over the last 10 s of visually uncontrolled movements relative to the mean position of extreme points under visual control is estimated. Negative values indicate greater displacement towards flexion, positive values attest to greater shift in the extensor phase of movement; VC—vital capacity; maximum exhaled volume after the deepest inhalation, liter; Tiffeneau—Tiffeneau index = Forced expiratory volume in 1 s/VC, dimensionless parameter, ALS, ILS, CLS—allergic-like, intoxication-like and catabolic-like shifts-initial, moderate and pronounced shifts in homeostasis.

This necessitates creation of an expert algorithm that takes into account the interaction between the parameters and their mutual influence under conditions of a weakly structured task. If this relationship is not taken into consideration, the deviations from the population mean in some cases can be erroneously interpreted as stress state of the studied physiological system, and vice versa, other parameters can be viewed as balanced, though they are not. A way of solving this problem during analysis of multivariate measurements in the absence of physiologically based equations of functional interrelationship is the use of multiple regression analysis that allows computing partial correlations. Analysis of interdependencies of these coupled processes is of particular importance when studying regulatory processes constantly inducing dynamic restructuring of physiological systems. To avoid errors in attribution of changed parameters of physiological systems to a particular centile, the most probable functional relationship between the parameters in conditionally healthy population should be deduced and mutual deviations of the parameters from the population mean should be computed; this will serve as the basis for constructing "dynamic individual norm". Unfortunately, physiologically-based equations (especially multivariate equations) of functional interrelationship for most parameters of body systems are unknown, so they should be deduced using statistical methods. When interpreting interdependence, apparent correlation between two variables can be just a reflection of the fact that they are both correlated with some third variable or a set of variables not included in the model. The situation can be clarified by computation of partial correlations between the two variables, while other variables are removed (partial correlations). It should be emphasized that even in case of high partial correlation coefficients, the causality between the analyzed variables should always be based on knowledge of physiology, but not exclusively on statistical relationships.

Considering the above, the algorithm for calculation of the balance between the parameters of the body systems includes the following steps:

1. Calculation of the matrix of partial correlations for scores of all parameters and for the entire population;
2. Selection of pairs of parameters with significant partial correlations;
3. Expert evaluation of significant correlations for their physiological feasibility and construction of a subgroup with physiologically based correlations;
4. Calculation of the matrix of multiple linear regression coefficients;
5. Calculation of the scores of "dynamic individual norm" indicators based on the scores of measured parameters.

The above calculations yield a vector that contains the scores b_j^* for each measured parameter obtained with considerations for the individual characteristics of the body; the score $k_2 \leq b_j^* \leq k3$ indicates that body condition by this parameter is satisfactory and correspond to the most common values for the conditionally healthy population, while at $k_1 \leq b_j^* < k_2$ and $k_3 < b_j^* \leq k_4$ it corresponds to the initial stress, and at $b_j^* < k_1$ and $b_j^* > k_4$—to pre-pathological and, possibly, pathological stress (Figure 3b). It should be taken into account that the same value of the studied parameter can be normal for one individual and a pathological symptom for another.

Functional sufficiency for each system was scored using a three-point scale, where 1 corresponded to balanced, 2 to sufficient, and 3 to strained state of the system. In a population not burdened by verified pathologies and intoxications, the proportion between these groups is 50%—40%—10% [15].

The integral functional balance was determined by the summary score for seven regulatory systems (each evaluated using a three-point scale):

1. Balanced (score from 7 to 10);
2. Sufficient (score from 11 to 12);
3. Strained (score $\geq$13);

The integral value of functional strain (38%) attests to a high burden of the studied population. The decisive contribution of the psychomotor regulation system is worthy of note: this system worked in a strained mode in 75% workers. In groups 1, 2, and 3, the functional state of the psychomotor regulation was interpreted as strained in 42%, 70%, and 96% cases, respectively. Taking into account occupational specific of these groups, we can assume that working conditions in group 3 promote aggravation of functional strain in the psychomotor system.

The use of the described algorithm made it possible to identify risk groups in mass surveys of employees exposed to potentially dangerous climate, geographical and occupational factors and to determine the degree of proximity to the tipping points for body systems in participants of an Arctic expedition [17].

An example of an effective application of this approach is described previously [12]. The development of methods for predicting catastrophic changes at the cellular level is in progress [18,19]. This trend seems to be very promising, because it allows identifying the patterns of interaction between the systems. This, in turn, makes it possible to use empirical data on the interaction between body systems during modeling and prediction of physiological balance level. Monitoring of the main body systems by using functional tests and applying the developed algorithms for detection of approaching the tipping points, followed by correction of the detected deviations, will lead to an increase in the duration of active life.

Author Contributions: M.K. formulated the idea of the paper. I.A. analyzed the experimental data. M.K. and I.A. participated in preparation of the paper text. All authors have read and agreed to the published version of the manuscript.

Funding: The research was supported by an RFBR grant 19-29-14104 "Instrumental assessment of the impact of digitalization of education on the physiological balance of the body".

Institutional Review Board Statement: The study was conducted according to the guidelines of the Declaration of Helsinki, and approved by the Institutional Ethics Committee of Institute of General Pathology and Pathophysiology (protocol No. 4, 02.09.2019).

Informed Consent Statement: Informed consent was obtained from all subjects involved in the study. Written informed consent has been obtained from the patients to publish this paper.

Data Availability Statement: https://cloud.mail.ru/public/1VaZ/NmK9au3HL.

Conflicts of Interest: The authors declare no conflict of interest.

References

1. Kryzhanovsky, G.N. Some categories of general pathology and biology: Health, disease, homeostasis, sanogenesis, adaptation, immunity. new approaches and notions. *Pathophysiology* **2004**, *11*, 135–138. [CrossRef]
2. Baggio, J.A.; Brown, K.; Hellebrandt, D. Boundary object or bridging concept? A citation network analysis of resilience. *Ecol. Soc.* **2015**, *20*, 2. [CrossRef]
3. Bak, P.; Sneppen, K. Punctuated equilibrium and criticality in a simple model of evolution. *Phys. Rev. Lett* **1993**, *71*, 4083–4086. [CrossRef]
4. Miller, B.F.; Seals, D.R.; Hamilton, K.L. A viewpoint on considering physiological principles to study stress resistance and resilience with aging. *Ageing Res. Rev.* **2017**, *38*, 1–5. [CrossRef] [PubMed]
5. Fleshner, M.; Maier, S.F.; Lyons, D.M.; Raskind, M.A. The neurobiology of the stress-resistant brain. *Stress* **2011**, *14*, 498–502. [CrossRef] [PubMed]
6. Karganov, M.Y.; Alchinova, I.B.; Yakovenko, E.N.; Kushin, V.V.; Inozemtsev, K.O.; Strádi, A.; Szabó, J.; Shurshakov, V.A.; Tolochek, R.V. The "PHOENIX" space experiment: Study of space radiation impact on cells genetic apparatus on board the international space station. *J. Phys. Conf. Ser.* **2017**, *784*, 012024. [CrossRef]
7. Karganov, M.Y.; Alchinova, I.B.; Polyakova, M.V.; Feldman, V.I.; Gorbunov, S.A.; Ivanov, O.M.; Rymzhanov, R.A.; Skuratov, V.A.; Volkov, A.E. Stability of dry phage Lambda DNA irradiated with swift heavy ions. *Radiat. Phys. Chem.* **2019**, *162*, 194–198. [CrossRef]
8. Alchinova, I.B.; Khlebnikova, N.N.; Karganov, M.Y. Information value of the test with radioadaptive response upon functional evaluation of pilots. *Hum. Physiol.* **2015**, *41*, 767–775. [CrossRef]
9. Thom, R. Topological models in biology. *Topology* **1969**, *8*, 313–335. [CrossRef]
10. Scheffer, M. Early-warning signals for critical transitions. *Nature* **2009**, *461*, 53–59. [CrossRef] [PubMed]
11. Dakos, V.; Carpenter, S.R.; Brock, W.A.; Ellison, A.M.; Guttal, V.; Ives, A.R.; Kéfi, S.; Livina, V.; Seekell, D.A.; van Nes, E.H.; et al. Methods for detecting early warnings of critical transitions in time series illustrated using simulated ecological data. *PLoS ONE* **2012**, *7*, e41010. [CrossRef] [PubMed]
12. Martin, C.M.; Sturmberg, J.P.; Stockman, K.; Hinkley, N.; Campbell, D. Anticipatory care in potentially preventable hospitalizations: Making data sense of complex health journeys. *Front. Public Health* **2019**, *6*, 376. [CrossRef] [PubMed]
13. Dakos, V.; van Nes, E.H.; Donangelo, R.; Fort, H.; Scheffer, M. Spatial correlation as leading indicator of catastrophic shifts. *Theor. Ecol.* **2010**, *3*, 163–174. [CrossRef]

14. Karganov, M.; Skalny, A.; Alchinova, I.; Khlebnikova, N.; Grabeklis, A.; Lakarova, E.; Eisazadeh, S. Combined use of laser correlation spectroscopy and ICP-AES, ICP-MS determination of macro- and trace elements in human biosubstrates for intoxication risk assessment. *Trace Elem. Electrolytes* **2011**, *28*, 124–127. [CrossRef]
15. Noskin, L.; Pivovarov, V.; Landa, S. Methodology, hard- and software of polysystemic monitoring. In *Polysystemic Approach to School, Sport and Environment Medicine*; OMICS Group Incorporation: Hyderabad, India, 2014; pp. 13–22. [CrossRef]
16. Bezrukih, M.M.; Kiselev, M.F.; Komarov, G.D.; Kozlov, A.P.; Kurneshova, L.E.; Landa, S.B.; Noskin, L.A.; Noskin, V.A.; Pivovarov, V.V. Age-related features of the organization of motor activity in 6-to 16-year-old children. *Hum. Physiol.* **2000**, *26*, 337. [CrossRef]
17. Pankova, N.B.; Alchinova, I.B.; Cherepov, A.B.; Yakovenko, E.N.; Karganov, M.Y. Cardiovascular system parameters in participants of Arctic expeditions. *Int. J. Occup. Med. Environ. Health* **2020**, *33*, 819–828. [CrossRef]
18. Meisel, C.; Klaus, A.; Kuehn, C.; Plenz, D. Critical slowing down governs the transition to neuron spiking. *PLoS Comput. Biol.* **2015**, *11*, e1004097. [CrossRef] [PubMed]
19. Smirnova, L.; Harris, G.; Leist, M.; Hartung, T. Cellular resilience. *Altern. Anim. Exp. Altex* **2015**, *32*, 247–260. [CrossRef]

MDPI
St. Alban-Anlage 66
4052 Basel
Switzerland
Tel. +41 61 683 77 34
Fax +41 61 302 89 18
www.mdpi.com

Mathematics Editorial Office
E-mail: mathematics@mdpi.com
www.mdpi.com/journal/mathematics

www.ingramcontent.com/pod-product-compliance
Lightning Source LLC
LaVergne TN
LVHW070717100526
838202LV00013B/1114
* 9 7 8 3 0 3 6 5 0 7 7 8 1 *